NATO ASI Series

Advanced Science Institutes Series

A series presenting the results of activities sponsored by the NATO Science Committee, which aims at the dissemination of advanced scientific and technological knowledge, with a view to strengthening links between scientific communities.

The Series is published by an international board of publishers in conjunction with the NATO Scientific Affairs Division

A Life Sciences B Physics	Plenum Publishing Corporation London and New York
C Mathematical and Physical Sciences D Behavioural and Social Sciences E Applied Sciences	Kluwer Academic Publishers Dordrecht, Boston and London
F Computer and Systems Sciences G Ecological Sciences H Cell Biology I Global Environmental Change	Springer-Verlag Berlin Heidelberg New York London Paris Tokyo Hong Kong Barcelona Budapest

PARTNERSHIP SUB-SERIES

1. Disarmament Technologies	Kluwer Academic Publishers
2. Environment	Springer-Verlag
3. High Technology	Kluwer Academic Publishers
4. Science and Technology Policy	Kluwer Academic Publishers
5. Computer Networking	Kluwer Academic Publishers

The Partnership Sub-Series incorporates activities undertaken in collaboration with NATO's Cooperation Partners, the countries of the CIS and Central and Eastern Europe, in Priority Areas of concern to those countries.

NATO-PCO DATABASE

The electronic index to the NATO ASI Series provides full bibliographical references (with keywords and/or abstracts) to about 50000 contributions from international scientists published in all sections of the NATO ASI Series. Access to the NATO-PCO DATABASE compiled by the NATO Publication Coordination Office is possible in two ways:

- via online FILE 128 (NATO-PCO DATABASE) hosted by ESRIN, Via Galileo Galilei, I-00044 Frascati, Italy.

- via CD-ROM "NATO Science & Technology Disk" with user-friendly retrieval software in English, French and German (© WTV GmbH and DATAWARE Technologies Inc. 1992).

The CD-ROM can be ordered through any member of the Board of Publishers or through NATO-PCO, Overijse, Belgium.

Series F: Computer and Systems Sciences, Vol. 141

The NATO ASI Series F
Special Programme on
ADVANCED EDUCATIONAL TECHNOLOGY

This book contains the proceedings of a NATO Advanced Research Workshop held within the activities of the NATO Special Programme on Advanced Educational Technology, running from 1988 to 1993 under the auspices of the NATO Science Committee. The books published so far in the Special Programme are listed briefly, as well as in detail together with the other volumes in NATO ASI Series F, at the end of this volume.

Springer
Berlin
Heidelberg
New York
Barcelona
Budapest
Hong Kong
London
Milan
Paris
Tokyo

Organizational Learning and Technological Change

Edited by

Cristina Zucchermaglio

Department of Developmental and Social Psychology
University La Sapienza
Via dei Marsi 78, I-00185 Rome, Italy

Sebastiano Bagnara

Department of Communication
University of Siena
Via del Giglio 14, I-53100 Siena, Italy

Susan U. Stucky

Institute for Research on Learning
2550 Hanover Street, Palo Alto, CA 94304, USA

Springer
Published in cooperation with NATO Scientific Affairs Division

Proceedings of the NATO Advanced Research Workshop on Organizational Learning and Technological Change, held in Certosa di Pontignano, Siena, Italy, September 22–26, 1992

Library of Congress Cataloging-in-Publication Data

NATO Advanced Research Workshop on Organizational Learning and Technological Change (1992 : Siena, Italy)
Organizational learning and technological change / edited by Cristina Zucchermaglio, Sebastiano Bagnara, Susan U. Stucky.
p. cm. -- (NATO ASI series. Series F, Computer and systems sciences ; vol. 141)
"Proceedings of the NATO Advanced Research Workshop on Organizational Learning and Technological Change, held in Certosa di Pontignano, Siena, Italy, September 22-26, 1992"--T.p. verso.
"Published in cooperation with NATO Scientific Affairs Division."
Includes bibliographical references and index.
ISBN 3-540-58917-1 (hardcover : alk. paper)
1. Organizational effectiveness--Congresses. 2. Learning--Congresses. 3. Technological innovations--Congresses. 4. Employees--Training of--Congresses. 5. Continuing education--Congresses. I. Zucchermaglio, Cristina. II. Bagnara, Sebastiano. III. Stucky, Susan U., 1949- . IV. North Atlantic Treaty Organization. Scientific Affairs Divison. V. Title. VI. Series: NATO ASI series. Series F, Computer and systems sciences ; no. 141.
HD58.9.N38 1992
658.3'1243--dc20 95-21988
CIP

CR Subject Classification (1991): H.5, H.1.2, I.2.6, K.3–4

ISBN 3-540-58917-1 Springer-Verlag Berlin Heidelberg New York

Typesetting: Camera-ready by editor
Printed on acid-free paper
SPIN: 10486630 45/3142 – 5 4 3 2 1 0

Preface

What the Book Is About

This book is about the problem of organizational learning, that is the analysis of organizations conceived as learning systems. In order to survive in a period of a rapid change, organizations must innovate and than to develop and exploit their abilities to learn. The most innovative organizations are those that can respond with great efficiency to internal and external changes. They respond to and generate technological change by acting as effective learning systems. They maximize the learning potential of ongoing and "normal" work activities. The organizational structure and the technology allow members to learn while the organizations itself learns from its members. So organizations reach high levels of innovation when structured to take advantage of the social, distributed, participative, situated processes of learning developed by its members in interaction with the technological environment. Organizations should consider learning as an explicit "productive" objective. They must create integrated learning mechanisms, that encompass technological tools, reward and incentive systems, human resource practices, belief systems, access to information, communication and mobility patterns, performance appraisal systems, organizational practices and structures. The design of efficient learning organizations requires cognitive, technological and social analyses. All the computer-based technologies (e.g. office automation, communication and group decision support) not only those devoted to and used in training activities, have to be considered as tools for organizational learning and innovation. The level of effectiveness highly depends on whether they are coherently and jointly designed with the organizations they are to be used in. This organizational analysis combined with a cognitive approach can also improve the design, implementation and use of new technologies in the workplaces.

To reach such a design aim the book brought together researchers from a variety of perspectives: cognitive psychology, educational psychology, anthropology, information technology and artificial intelligence. The interaction among the differential points of view allowed us to clarify this issue and let elements for a new synthesis appear based on the merge of ideas from the different disciplinary backgrounds represented in the book.

There is no pre-existing common pool of knowledge about organizational learning nor even a clear shared definition of words such as technology or learning.

Differences of perspective come from different disciplines, and from different perspectives of practitioners, scholars and theoreticians - though many of the authors belong to several camps.

The Workshop

The book is based on papers presented at an Advanced Research Workshop held in Certosa di Pontignano, Siena, September 22-26, 1992, entitled "Organizational Learning and Technological Change". The workshop was sponsored by the NATO Science Committee as part of its Advanced Educational Technology Programme.

The majority of the people, acting within a general cognitive framework, had never meet before and this special opportunity was a unique chance that has been fully exploited. Twenty-five speakers were asked to submit short papers in advance of the workshop and these were distributed to the participants. Each participant then gave a full presentation on the basis of the paper. All of the presenters were asked to write up their presentation in a paper. Twenty presenters agreed to write a chapter for the book and the editors will write a general introduction summarizing also the best parts of the discussions.

The debate was framed around three major themes: organization, technology and theory of learning. Inside this general framework, each contribution focused on different aspects of the general problem of organizational learning and technological change.

A first key concept that emerged during the work was that of community of practice: "real" emergent organizations (as opposed to the "artificial" designed ones) are to be considered as a socio-communicative system, not only as a set of information flows and execution procedures. The concept of "community of practice" allows the analysis of learning considered as a cultural development, a process of enculturation, of entry into a culture of practitioners; learning considered as a complex social action embedded in historically and culturally defined frames of activity

A second key concept, strictly related with the first one, was that of co-operation: social processes of knowledge construction were analyzed not as particular or special conditions for learning, but as an essential and integral feature of efficient learning and working contexts, that should be allowed with appropriate technological tools. Some cooperative technological tools were presented and the design principles of such learning/working tools were discussed.

A third key concept was that of learning strategies in organizations. In particular both the positive function of errors in the learning processes and the "nature" of the concept of error were analyzed. It has been stressed that the organizational norms, that tend to put the responsibility for errors on the individual, are the societal and institutional consequences of a very general cognitive bias. In assessing the causes of error, people are biased toward violations that have immediate and visible negative effects, whereas latent failures (for example organizational decisions) that create the conditions of errors are overlooked. Moreover, technology as a cognitive and organizational artifact was analyzed. In this perspective, the design of a user interface becomes a very complex and strategic task. Both the functionality provided to the user and the interaction tools among the user, the task, the system, and other users should be considered.

This book is part of the process of organizing the new ideas, developments, and claims, and laying them open to the scrutiny of all interested parties. The book offers a cross-section of the concerns of researchers at the time of the workshop. It does not claim to be a comprehensive report on the field. It does, however, represent an illuminating and stimulating selection of papers, drawing on a representative variety of disciplines, approaches, and issues.

Overall, we believe that the workshop and the book will substantially enhance the understanding of organizational learning and technology and will have an impact on related areas such as analysis of work practices and learning processes, organization design, human-computer interaction, and technology design.

April 1995

Cristina Zucchermaglio
Sebastiano Bagnara
Susan U. Stucky

Table of Contents

Part I
Organizations as Learning Bodies

Introduction

Susan U. Stucky

Institute for Research on Learning, 2550 Hanover Street, Palo Alto, CA 94304

The participants of the workshop on organizational learning and technological change came from overlapping communities of researchers for whom organizations and organizational life have become newly important to understand. As long as learning and its attendant technologies were viewed primarily as something done or used by an individual, the fact that individuals live and work in organizations was seen chiefly as context. However, both learning theory and real-world contexts have changed in such a way that organizations have become a focus of inquiry. Learning theory, for its part, has been gradually showing the explanatory value of the social. Which means taking the social world to be not just the context of action, but constitutive of it. Also, the technological advances that were supposed to have enhanced learning have turned out to be more problematic than anticipated. And finally, the scale and scope of human interaction has increased dramatically in such a way as to put real pressure on our organizations to change and adapt, in short to learn.

The papers in this volume reflect this concern. Five of them, those by Clancey, Elkjaer, Orr, Stucky, and Zucchermaglio, bring insights from recent thinking in artificial intelligence, educational psychology, anthropology, cognitive science and cognitive psychology to the organizational realm.

Clancey, in his paper, "Practice Cannot be Reduced to Theory" takes the reader through a representation of central techniques of artificial intelligence. By taking a new view of knowledge, Clancey shows how the traditional representations of AI when viewed as a part of a cycle of behavior - as created in the course *of* activity not as instructions *for* activity - can become part of an organizational cycle of learning. He uses as an example the role that representations of work practice can play in work re-design. This use not only holds promise for more effective work, but also for double-loop learning at the level of the group. And this attention to group-level learning is a first step toward getting organizations to learn as organizations.

Elkjaer, in her paper, "Learning is an Active and Personal Process" brings forward a study of the individual learner in the organizational context that argues that viewing organizations merely as aggregates of individuals obscures the fact that the individuals are particular persons. And she demonstrates that valuing learning is dependent on the individuals' finding meaning both as persons and as employees of an enterprise. This negotiation of meaning is where the fit of organization to individual is worked out. As such, it is also at the crux of organizational learning. Failure to negotiate this boundary successfully will ultimately fail not only the individual but the organization as well. Elkjaer concludes with a number of suggestions for how to create more useful contexts for training in the corporate context.

Orr, in his paper, "Ethnography and Organizational Learning: In Pursuit of Learning at Work" is concerned to show us what learning at work looks like on the ground in the context of real work in an organization. It happens, Orr says, "primarily through diagnosis, the struggle of one or more technicians to understand a particular problem." The learning is then embedded in this work community's group memory by a practice of story-telling and re-telling. The most striking finding of this study is of the larger organization's ignorance of this learning and its failure to "learn" from the service technician's experience to provide this support more broadly. Part of the reason is that learning is local. And, Orr's account warns us, unless we all find ways to see the local, we will continue to miss much of the learning that goes on in organizations. And then we will be much less effective at providing support for this kind of activity, which fundamentally, we want more of.

Stucky, in her paper, "Technology in Support of Organizational Learning" proposes that there are two kinds of organization in any organization - the designed and the emergent. The emergent, far from being merely informal, is a robust kind of organization that, when seen clearly, forms the explanatory basis for understanding how a corporation (a kind of "designed organization") actually works and learns. She then proposes that technology to support learning should be conceived of as supporting participation rather than for transferring information. Clancey's work design tools are an instance. So are the radios for Orr's service technicians. Like Clancey, Stucky suggests re-thinking the use of the tools we already have while opening up the possibility of new ones. Participation too, is essential to Elkyaer's "meaning negotiation".

Zucchermaglio, in her paper, "Organizational and Cognitive Design of (Technological) Learning Environments" argues (in agreement with Clancey) that the work context should be designed "to take advantage of the learning potential" that is inherent in any productive activity. Taking learning to be a social activity, she proposes a set of basic principles of a "design science" of educational contexts. In so doing, she honors many of the points made by her co-authors (that learning is personal, that representations can and should be designed for collaborative use, that learning happens on the job in normal everyday life). Continuing to evolve these principles will be a very useful activity for the educational technology community.

Altogether these papers rely on a picture of organizations that is not as rational or as orderly as the fields of organizational theory and organizational design might have us believe. Indeed, each of these papers suggests that the level of description of organization that will be relevant for making progress on the general problem of organizational learning, will be smaller, more local and more particular. Organization looked at this way is not, however, unsystematic. This new kind of systematicity presents new challenges, and, we hope a fruitful way to go about taking next steps in developing technology for learning.

Technology in Support of Organizational Learning

Susan U. Stucky

Institute for Research on Learning, 2550 Hanover Street, Palo Alto, CA 94304

Abstract. This paper presents a position on the use of technology in support of organizational learning. In bringing together some recent work on learning it is argued that to enable learning in the organizational context, computational technology - software, hardware, as well as networks - should be designed for social participation. This approach stands in stark contrast to other purposes for technology, e.g., technology as productivity amplifier, as agent of information transfer, as expert system or even as a knowledge delivery engine. This participatory position is based on two findings:

(i) that organizations function, i.e., accomplish their work and learning, in virtue of their *emergent* organization rather than in terms of their *authorized* organization;

and

(ii) that the social negotiation of understanding is a primary way in which organizations can be understood to learn.

From these findings, the conclusion follows: technology designed to support social participation within and across *emergent* organization will have a better chance at supporting organizational learning than technology that is not explicitly designed with this kind of organization in mind.

Keywords. Computer-supported collaborative work, human-computer interaction, learning technology, organizational learning, participatory design, technology development, training.

1. Introduction

In organizational contexts, computing technology has played to mixed reviews. This is hardly surprising, given the complexities involved. The original inventors often have speculative ideas on what their products can do technologically, but little sense of how they would actually function in real-world settings. Development and marketing departments, each from their own perspective, struggle with technical details and with how to position the new systems. Customer organizations have their own points of view; what matters to them is not the technology *per se*, but how to improve their own businesses (send out invoices faster, manage more inventory, develop drugs more efficiently, get more kids to add and subtract correctly). Producers and consumers construct elaborate points of contact, chiefly around the idea of increased productivity. Of course, no one would ever suggest starting over completely, but neither are efforts so far counted as unqualified successes.

What lesson should we learn from this checkered history? That if we wish to embark on any large-scale development of technology to support organizational learning we urgently need an integrated understanding of the role of technology in organizations. If we do not, we in the learning business risk getting the same mixed reviews as our colleagues in information technology have.

The challenge we all face then is to understand how to integrate technology in organizations in order to enhance learning. Technology in organizations, whether it is used in the workplace, (say a system to aid in the processing of insurance claims) or in the school context (say a CAD system for learning and using mathematics) or in everyday life (say ABS brakes) needs to satisfy multiple perspectives. The two most familiar are the perspective of the technologists (for instance, whether the system in question is viewed as a database, or as an expert system, or as word processing software) and the perspective lent by the needs of the users (whether they can meet payroll on time, or learn math or, even, get their school assignments done in a helpful way). In the context of organizational learning, I wish to add a third player: the organization itself. This integrative agenda can be helped, I will argue, by a different view of organization than the one that informs most large-scale technology development in organizational contexts, and by a correspondingly different view of the role of technology in those contexts.

This view of organization, technology, and learning is emerging. It is suggested by a body of theory highlighting social participation as central to human cognition. This body of theory, together with its accompanying methods of research and participation comes from many quarters, though it is not so widespread as to be considered in the mainstream of any of the relevant disciplines (i.e., cognitive science, sociology, anthropology, artificial intelligence, or even organizational theory). In this paper I will be relying chiefly on the work at The Institute for Research on Learning because it has an uniquely encompassing theoretical stance.

IRL's approach to organizational learning and technology is guided by the notion of *design for participation.* This approach draws deeply from work in human-centered design, user-centered design, human-computer interaction, cscw, participatory design and the like (see, for instance, Greenbaum and Kyng, 1991; Norman and Draper, 1986, and references there). As with these other approaches to design, there is an emphasis on work practice, the tacit habits of groups of people that have developed in the course of their work. *Design for participation* likewise makes use of research methods that help make visible the kind of social organization which emerges from the processes of participation with the same practice (see for instance Jordan, 1994 and Jordan and Henderson, 1994). But *design for participation* also accommodates new insights into the work of organizations: specifically, it appeals to a of level of organization which arises out of members' participation in the workplace. Because this particular kind of social cohesion has been shown to be useful in understanding how work and learning take place together in situation, it is particularly appropriate to underlay the design of technology that has the purpose of enhancing learning.

The demands on technology in the organizational context are twofold: first, technology itself must be able to be used. And that takes learning on the part of not only the individuals in the organization, but of the organization itself. We shouldn't and we can't relegate learning to user manuals or to training.

Secondly, technology needs to be designed so as to support emergent social interaction for it is through those that organizational learning happens. Technology must form a resource for this kind of social participation. It is a necessary pre-condition for the social construction of understanding and, we infer, for organizational learning. *Design for participation* is explicitly set up to enable learning at work, to attend to learning from the very beginning of any design enterprise. It is a pro-active stance.

The structure of the paper is this: it begins with a description of the problems encountered by organizations in the use of technology. This is followed by revised pictures of learning, of organization, of learning and organizations together, of technology in support of organization and, finally, of technology in support of learning in and by organizations.

2. Two Perspectives on Technology in Organizations

Let's take up the motivating context from two perspectives: from the side of the technology providers and from that of the organizational implementers. First, from the side of the technology providers. It is hardly news that computer software systems have been harder to deliver than had been anticipated several decades ago. I take it that no documentation is really needed to substantiate this claim. The popular and academic press are full of reports of failed expectations around large systems. Artificial intelligence has had its bubble burst. So have expert systems.

And we can still find large-scale efforts aborted after hundreds of thousands of dollars of expenditure. Two recent (1993) examples include abandoning an effort to automate the London Stock Exchange and the complete re-negotiation of the contract to provide a new computer system for U.S. air traffic control only a few years from release.

The problem from the point of view of the organizations is that technology has been harder to integrate into organizations than expected. In other words, organizations haven't learned to use the technology that has been delivered. Here I will relate a number of tales from IRL's experience in the field, though to protect the actual corporations, I will not reveal the identities of our research partners. One large computer-maker reported the absolute rejection of a tool by its own sales force which was conceived of and implemented by the company's own AI department for them. A chief scientist in an expert systems lab of a large communications company finds that many of the expert systems they build are simply not used. The department of organizational excellence in a large health-maintenance organization in looking into the development of electronic medical records concludes that the organizational implementation of the systems must happen in parallel to technology development because they have seen too many systems fail at the stage of organizational adoption. Though these are recent experiences we at IRL have heard from our partners, they are not atypical; they serve as a demonstration of the widespread perception of the difficulty of fit between organizations and technology, particularly of large software systems.

Seeing the connection between organizational learning and technological fit is not as pervasive, but the recognition of this problem has been around for a long time. Typical of the academic business press (which has been calling attention to these issues for nearly two decades) is the following by Calvin Pava (1983) in a book entitled *Managing New Office Technology*:

> "The purchase of exotic technology does not automatically result in its productive utilization (Ackoff, 1967). There is a vital distinction between equipment functionality and tangible user benefits. One device may offer more memory, faster processing speed, and better quality graphics than another, but these attributes belong to the machinery itself and the relative efficacy of technologies around which it is built. Technically enhanced functionality, however, is not the same as tangible benefits from a user's point of view. Concrete advantages to specific operations can be realized only if changes are made that take advantage of the equipment's capabilities. In most cases, the translation of enhanced technical functionality into substantive benefits requires both learning and changing on the part of the user organization, ..."

The second problem is that technology, often requires people to learn to use it. As the nearly apocryphal VCR interfaces (yes, mine blinks '12:00' all the time, how about yours) have demonstrated and as one of our research partners in consumer electronics has claimed, it is manifestly easy to design technology that is impossible to learn or to use. What's more, not only is there often a steep learning curve in adopting technology inside an organization, but there may also be concomitant change needed in work processes in order to accommodate the new technology. Or, alternatively, in business process re-engineering contexts, information technology is often introduced as part of the solution to more productive work instead of understanding that the technology actually changes the situation, often complicating work for a time. Thus, there is a push towards learning both the new technology and the new work processes in the organizational context. Often, in re-engineering contexts, these challenges come at the same time.

In contrast, learning (at least in organizations) has traditionally been relegated to the training classroom. Only recently has learning, in contrast to training, been seen as a natural response to a growing sense of complexity and change. Organizations are being asked to acquire new capabilities of, what seems to them, both increasing and widening scope. Whole corporations seek to change their business. Xerox, for example, has begun a shift from "the copier company" to "the document company". Nynex, one of the seven "Baby Bells" in the US., recently had to redesign its T-1 Provisioning Process to stay ahead of competitors in the installation of fiber cable. Schools are being asked to restructure dramatically. Nation-wide efforts (e.g., by the New American Schools Development Corporation) to start "break the mold schools" require huge expenditures of public time and energy in service of reform. Organizations are also being asked to learn to work in new kinds of configuration. As the movement away from hierarchically structured command and control organizations and to a much flatter organization with teams of all sorts - autonomous work groups, high-performance work teams and cross-functional activities - people in organizations are asked to learn how to work all over again. Typical of the rhetoric that surrounds the arguments linking learning and change is the following from the introduction by Robert Howard (1990) to *The Learning Imperative: Managing People for Continuous Innovation*:

> "Every manager knows that the competitive environment is changing. Competition is getting tougher. Technological innovation is happening faster. Both customers and business partners are becoming more demanding. In so dynamic and volatile an economy, the only sure source of competitive advantage is an organization's people - in particular, their ability to anticipate change, adapt to new circumstances, and innovate new kinds of business practices.

To achieve these goals, learning at all levels of the organization is not just an advantage. It is an imperative for long-term competitive success. But this learning imperative poses new kinds of challenges for managers. They need to know how to design, build, and lead organizations that are equal to the more stringent demands of the new economy. They need to make continuous 'organizational innovation' the cornerstone of their company's competitive strategy."

What is important about these demands for organizational learning, is that they are forced upon us the real-world context into which these organizations are situated. Yet though these demands are real enough, they are not the terms in which anyone does his or her everyday work. This discrepancy also exists for learning. Knowing that you have to make continuous innovation is not the same as doing something that can be regarded as an innovation. Learning, as we will come to see, is local. It can't be seen from headquarters. (It follows, of course, that it can't be managed from there either.)

3. Revising Our Picture of Organizations

There are many definitions of 'organization' in the literature, of course, just as there are many kinds of organizations in real life. What I want to do here, however, is point out the usefulness of seeing two very different forms of organization at play in our so-called "organizations", in our corporations, schools, nations, football teams and the like. One, the kind of organization reflected in an org chart, or the official make-up of a team, is often intentionally designed. I'll call this 'authorized organization'. Authorized organization is basically reified social organization. The official membership and reporting structure of a team is authorized, for instance.

The second kind of organization is organization that emerges, or 'emergent organization'. A quintessential form of emergent organization is a *community of practice* (a concept developed by Lave and Wenger, 1991 to explain how learning takes place socially.) A *community of practice* is an aggregate of people who exhibit a kind of social cohesiveness. They come to share a common view of the world; they have a way of doing things. Often, they come to look like each other or talk like each other. They are a kind of community and they have a practice. We all belong to lots of them. Some teams are cohesive communities of practice. Some aren't. Sometimes there is the "real" manager and sometimes there is also a "de facto" manager. Then the authorized and the emergent start to pull apart; sometimes they come so far apart that there is almost no overlap. I'm not going to judge what is more desirable but simply describe the difference and then use it to show how we can think of technology differently in virtue of this revised picture of organizations.

Figure 1 below is a picture of how these concepts are meant to relate to each other. If you start with authorized organization, the job is to enact it (in the sense of Weick, 1979). So, for instance, if you put together a software team where there is no prior history of working together, no existing emergent patterns of work and play, it may take some work for it to become a team. To enact teaming will require a lot of participation through the real work of the team. Usually, we have a bit of both, of emergent and of authorized going on at the same time. At the other extreme, suppose you have in your organization an extremely talented group of folks who have, over time, emerged as very effective at planning conferences. Similarly, not authorizing the group at all may mean that a new manager not "see" the expertise that has been developed right away. A newcomer, also good at planning conferences, may find it almost impossible to "get into" this group. Practices may be different, or the community may be closed. This takes real managerial talent to work through the ensuing mess. Just authorizing a role from the outside will probably not be enough to get that person accepted.

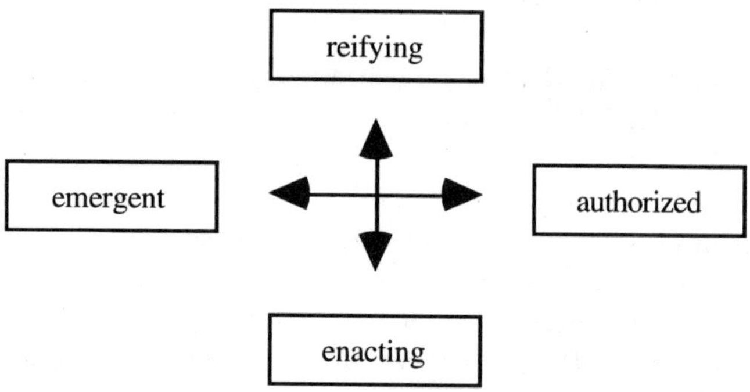

Fig. 1. Types of organization

For the moment, you can think of 'authorized' and 'emergent' as analogous to formal and informal organization. However, as I have conceived of these two forms of organization, emergent organization isn't without form; indeed, it has a lot of form. However, its form is constituted by participation, not by declaration (or design or authorization). If the participation ceases, the community of practice ceases to exist. Conversation has this property too. You stop talking, your partner stops replying and soon the conversation stops existing. But conversations have a lot of form, as any conversation analyst can tell you. Or as you can verify on your own by trying to get into a conversation which is already well under way. Authorized organization, on the other hand can stand up on its own whether or not it is instantiated by a particular person or persons. For it to continue existing, the reified position has to get its own practice going. As a consequence of these differences, the size of authorized organization is not limited in quite the same way as emergent authorization is. It all depends on how much other organization you are willing to put in place in order to keep the authorization, i.e., the reification in place.

The relevance of this distinction in the context of organizational learning is this: learning happens as people come to be members of a community of practice. Correspondingly, organizations learn as people are changing their membership within the communities of practice that make up the organization, as new communities of practice are emerging and old ones dissolving and as the configuration of the communities of practice themselves are shifting. Put simply: becoming a student in a class does not ensure learning what the course is about. Becoming a member of a community of practice however *always* involves learning what that community is about. So if we want to understand learning, it is clear where we need to look.

Our general experience at IRL is that if you want to understand who learns what, you need to look at the patterns of participation and the emergent organization which these patterns reveal. Perhaps the most dramatic example in IRL's learly work came from the work of Wenger in an insurance company (Wenger, 1990). Wenger, a computer scientist and budding ethnographer, spent a year becoming a claims processor and working in the company himself. First of all he found a fairly big mismatch between what the training provided and what the work required. Secondly, he noted that there was a high degree of turnover on the job. This provided a very good naturally occurring experiment to find out how learning takes place on the job. The question became: how did the claims processors who did stay on the job, actually come to learn the work that it took? And the answer was, that those who stayed somehow became members of the existing community of practice of claims processors.

We have found these dynamics at work in many settings - in schools as well as in the workplace. It turns out that who learns what (who learns to take tests or to do drugs or to solve math problems) depends very greatly on which social group kids hang out with (Eckert, 1989). These social groupings, while correlated in some ways with other societal indicators such as social class or ethnicity exhibit their own kind of social cohesion and form rich environments in which the world is made sense of.

This picture of emergent organization also helps close the gap between our understanding of individual learning and organizational learning. As Cook and Yanow (1993) have pointed out, most work on organizational leaning has explored how individuals learn in organizational contexts or has explored ways theories of individual learning can be applied to organizations as organizations. But as they point out and as is evident here, there is a sense in which inside our formal or authorized organizations, there is another kind of organization at work. This kind of organization is formed out of participation in work. It is more like a biological organism, if it is like any "thing" at all. This body, this multi-faceted participatory form, can change its practice; it is in this sense that organization can be said to learn.

Though it is very much outside the scope of this paper, it seems worth pointing out that we can also see that with the concept of emergent organization there is room for another picture of the individual to be formulated. Though it is natural to see authorized organization as made up of individuals, it is less obvious that this is the way to think of the relation of individuals and the emergent. Rather, as Wenger (in preparation) and Wenger and Eckert (personal communication) suggest, individual identity and what I am calling the emergent is in some ways mutually constitutive. They each define the other. (For a preliminary theoretical position from which to understand individual cognition in the emergent context, see Stucky, 1992).

In short, the claim is that in order to understand how work and learning take place, looking in the emergent realm will provide a more satisfying explanation than looking only at the authorized forms. If this claim turns out to be substantiated, we need to try to figure out how to enable the emergent realm thereby enhancing the ability of our organizations to learn. There are many areas in which this perspective can help improve the chances for learning: the design of work of organization (the authorized kind), and, importantly for our purposes in this paper, of technology.

4. Changing Views of Technology

Two views of technology have driven the development of information technology at different times in its short history. One sees technology to be a prosthetic device (a tool); the other sees technology as intelligent/cognitive agent. There is the same choice here: technology can be seen as a social prosthetic (extending our social reach in space and time) or technology can itself be seen as a social participant. However interesting (and important!) it may be to understand the implications of each of these views, there are several points that can be made without taking a stand on this.

Once one understands that there are two types of social order, and that one of them is home to the activities of work and learning, it becomes clear that to support that kind of social order - here called the emergent - is imperative. This is what is meant by the phrase "designing for participation". We will need to change fundamentally how we view the role of technology in the organization. Here are three major changes that need to be made:

Think participation, not information transfer

Learning requires that information received has to be made meaningful in the local context. This takes participation of people with each other. One place where meaning-making happens naturally is in the local communities of practice. Indeed, information, IRL hypothesizes, travels very quickly within an emergent organization, such as communities of practice, but much more slowly over the boundaries of them. This suggests that attention needs to be paid to who is on the receiving end and whether what is there, is socially cohesive. Without any cohesion, chances are that the information will fall on barren ground. This is particularly important for corporate training organizations which are increasingly moving toward distance learning. In brief, if technology is seen primarily as enabling information transfer, rather than enabling social participation, then the risk is that more and more will be sent out and less and less will get learned.

Design technology with work practice (not just work process) in mind

Clancey's paper (this volume) makes a strong case for designing work that reflects how work actually gets done in group settings, not just how it is described in the work process or how it is supposed to be done. Technology could be more effectively designed if it were designed with practice in mind too. However, making practice visible in the design process is not easy. It's not just a matter of asking people what they do. For one thing, most practice is tacit and when it is tacit in a group it takes additional work to figure out what is going on. Ethnographic research techniques hold great promise for helping us to do this because they make use of participant observation (Jordan and Henderson, 1994). Also, many of the methods designed to get at interaction (conversation analysis, interactive video analysis) are very helpful in getting at this sort of data.

Connections, connections, connections

People being able to connect with each other sustain participation. Indeed, if the communities of practice that sustain knowledge and learning are valuable to an organization, then there need to be ways of supporting it. And the authorized organization needs to not get in the way. On the technological front, it means not only connectivity but increased attention to programs that enable group work. That number crunching is so much better supported both by organizations and by the technology itself is a very strong indication that connecting people has not been the highest priority. That groupware is a relatively new phenomenon is likewise an indication of the lack of this mindset.

5. Conclusion

There is a social aspect of learning that is only now being understood fully. With that understanding in hand, we can see that our technologies can be designed to enhance learning within the organizational context and that our technologies can be designed for learnability as well. We at IRL suspect that the failure to appreciate the social nature of learning lies at the bottom of much of the early failure of technological innovation in the workplace and we are developing ways of approaching work and learning in a whole variety of settings with that in mind. Learning how to pay attention to that dimension of our organizations is the challenge that we have set for next steps in both research and development.

Acknowledgments

Thanks to the NATO workshop participants for engaging discussion; to my colleagues at the Institute for Research on Learning especially Bill Clancey, Gitti Jordan, Charlotte Linde and Etienne Wenger; to Scott Cook, Estee Solomon-Gray, and Alan Perey of the emerging community on Competence in Practice at Xerox PARC. Finally, thanks too to colleague Helga Wild for very helpful comments on the ideas as well as the construction of the paper at the very last minute. Thanks for support to the Xerox Foundation for its grant to the Institute for Research on Learning. Part of this work was made possible by a gift from the Systems Development Foundation to The Center for the Study of Language and Information at Stanford University.

References

Argyris, C., & Schon, D. (1978). Organizational learning. London: Addison-Wesley

Cook, S.D.N., & Yanow D. (1993). Culture and organizational learning. Journal of Management Inquiry, 2:4, 373-390

Eckert, P. (1989). Jocks and burnouts: social categories and identity in the high school. New York: Teachers College Press

Greenbaum, J., & Kyng. M. (Eds.) (1991). Design at work: cooperative design of computer systems. Hillsdale, NY: Lawrence Erlbaum Associates

Howard, R, & Haas, R.D. (1990). The learning imperative: managing people for continuous innovation. Cambridge, MA: Harvard Business School Press

Jordan, B. (1994). Ethnographic workplace studies and computer supported cooperative work. IRL Report No. IRL94-0026. Palo Alto, CA: Institute for Research on Learning

Jordan, B., & Henderson, A. (1994). Interaction analysis: foundations and practice. IRL Report No. IRL94-0027. Palo Alto, CA: Institute for Research on Learning

Lave, J., & Wenger, E. (1991). Situated learning: legitimate peripheral participation. Cambridge: Cambridge University Press

Norman, D.A., & Draper, S.W. (1986). User centered system design: New perspectives on human-computer interaction. Hillsdale, NJ: Lawrence Erlbaum Associates

Pava, C. (1983). Managing new office technology: An organizational strategy. Cambridge, MA: Harvard Business School Press.

Schuler, D., & Namioka, A. (Eds.) (1992). Participatory design: Perspectives on system design. Hillsdale, NJ: Lawrence Erlbaum Associates

Stucky, S.U. (1992). Situated cognition: A strong hypothesis. In: F.S. Engel, D.G. Bouwhuis, T. Bosser, & G. d'Ydewalle (Eds.) Cognitive modelling and interactive environments in language learning. NATO ASI Series F, Vol. 87, pp. 27-34. Berlin: Springer

Weick, K.E. (1979). The social psychology of organizing, London: Addison Wesley

Wenger, E. (in preparation). The production of meaningfulness

Wenger, E. (1990). Toward a theory of cultural transparency: elements of a social discourse of the visible and the Iinvisible. Unpublished Ph.D. dissertation, University of California, Irvine

Practice Cannot be Reduced to Theory: Knowledge, Representations, and Change in the Workplace

William J. Clancey

Institute for Research on Learning, 2550 Hanover Street, Palo Alto, CA 94304

Abstract. Changing views of the nature of human knowledge change how we design organizations, facilities, and technology to promote learning: Learning is not transfer; using a plan is not executing a program; explanation is not reciting rules from memory. Such rationalist views of knowledge inhibit change and stifle innovate uses of technology. Representations of work (plans, policies, procedures) and their meaning develop in work itself. Representations guide, but do not strictly control human behavior. Every perception and action involves new, nonlinguistic conceptualizations that reground organizational goals and values. This essay explores how the epistemology of situated cognition guides business redesign.

Keywords. Organizational learning, situated cognition, situated learning, rational agent, business process design, re-engineering, ethnography of work

1. Introduction

Change is always occurring in the workplace, often without a theoretical understanding of why one technology or organizational design works and another doesn't. Many programs for corporate change seek to impose policies, failing to acknowledge how people make sense of plans and standards in everyday work (Kling, 1991; Suchman, 1987; Wynn, 1991; Zuboff, 1988). On the other hand, focusing on the social nature of cognition may devalue how individual experiences shape new organizations and technologies.

This essay explores how new views of knowledge can be applied more systematically to organizational learning and use of technology. How can we help businesses create policies and standards that will guide creativity rather than stifle it? The central idea is to avoid equating knowledge with representations of knowledge. We want to shift from the "capture and disseminate" view of managing work to designing workplace processes that facilitate learning. In particular, explanations of delays and exceptions to standards can be exploited as opportunities for articulating new patterns and theories, instead of making an employee feel guilty for not rotely adhering to bureaucratic procedures.

Building on the idea of situated cognition, I broadly articulate the nature of knowledge, representations, and change in the workplace. I set the stage by giving examples of misconceptions about knowledge (Sect. 2) and summarize the tenets of the rationalist view of learning (Sect. 3). I illustrate the relevance of situated cognition to business process redesign by an example of a recent organizational change in a large corporation (Sect. 4). Then I elaborate on the relation of practice, plans, and justifications, giving examples from software engineering and expert systems (Sect. 5). I present a model of workplace change in terms of intersecting communities of practice, using as an example the development of a medical information system (Sect. 6). Finally, I examine how representational change occurs in the course of reflecting on experience and re-using models in different settings over time (Sect. 7).

2. Common Misconceptions

At the NATO workshop at which this paper was first presented, I heard several statements about knowledge, perception, information, learning, memory, representations, consistent with the dominant views in cognitive science, sometimes called the "rationalist position" (Winograd and Flores, 1986). Such views have been increasingly criticized since the mid-1980s (e.g., Bickhard and Terveen, in preparation; Clancey, in press b; Dreyfus and Dreyfus, 1986; Edelman, 1992; Gasser, 1991; Lave, 1988; Smith, 1991; Suchman, 1987), signaling a shift within cognitive science, sometimes called the "situated cognition position." The controversy is in fact quite old; most of the situated cognition ideas were clearly articulated by Dewey, Collingwood, Bartlett, Bateson, Ryle, among others (Clancey in press b). The point of this paper is to develop implications for organizational learning, rather than redevelop the situated cognition argument. Nevertheless, it may be useful to illustrate the situated cognition point of view, particularly since it is controversial to many participants of the workshop itself. I quote seven statements made during the workshop, explain them according to what the participant probably intended, and use the opportunity to summarize situated cognition ideas.

"They have knowledge but cannot act."

What is meant is that a person can have a relevant, but incomplete model in hand, know that it is incomplete, and still not know what to do. This remark suggests an apparent paradox: A person can "have knowledge" but not be able to use it. This equates knowledge with representations and acting with applying representations. Knowledge is not a thing but a capacity to interact (Dewey, 1938; Ryle, 1949). Knowledge is more analogous to energy than a substance. Knowledge and representations are not interchangeable: We can represent what someone knows, but the representations are not the knowledge itself (the map is not the territory) (Clancey, 1992b). Conversely, we can act without manipulating linguistic representations at all (but indeed, we often use maps and models to orient new behaviors).

"We have too much information."

What is meant is that we have too much data, more descriptions than we know what to do with. We lack abstractions, trend analyses, ordering, causal concepts, and theory. That is, we have too much data and not enough information. As Bateson (1972) would put it, we don't know what differences make a difference.

"Knowledge should be stored before it gets lost."

What is meant is that a representation should be recorded before it gets lost. In equating knowledge and representations, this remark views human memory as a place, so learning is like storing representations in a knowledge base (Rosenfield, 1988; Clancey, 1991b). This also suggests that using representations involves merely accessing them and applying them, leaving out the re-conceptualization that always occurs when we read and comprehend text (Tyler, 1978). The idea of getting "lost" focuses on how to retrieve representations when they are relevant. In contrast, human reminding is always relevant when it occurs (though the memory might not be useful) (Bartlett, 1932). In effect, human learning and modeling are conflated in this remark, distorting how knowledge bases are actually created and used.

"Expert systems don't learn from experience."

What is meant is that the program is not revising its representations (knowledge base) as a result of feedback from incorrect behavior. But the deficiency is more stark: People learn with every motion, for example, in the process of conceiving questions, in the process of formulating rules - within a problematic situation, not just afterwards. The problem is that expert system reasoning is timeless and disembodied. Programs generate questions by instantiating plans, which are effectively grammatical descriptions of interactional strategies. In computers, every perception and action is linked by deliberation, rather than automatically adapted by processes of re-categorization and re-coordination at a prelinguistic level, as in people (Dewey, 1896; Clancey, in press). The learning deficiency of computer programs exists because human knowledge, memory, and perception-action coordination have been inadequately characterized.

"Mental models are in the head of the user of the computer system."

What is meant is that the person may have an experience of visualizing something or silently saying something. In this sense the experience is internal. But such experiences are not in the head in the same sense that representations are in a computer system. Relative to scientific models in general and knowledge bases in particular, mental models are not models at all - they are experiences, not substances. We talk about "running a mental model" But what is the "it" that is being run? A metaphor is taken literally here, as if a mental model is something stored and applied like a computer program. Experiences of creating and interpreting representations must not be equated with manipulation of a symbolic calculus (as in qualitative models of expert systems (Clancey, 1989)).

"Learning is an individual, knowledge-based process and a social communication process."

What is meant is that learning has both individual and social aspects. To preserve an objective, given world, the remark suggests that an individual determines the truth and communicates it. But viewing learning as inherently individual misses how it occurs in activity, in cycles of perceiving and acting over time (Bamberger and Schön 1983). Articulating representations is inherently interactional, involving speaking or writing, or imagining. Furthermore, in social activity, individual perceptions and statements are occurring as part of coordinated discourse whose (social) forms shape goals, interests, and argumentation (Roschelle and Clancey, in press). Put another way, creating models inherently involves communication (always with oneself, often with others); communication is not something that occurs after the models are in hand.

"Perception is controlled by theory"

What is meant is that representations (maps, instructions, goal statements, beliefs, causal relations) are interpreted to influence where we look next and what we see. But interest, a sense of similarity or discord, and a sense of value in general precede creation of theoretical representations (Schön, 1979). Consider how you find items of interest in an museum before you know what you are looking at. Perception is always value-laden and occurs at a pre-linguistic level, grounded in prior coordinations and perceptions (Dewey, 1896). How descriptions (e.g., reading instructions) change behavior becomes an important area to reconsider; theory does not control human perceptions and behavior in the sense that rules and programs control a computer's data processing (Suchman, 1987).

3. The Rationalist Argument

Understanding the situated cognition perspective may be difficult at first because our concepts of knowledge, information, memory, perception, representations, etc. are intertwined. Change one and you must change them all. Figure 1 summarizes the rationalist line of argumentation, which I claim is implicit in the seven statements we just analyzed.

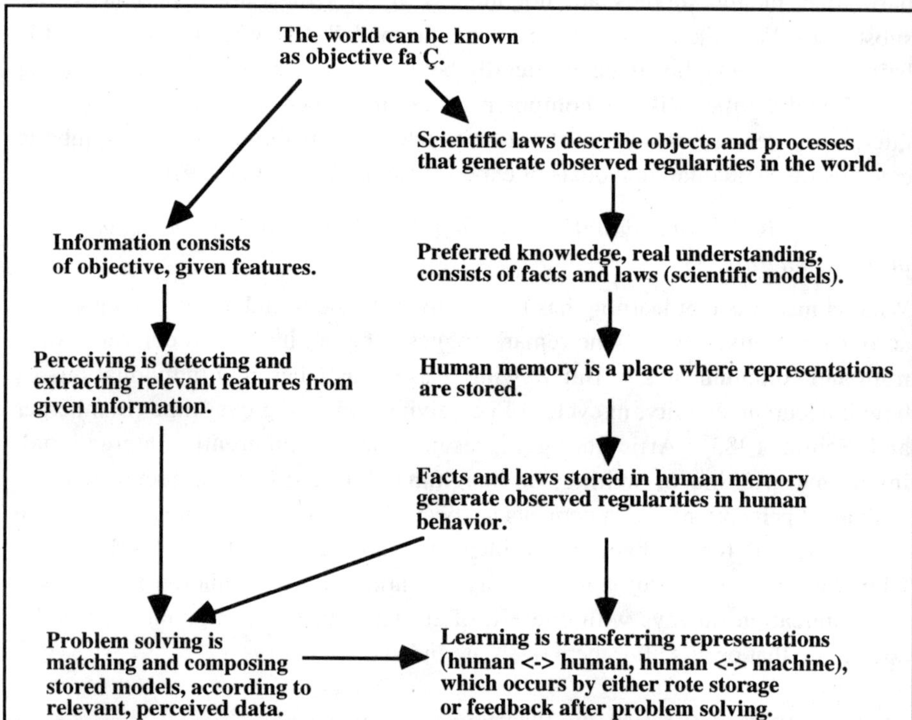

Fig. 1. Rationalist view of knowledge and representations

The strong relation between normal science and the idea that knowledge consists of representations is partly responsible for prevalent views of memory and learning in cognitive science. Changing this view sometimes appears to threaten the scientific enterprise itself, or it may appear contradictory, suggesting that all theories are relative and it doesn't matter what we say (Slezak, 1989; 1992). The key idea is to not equate phenomena being studied with models or equivalently, in the realm of human endeavor, not equate practice (what people do) with theory (what people say is true). The view that natural laws create regularities in the physical domain has been translated to the view that corporate policies and procedures create regularities in the workplace. Such causality between representations and behavior holds in computer systems, but not in human beings. Indeed, the presumption is that violation of business policies can only mean anarchy, when in practice anarchy is what would result if policies were literally followed.

It may appear at first strange to bring up issues of human memory in discourse about organizational learning. But we must root out fundamental assumptions about the nature of knowledge and representations if we are to understand how organizations actually generate and use policies, and what designs for organizations and technology would be more productive (Winograd and Flores, 1986; Ehn, 1988; Hirschorn, 1984; Kukla et al., in press; Nonaka, 1991). As the next example illustrates, change can occur without such theorizing, but at the risk of being ineffective or nonsystematic.

4. Organizational Design Example

Table 1 illustrates a shift that is occurring today in an American corporation. The shift is described in terms of changes to the organization:

- The changed organization is intended to build on internal diversity among existing employees, enhance teamwork and communication, promote learning through story telling, etc.
- The changed organization is intended to provide stable interactions with customers, so that client needs are "known deeply," while providing a single face or view of the corporation.

These changes appear innovative and intuitively appealing. But corporate management doesn't articulate the view of knowledge and learning implicit in this design. Hence the corporation is a poor position to compare alternatives and guide concomitant design of facilities and technologies in a principled way. Within the corporation, the lack of a principled theory might impede change by making this design appear arbitrary and potentially no better than any competing approach.

In effect, the corporation is shifting from a transfer view of learning to an interactive view. The old view suggests that knowledge resides in individuals and involves technical details; furthermore, the company's products and services are conceived internally, a priori, and disseminated to customers. One sales person serves many customers. Because sales people are specialists, each customer must potentially deal with many employees. Each sales person works for him or herself, seeking to move to the best geographic areas and to sell the most profitable products, in order to maximize personal income and promote his or her career.

In the new view, knowledge is about customer interactions. The view is inherently interpersonal and relational. Employees are generalists responsible for many products and services. Employees necessarily need to interact with each other, working as a team, to sell coherent systems specialized for customer needs. That is, each employee has an inherent need to get technical help from his or her colleagues. Communication within the corporation is encouraged and formalized by formation of functional workgroups and customer-liaison teams. Knowledge, as a capacity to interact with customers to define products and services, develops during interactions with customers (as opposed to being transferred in classes). The anticipated result is a more stable, single-face interaction with customers. Internally, there should be less churning of sales representatives into more lucrative markets, and promoting loyalty instead to the customer-team relation.

Table 1. Epistemological shift implicit in an organizational redesign

	Knowledge is about	**Knowledge resides in**	**Knowledge is developed by**
Individual View: Reify the individual employee, a constant player who moves in the corporation	Technical details of products and services (internal capacity)	Specialized employees (stored in individual heads)	Training given to individual
Interactional View: Reify company-customer relations as stable and responsive	Customer relations (interactive capacity)	Cross-functional team (manifest in activity)	Project activity of functional workgroup and teams

My point is not to argue specifically for this organization, although it does fit the theories of knowledge and learning I support. Rather, this example illustrates how explicating underlying principles of knowledge, memory, and learning can help us understand organizational alternatives. Furthermore, the example illustrates that theories of knowledge that might appear esoteric to business managers have direct implication for organizational design. In particular, the example illustrates that the jargon of today's organizations (e.g., "cross-functional workgroups") can be usefully related to situated cognition theories.

5. Practice, Plans, and Justification

In this section, I will consider in more depth the nature of plans from the perspective of software engineering and expert systems. We will examine how plans are created and used, and what happens when a plan is violated. In particular, we must change our understanding of how people justify their work. I will argue that organizational change is impaired if plans are not appropriately understood with respect to the inherent adaptiveness of human knowledge and activity. The rational view of stored knowledge and planned action inhibits organizational learning: It distorts descriptions of how work actually gets done, inhibiting reflection and preventing learning from past actions.

5.1 The Rational View of Work and Plans

Many software managers believe that the problem of software development is to plan, control, and predict software programming throughout the development and maintenance cycle (Fig. 2). Design is framed as an activity that is rational, that is, an activity whose *every step is a deliberate choice* among alternative actions, justified by appeal to scientifically-derived standards and metrics. The problem of improving software development is conceived as defining appropriate standards for the development process and metrics for evaluating deviation from the ideal. Learning is conceived as a feedback process by which deviations are translated into issues that guide reorganization, development of better tools, and new standards of behavior. From this perspective, organizational learning is often conceived as a process of continuous improvement and *dissemination*, by which successes and failures are continuously analyzed and formalized as methods and tools to be transferred throughout the company.

This rational view of work has merits, for it is true that progress depends to some extent on reflecting on past performance and formalizing procedures to improve future outcomes. But this view of human activity over emphasizes the planful nature of cognition. When people behave - for example when programmers write a line of code - they are not merely executing procedures or applying rules in the manner of a computer. Except when people are deliberately behaving like automatons, as in reciting a poem or reading a list of numbers, no human behavior has such a mechanical, template-driven nature. As we behave, the brain is not merely applying a stored plan, but is *constructing new ways of coordinating*

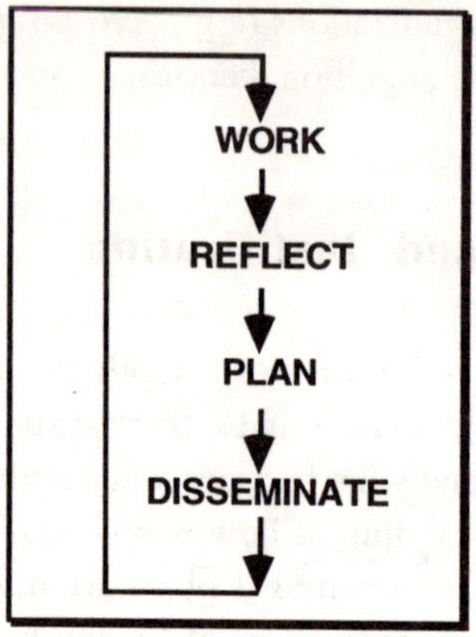

Fig. 2. Rational ordering of work, learning, and policies

what we see and do. Our moment-by-moment processes of interacting with our environment always result in new, adapted behaviors. Learning is occurring with every perception and movement.

In contrast, by the rationalist view intelligent behavior is always justifiable. Everything we do must have a reason, *which we have considered before acting*. This supposedly distinguishes us from animals, who don't represent their world and don't plan what to do. Supposedly people don't behave instinctively; they don't simply react. As civilized human beings, we are always supposed to follow rules, to behave according to scientific principles, to be methodical, to be lawful.

As an attitude for socially orienting behavior, this view has merits. But it is a poor description of how the human brain works. There are no stored plans, no procedures put away that we simply execute. Knowledge is not stored away like tools in a shed, which remain unchanged between uses. Learning is not just something that occurs on reflection, after we have acted. Every human action is unique, adapted, and hence process of learning. When I speak, I am not translating from an internal, hidden description of what I planned to say. I am conceiving. I may tell myself silently what I plan to say, but this silent telling is itself a form of reconceiving.

According to formal models of work, each step is viewed as manipulation of representations according to other representations (facts, rules, policies). In practice, each step involves action grounded on non-linguistic, value-oriented conceptualization and perceptual categorization. That is, action is not grounded on representations like written policies and standards. Following a recipe or implementing a corporate policy involves improvising and creating new causal stories and theories. Hence, rationalization is not reciting a plan that was followed in the manner a computer program is applied. Reflection is a time for reconceiving and theorizing, looking toward the future, not merely justifying with respect to rules.

5.2 The Symbolic View of Explanation

The example of "consultation program" explanation systems (Scott, et al. 1984, p. 338) reveals how our view of explanation becomes embodied in technologies and interactions between people and machines. Consider this typical example from the Mycin program, developed circa 1975 (Fig. 3).

```
**  How did you determine that the aerobicity
    of ORGANISM-1 is facultative?
The following were used:
RULE027 indicated there is strongly suggestive evidence
     (.8) that the aerobicity of ORGANISM-1 is
     facultative
RULE027 indicated there is weakly suggestive evidence
     (.2) that the aerobicity of ORGANISM-1 is
     anaerobic
Since this gave a cumulative CF of (.8) for
     facultative, and (.2) for anaerobic, it has been
     established that the aerobicity of ORGANISM-1 is
     facultative.
```

Fig. 3. Example of MYCIN's explanations (from Scott et al., 1984, p. 347)

Table 2 summarizes the symbolic view manifest in the design of Mycin's explanation system. The second column summarizes how we view the relation of knowledge and representations today.

As pointed out by Lave (Lave, 1988; Lave & Wenger, 1991), the view of knowledge as consisting of stable, stored representations suggests a transfer view of teaching, advice giving, and explanation. Indeed, in the heyday of the Knowledge Systems Laboratory at Stanford, some of us wrote a manifesto summarizing our discovery that creating computer programs, giving advice to people in the workplace, and teaching students could all be accomplished by transferring representations from computer memory to user memory to student memory (Barr, Bennett, & Clancey, 1979). Of course, nobody in the 1970s would have argued that MYCIN's advice was intended to be followed by rote. But aside from providing access to the program's deductions - an important innovation at the time - nothing in the design of the system related to the user's sense-making capabilities (see Clancey, in press b, for further discussion).

Sometimes the deficiency of the transfer view of learning is described in terms of the context dependence of knowledge. But the issue is subtle. AI researchers know that knowledge base representations are not context independent. But they generally believe that the contexts to which knowledge base rules apply can be represented as conditions (features of the environment) that are themselves non-problematically given (objective) and stable. Indeed, the basic components of expert knowledge are called "situation-action rules." Similarly, the justifications for rules themselves are assumed to rest on a stable, objective, representational bedrock (e.g., see Wallace and Shortliffe, 1984). In this way, all information about the world, all knowledge of the world, and all justifications for actions rest on representations - what I have characterized as "representational flatland" (Clancey, 1991a). Hence, all action can be effectively controlled by "acquiring" the right representations. To understand what is wrong with this position, we need to consider in more detail the nature of human creativity and emergent aspects of behavior.

Table 2. Shift in view of explanation

	Exclusively symbolic view of knowledge	**Situated Cognition View**
Human knowledge	Facts and rules	Capability to interact; inherently unformalized; cannot be inventoried, grounded in non-linguistic perceptual categorizations, conceptualizations, and coordinations.
Knowledge base	Equivalent to human knowledge	A qualitative model of causal, temporal, spatial and subtype relations between objects and events in some domain of inquiry.
Explanation	Reciting facts and rules previously used to solve problems.	Reperceiving, reconceiving situations, theorizing freshly in a way that relates previously articulated models to non-theoretically controlled human actions.
	Looks backward to justify what happened.	Looks forward to improve future interactions.
Use of a consultation program	A mechanism to which people supply data and receive advice to be followed by rote.	A tool that facilitates perception of relevant data, reinterpretation of data categories, and reconception of previously articulated rules in each unique situation.

5.3 Emergent Effects: How Representations are Created and Used

A dynamic, *interactional view of behavior* can be difficult to understand, because the view that knowledge is stored and that speaking is a process of translating ideas into words is so ingrained in our society. A better psychological understanding requires turning the rationalist view inside out.

Representations - such as plans of what to do and reflections on our own behavior or what other people have said - are created in our activity, as we speak. Representations do not exist until we say something, or write something down, or imagine something in our head. Representations are created in the course of activity; they are not part of a hidden mechanism that drives our moving and speaking and seeing. Representations play an essential role in human behavior, but only in *cycles of behavior*, as we perceive what someone said, and say something else in return. As we perceive representations, we reorient our activity: We look somewhere else, we reperceive our partially completed work, and we organize our activity in a new way (Bamberger and Schön, 1983).

An architect designing a building, a scientist writing an article, and a computer programmer all have in common the broad pattern of putting something into the environment - a line, a statement - and periodically stepping back to reconceive the nature of the whole. We are always interacting, putting some *thing* into the environment, reperceiving what we have done, and replanning (and this interacting can go on privately, in our imagination). Plans orient this process, but how the plans are interpreted is itself an interactive, non-predictable process. We cannot fully predict what the completed work product will look like because it is not generated by mechanically applying formulas. The architect, the writer, and the programmer don't know what they will produce until after they put their representations down on paper. Periodically, they will reperceive the whole and observe unanticipated effects. They observe trends, flows, and directions that they didn't expect. New opportunities arise for reshaping the project. New plans are formulated to account for unanticipated difficulties or interesting new directions. Every behavior is an improvisation.

How can this interactional view improve organizational learning? We must begin by realizing that holding people accountable to representations, such as standard plans and procedures, must *take into account how people use representations*. In particular, we must remember that the very essence of human behavior is novelty, flexibility, and adjustment. We are always, at the root level, innovating. Reperceiving, readjusting, and replanning are integral to intelligent behavior.

Reflecting is necessary because we are creating something new when we design a building, an article, or a computer program. Each statement is a creative act. We must regularly stop and look at what we are doing. But also interactions between the parts can produce unanticipated effects. Both positive and negative properties can emerge, which are perceived as we reflect on what we have done so far. The building may become too large for its site, the article may become rambling and unevenly developed, the computer program may bog down against resource limits. Allowing for refocusing and replanning is therefore necessary for several reasons: creating something new can't be predicted, we want to benefit from serendipitous effects, and we must cope with undesirable interactions.

5.4 Coping with Rationalist Standards in a Chaotic World

Where does this leave the programmer and software manager? They have been told that their activity must be planned, predictable, and controlled. But unless they are engaged in some rote, mechanical process, they cannot strictly predict or control what they will create. They won't know what they have or even what they can do until something of substance is on paper and they can reflect on where they are headed.

Pity the manager and worker who are held accountable to plans, but are expected to be creative and flexible. Programming manuals describe a highly-structured process full of documentation, obligatory liaisons between teams, approval meetings, checkpoints, and product specifications. The organization, in putting out these standards and procedures, seeks predictability and uniformity. Yet, in all this planning and proceduralization, there is no sense of what the manager or programmer should do first and what after that. How does the programmer know when he or she is making progress? New interactions are continuously being discovered, new opportunities and difficulties are emerging. We can see the date and where we stand in the schedule. But who knows what shortcuts or detours will occur tomorrow?

Human behavior is inherently *ad hoc*, inventive, and unique. But Western culture has biased us to believe that intelligent behavior is planned and under control. Every action is supposedly governed by representations. As adults we are supposed to be conscious of why we are doing what we are doing; we must have reasons. All of us, as members of this society, are influenced by this view of human nature. Unfortunately, since human behavior is not governed by representations in this manner, we find ourselves in a double bind.

Examining our own behavior, we find that we are simply doing things. We are always at some level being intuitive. Again, at some point, even the plans we generate can't be accounted for or strictly justified. But we must respect organizational authority, through our tacit agreement as participants of this society to be rational, to follow the regulations. We are caught: We don't reveal to other people or even to ourselves the true nature of how we do our work. Instead, we put up good appearances. We learn how to justify our behavior, and such representations of what we have done are always *post hoc*. Like mathematicians, we learn to tidy up our reports into nicely formatted proofs. It always appears as if we did things according to the rules: We made observations, we gathered evidence, we deliberated, we drew conclusions. We followed a plan, we were scientific. We were intelligent. We were good members of the organization.

Endorsing this rationalist view of human behavior inhibits learning. We don't reveal to ourselves or others how our work actually gets done. Indeed, working within the rationalist conception of science, we do not even have the concepts for describing how our mind works and how ideas develop in a community. We feel obligated to say that we followed the template, that our knowledge really did consist of schemas and rules, and we did what we were supposed to do. Rather than reflecting and seeing how we are really interacting with people and our environment, we (usually unknowingly) say what people want to hear: We endorse not only the rationalist view of knowledge, but the rationalist genre of justification. We get good at telling the rationalist story.

With this vested interest in a command and control mentality, neither the workers nor the managers are able to see how work really gets done. Indeed, we may be momentarily stopped in our tracks when "called to task" for not following a plan. Confusion becomes acute when we realize that no thing, no representation, underlies our behavior. There is ultimately no deeper plan or better understanding to refer to. Even when we realize that influences are beyond our control, we make up names like "tacit knowledge", as if to say that there is something written down and stored inside that properly justified our behavior, we just weren't thinking about it explicitly.

It takes courage to put forth the interactional view, to realize that there are no well-thought-out rules or tidy laws inside that make us the rational machines we hoped to be. We must accept that story-telling is all we have ever done or ever could do in justifying our behavior. Indeed, our integrity is called into question as we are called to justify ourselves according to the rationalist standard and cannot find the plans or schemas inside. There are no deeper facts to appeal to, we simply behaved.

5.5 Giving People the Right to Reflect

Improving learning, particularly within business, requires granting people the uniqueness and openness of experience. People must have the freedom, when appropriate, of saying, "I am not a machine, I am always doing something new. I am creating. I can't tell you what I will produce for sure until it is done. I can't say when the whole project will be done because I don't know what it will look like or how it will interact with other parts until it is well along." We must *give people a different view of the self and what a justification is*:

- What we do is always coming into being in our interactions, it can't be strictly predicted or controlled.
- Representations, such as business standards, orient, but don't control human behavior in the manner of computer programs.
- Reflection is itself a creative interaction, a reperception of emergent properties of the whole and interactions beyond the context of our immediate focus.

We inhibit learning when we view people as machine-like, suggesting that they follow instructions like a machine, and force them to justify behavior exclusively in terms of previously articulated plans. In one sense we have a poor view of what standards are for, how people use them, and how to relate standards to behavior. We have a poor view of accountability. We don't properly evaluate human work because we have obscured the inherent unplanfulness and inventiveness of every action.

Of course, in practice explanation does focus on understanding the relation of what we said we would do and what happened. What I am suggesting is an attitude and an understanding between employees and managers that *acknowledges the inevitability of diversions and discoveries*, and seeks to work with them, rather than apologizing or obscuring what is happening.

We must give people the right to acknowledge that they are constantly reperceiving and reorganizing their work. We must emphasize their crucial power to resee and reappraise what they have done so far. We must give them the right to be assessors of their own work, giving them time and tools to reflect and reconceive what they are accomplishing. We will do this not just because it is good, democratic policy, but because it is inherently the nature of human activity. Without the right to reassess and redirect, we are flying blind or idly, mechanically, doing what someone else conceived before current possibilities and problems emerged.

People do not simply plan and do. They continuously adjust and invent. Managing this process means *managing learning*, not managing application of a plan. Managing a creative process means *orienting the process of inventing new orientations*.

5.6 Disseminating Practice, not Theories

We do not deny the value of a technological view of work and transfer view of learning. Technological tools and methods exist. Certainly we desire other groups to know about them. But this means more than giving lectures, printing more manuals, or even seeding organizations with experienced players.

We must begin by better understanding organizational successes and failures. We must study what people do, understanding better how they interact with each other, their tools, and their physical environment. We must fundamentally respect the unplanned nature of behavior, the informal network in the community, and the novel, local contributions to standard company procedures. By looking more carefully, we can collect new ideas, respecting improvements, but especially respecting the climate of collaboration and exploration that allowed new contributions and adjustments to take hold and to spread.

We must view knowledge not as a storehouse of facts and procedures that can be inventoried, but as a capacity to interact, to reflect, to innovate. We must shift our view of assessment of productivity from "What have you done for me lately?" to "What surprises might you produce for me tomorrow?" Any investor knows that past performance is no guarantee of future results. The manager must be concerned not only with today's numbers, but tomorrow's promise. This requires sensitivity to trends and orientation, to what the organization is becoming.

Practice, what people actually do and how their communities evolve, is not reducible to descriptions of what they believe or what they do. The patterns we find are always a step removed from the continuously adapting and reconforming interactions. We might characterize a group's activity in terms of a certain vocabulary, a certain orientation, or a plan. But in the very next statement, the next moment of stopping to see what has been done so far, any member of the group is likely to surprise us, producing a new conceptualization, a new way of viewing the goals, and a new value for appraising work. Indeed, every theoretical description of the group is not just a snapshot, but a reflection of an observer's own interactions with the group, and the group itself is a moving target. What is needed is not better or more accurate theories, but an appreciation for the dynamics of the group's ongoing development, an understanding of where new contributions are manifest, and how they change the group's behavior.

6. Change

In this section, I consider broadly the implications of situated cognition for deliberately changing organizations and technology. The idea of participating in a workplace, as researchers, is considered from several perspectives, including the notions of participant observation and participatory design. Studying an ongoing software development effort, I illustrate how change can be understood in terms of interacting communities of practice. In this example, people with diverse roles and goals are brought together to construct formal descriptions of work that will be used as tools in the workplace. In section 7, I consider what this reveals about tools and representational change in general.

6.1 Two Spaces, two Commitments

The laboratory research of computer scientists can be characterized as exploring what computers can potentially do. Computer scientists involved in action-oriented research consider the problem of designing and changing social organizations, facilities, and technology in everyday life (Fig. 4) (Brown, 1991; Clancey, in press c). This effort naturally requires collaboration between people with diverse backgrounds, including workers, managers, graphic designers, organizational specialists, psychologists, programmers, facilities specialists, etc. Research focuses on inventing design processes that facilitate fruitful collaboration (Greenbaum & Kyng, 1991). Hence, we need to understand better how learning occurs in such interactions. Consistent with the situated cognition view of knowledge, note that our theories about work, training, and computer tools will be grounded in our experience in changing everyday life (Ehn, 1988; Bannon, 1991). Our knowledge about learning and tools for learning will first of all be an ability to participate in redesign communities, and only secondarily theories about that redesign (although we may easily create theories without having relevant knowledge and experience).

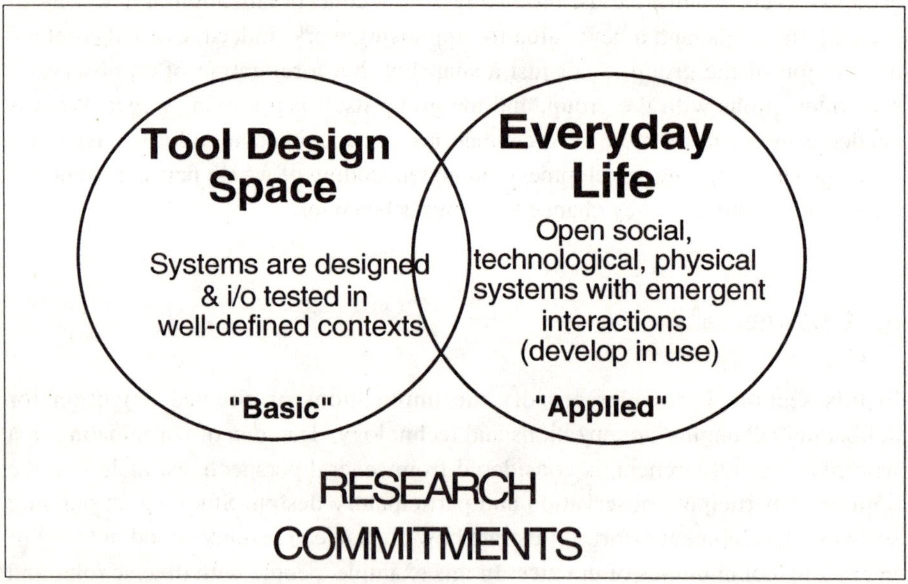

Fig. 4. "Applied" research concerns how to design systems that are part of everyday life

An overriding theme of our focus on everyday life is making social processes visible in the workplace. This concern appears at two levels: in our models of business processes and in our metatheories of how the modeling and change processes occur. At each of these two levels, we advance our understanding by analyzing current models and contrasting them with what actually occurs in practice. We promote social and technological designs in which we find that collaboration and learning have been successful in the past. In this perspective, we are taking the idea of "participant observation" from ethnography and applying it to business redesign. We are recognizing that understanding how successful change occurs is a research problem, and it is just as basic as what occurs in the laboratory (hence the use of quote marks in Fig. 4). To develop this idea further, I first present my perspective on the ethnographic approach, and then illustrate what we have learned about the social processes of "knowledge acquisition" in constructing expert system software.

6.2 The Ethnographic Method of Reflection

Ethnography is a method of observation based on participating in a community, so one can experience first-hand the habits, new ways of seeing, and evolving sense of values of the group (Jordan, 1992). Ethnography brings us closer to experience, to practice, so we can partially avoid the conflict between what people do and what they say they do.

Again, the disparity between saying and doing is not caused by people's inability to remember what representations they used when acting. It makes no sense to speak of accuracy of recall, *per se*, during rationalization because there are no stored recipes, rules, or past decisions to appeal to. What is usually meant is the adequacy of the model of what occurred. Indeed, people can lie, and they may forget what representations influenced their behavior (e.g., what directions they were given). But we must remember that every human action is, at a certain level, new and direct. Saying what we did is a primary activity of theorizing, *describing for the first time*, which is inherently apart from the activity it is about. As Schön (1979) tells us, "To read the later model back onto the beginning of the process would be to engage in a kind of historical revisionism."

In ethnographic observation, we aren't seeking better descriptions, and hence procedures that we want everyone in the company to rotely follow. Believing that work can be managed better by better describing the work and writing more accurate job descriptions obscures the nature of the work, how such descriptions are actually used, and indeed the kind of work process we should be aiming to identify, reward, and disseminate. We seek instead to appreciate the dynamics by which a given community is continuously reperceiving what they are doing, theorizing about what they have accomplished, and planning what they might do next. We learn to see that the workplace is not a strictly organized and controlled place. Neither knowledge nor *work* is strictly decomposable into pieces. Components of work products are not simply produced in a linear fashion and arranged on a table and assembled. Pieces of work (e.g., software modules) are coming into being along with the concepts for describing what the group is doing, values for appraising work, and theories that justify what is produced. We focus on how a community reflects, how it adjusts to changing conditions, how it develops day by day. That is, *we focus on how the community learns.*

In an ethnographic study, we might begin by considering how a community of practice interprets standard procedures - how their attitudes, experience, and sense of belonging to a larger community affect what they see in the procedures and how these perceptions organize their future work. Second, we might consider how the complex of interactions in the social, physical, and information processing environments allows for reflection, how individual reflections are composed into a group assessment of progress, and then again, how these representations influence their future work. Third, we consider how the group's ongoing assessments and restructuring interact with similar processes in other communities with which this group interacts. And then again, we consider how these intergroup interactions become manifest in further reflections and reorganizations of future work.

6.3 Example: Knowledge Acquisition as a Social Process

As an example of how we are making social processes of change visible, I present what we have learned about the process of constructing models to be used in a tool for medical clinics[1]. Our study focuses on the representation of a vocabulary of patient symptoms, diagnoses, and treatments that form the basis of an online patient database. In subsequent sections, I draw on this example to again contrast situated cognition with rational theories of change.

Our initial studies of the modeling process indicate that a multidisciplinary group draws from many sources and projects different uses of their model of work (Fig. 5). A clinician may be viewing the model through the advantages for better patient care; a public policy expert may be viewing the model as a means to standardize patient care; a research MD may be focusing on gathering more accurate data in order to test alternative treatments; and a programmer may be focusing on the completeness and consistency of a patient data-base. The negotiation of design sessions is therefore a complex interplay of previous and imagined future experiences, of conflicting constraints drawn from different communities of practice, of competing evidence, goals, and future uses.

---

[1]This study is the work of Kären Wieckert, in collaboration with the T-HELPER project at Stanford University, in the Department of Medical Computer Science.

This model of how representations are created and used can be compared first of all to the original "knowledge acquisition" process of interviewing an expert and codifying his or her knowledge. Of course, AI researchers realized early that experts have different opinions; it was even proposed as early as 1975 to use MYCIN as a means of bringing national experts together and developing a unified model of infectious disease diagnosis and therapy. However, this diagram illustrates a more profound cause of disagreement: *The participants may agree on the need to develop a common language and models, but they have different uses in mind.* It's not just a matter of having different theories (as if theories are just individual opinions), but theories that develop and function within different communities of practice. Again, models are not be understood with respect to a single, objectively correct view, but within the social processes in which they are created and used. This requires a new analysis of what the participants in this process share: if they share a "common goal" it is at the level of spending time together to develop a representation that has multiple uses and is grounded (perhaps) in some empirical phenomenon (e.g., a population of patients) or tied to a common legal standard.

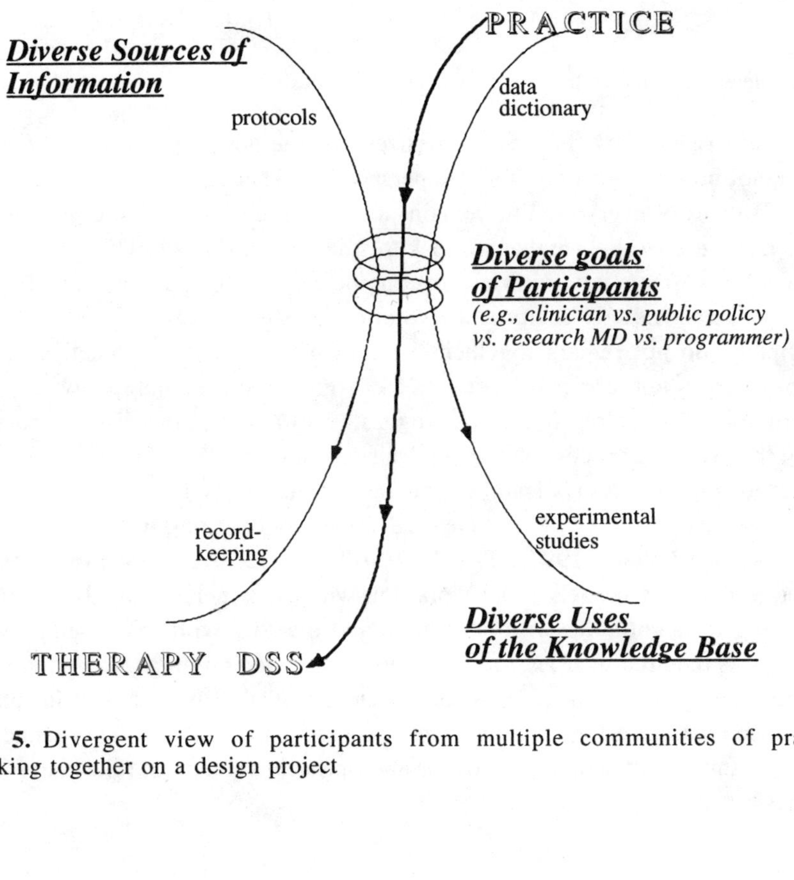

Fig. 5. Divergent view of participants from multiple communities of practice working together on a design project

6.4 The Interplay of Practice and Theory, Workers and Researchers

Our current perspective on representation creation and use is distinctly non-linear. It should be contrasted with the idea that a project has a start point, shared involvement, and a common end point (Fig. 6). For example, the idealistic view of participatory design is that of lockstep advancement, with everyone marching together toward a common goal - probably an idea that only those who have never worked in a group on a large project would imagine.

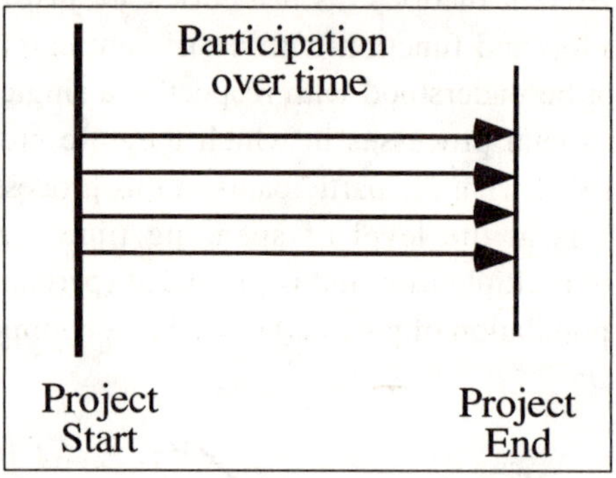

Fig. 6. Idealized, lock-step view of participatory design

The divergent view (Fig. 5) emphasizes that the design process is not linear and cannot be captured from a single perspective. That is, the practice of design itself, as a part of everyday life, remains as open to change and interpretation as the world of the workplace that we seek to study and influence. This points out a weakness in a simple iterative view, cycling between design and practice (Fig. 7), what has been called "reciprocal evolution" (Allen, 1991). In fact, project participants do not remain together in a single community of practice during design cycles, nor are influences and design effects contained within one community, as a cycling line might suggest. Furthermore, theorizing occurs as part of the everyday process of making decisions in work.

The idea of continuous learning emphasizes that the redesign process occurs within everyday work. Goals, policies and values are changing within every act in the workplace (Wynn, 1991). That is, learning is occurring not just when we reflect, articulating models about work, but with every perception about what is happening in negotiating with customers and meeting with fellow employees. Judgment is required in at least minor ways even in determining that a situation is "routine." Redesign is in some sense occurring all the time. A key insight of workplace redesign efforts is putting into practice a process by which change that occurs as a matter of course everyday is appropriately valued and related to formal policies.

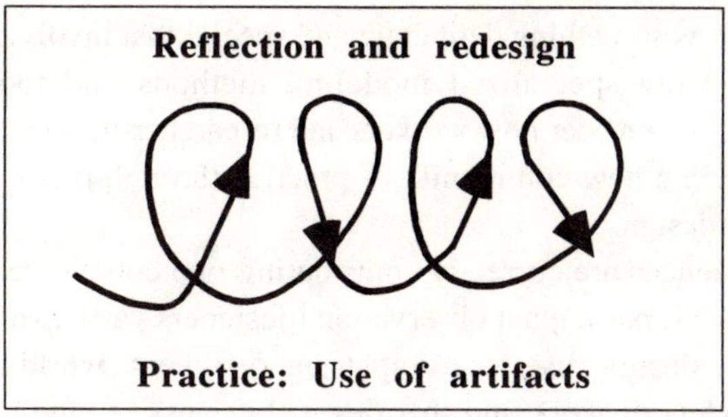

Fig. 7. Iterative model of reciprocal evolution, cycling between use of models and tools in practice and theorization with new design

A better view of the design process combines the "influences and effects" diagram (Fig. 5) with the iterative process of reciprocal evolution. This revised diagram (Fig. 8) shows time moving from left to right, with new theoretical influences coming in over time from outside (arrows coming in from above), as well as from the workplace (arrows pointing upwards). In addition, effects diverge from initial intentions, including serendipitous uses (arrows coming from below) and effects in outside communities (arrows pointing downwards).

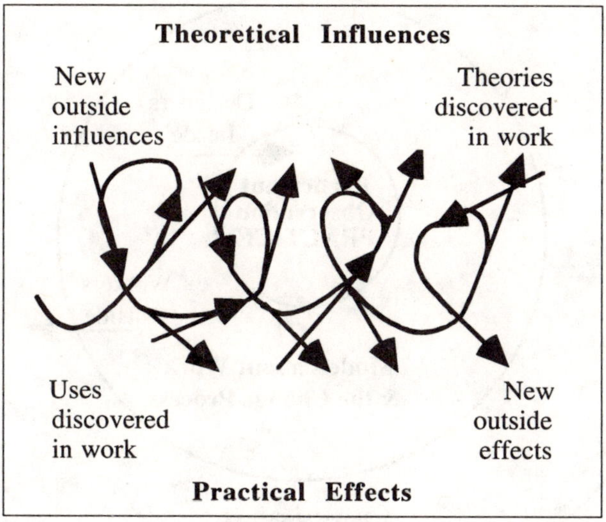

Fig. 8. Process map of participatory design involving multiple communities of practice

Stepping back, we need to acknowledge that there are different gradations of change. At a certain grainsize, we are not merely "doing our work," but changing the business. Despite acknowledging that change occurs all the time, we view business redesign as something deliberate and special that involves other players in new activities, using specialized modeling methods and tools. One way to understand this is to consider how workers and researcher-designers are deliberately brought together in a new community of practice through participant observation and participatory design.

The social sciences are currently contributing two complementary ideas about the redesign process: participant observation (designers participate in users' world) and participatory design (users participate in designers' world). The practice of constructing models of work and theories of business re-engineering (Hammer, 1990) mediates these two worlds of practice and redesign (Fig. 9). Crucially, there are two practices: the practice of the workplace we seek to change and the practice of redesign. The redesign practice is a metapractice in the sense that it seeks to describe the practice of work and change it. So we want to acknowledge that the

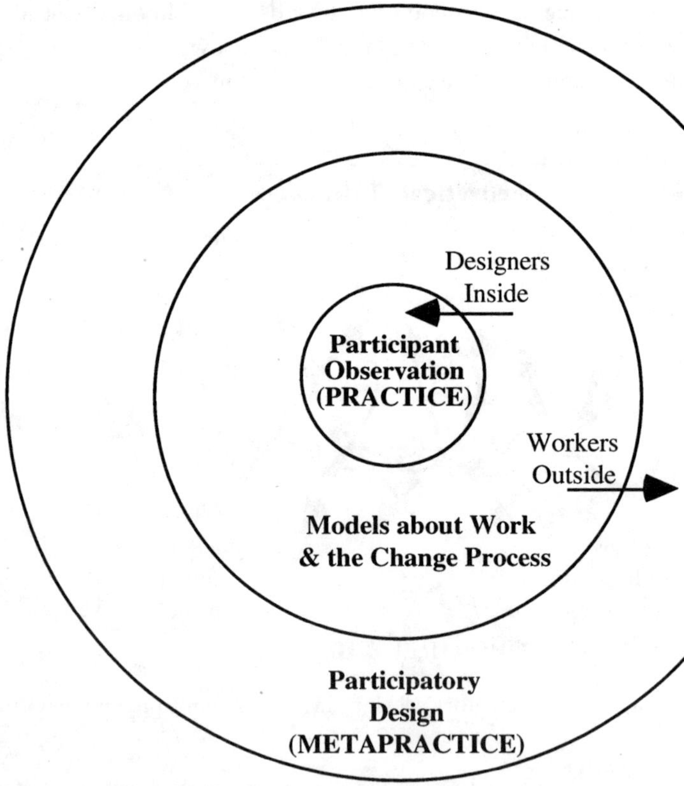

Fig. 9. Relation of participatory design to participant observation

worlds of workers and designers are different, and there is a need to develop new processes for working together, so that the workplace is included in the world of redesign.

6.5 Individuals in Changing Communities of Practice

To recap, we need to better understand work so we can understand how to bring about effective change. This understanding involves better understanding the nature and role of models, how identity is constructed through participation in communities of practice (Wenger in preparation), and how emergent understandings and interpersonal interactions can be encouraged and recognized. A key idea is that individuals contribute to group efforts by reflecting on what they say and do in *other communities* in which they participate. The experience of difference - which starts with a feeling of discord - is the source for articulating theories about past activities, as well as understanding emergent goals and values in the new group.

The example of vocabulary development from the T-Helper project, suggesting the divergent view of representational change (Fig. 5), is broadly applicable. First, we see that it fits different "capture and disseminate" activities:

- **standardization across communities of practice** (e.g., software reuse within divisions of a company, between research and manufacturing, and internationally);
- **curriculum design for school and workplace training** (e.g., collaboration between scientists, teachers, and educational researchers);
- **preserving corporate memory and intellectual property** (e.g., corporate efforts to "capture expertise" in materials and designs before people retire).

In short, this divergent picture describes what happens whenever an effort is made to bring together people with diverse roles to create a common vocabulary and models (e.g., curricula, knowledge bases, policies and procedures, national standards). Although project participants may create a single model as a result of a collaboration, the use of the model and its meaning will be prone to divergent interpretations in different communities of practice over time. Indeed, the modelling effort focuses articulation of past experience, as participants come to understand how constraints from different domains interact and generate new ideas for synergistic interactions between different communities. Ideally, such reflections will be facilitated by researcher-designers in a business re-engineering effort.

Furthermore, the diagrams (Figs. 5 and 7) show what happens *within* the workplace itself. Besides describing what is happening in Stanford's laboratories as clinicians, research MDs, and computer scientists collaborate, the diagrams show what is happening inside the medical clinic itself between nurses, physicians, interns, and patients. That is, it is helpful to realize that the home turf of a single participant in a multidisciplinary team appears to him or her like another multidisciplinary team with divergent experiences and divergent future-oriented intentions.

The thrust of Fig. 5 is to view individuals as the locus of overlapping interests, which have their origins and ultimate meanings in participation in diverse communities of practice. Consequently, the researcher community itself is multivocal, with diverse influences and effects. For example, we are well aware that the Institute for Research on Learning (IRL) is a multidisciplinary group (Clancey, 1992a). Each IRL researcher is always participating in many communities of practice (in academia, schools, and businesses). New ideas at IRL develop from individual experiences of "reseeing" their previous experiences and future activities through the evolving language and world models of the IRL community. Thus, another way to express the divergent view of Fig. 5. is that a community of practice involves individuals interacting together, but *each participant looks backwards and forwards into the realm of personal experience.*

To continue the application of this idea to IRL, the effects we produce will occur in different communities *outside*, through the articulations and participation of IRL individuals. Even this picture gets complicated when we realize that an IRL anthropologist and computer scientist may attend a single-discipline workshop together, changing the experience of everyone all around.

Finally, the diagram of divergent sources and effects is useful for understanding strategic partnerships between research and business organizations. Here the inner ring of interaction (the loops at the center of Fig. 5) can involve workshops, research papers, and focused redesign projects. In such partnerships, we share the experience of a need for change; we negotiate what tools and methodologies will be used during our activities together. We may share a vision for what kind of artifacts and social organizations we wish to see develop from our work together. But always the redesign process will be facilitated by becoming aware of our diverse past experiences, values, and imagined results of our joint work. This analysis shows why scenario depiction is especially valuable. Story telling is seen as a way of grounding design ideas in individual experiences – to convey what life is like in another place as well as the impact of social and technological change.

7. Representational Tools

To illustrate some additional, subtle aspects about representation creation and use, consider how Fig. 5 developed. I drew this first as a way of "holding in place" the many players and constraints that Kären Weickert observed in her study of T-Helper design sessions. (This idea of "holding in place" relations via an arrangement of physical things is developed by Jeanne Bamberger (1991) in her analysis of children's representations of tunes.) I later used the figure for a research project presentation at IRL. In the course of that meeting, a video about reciprocal evolution was shown (Fig. 7). The curving lines of Fig. 7 reminded me of my own Fig. 5, and I realized that they could be overlaid (Fig. 8), and then contrasted with other views (Fig. 6). In this respect, the ideas of Fig. 5 were developed as commentary on the shortcomings of other depictions of the design process. A participant at the IRL research presentation commented that Fig. 8 was a "process map," leading me to understand that it showed Fig. 5 over time. That is, the diverse sources and purposes of a representation aren't all known at once, but articulated over time, as the representational artifact develops and is used. The process of representational change is shown here *as an inherently social feedback process*. Notably, the evolving system is open, as new constraints are imposed from above by outside communities of practice (e.g., by funding or political agencies) and opportunities for new uses below become known.

In effect, the diagram serves as a way for focusing and reperceiving relations that were not originally represented in the diagram itself. The idea of outside communities of practice, serendipitous use, and feedback were added later. By bringing in the reciprocal evolution idea, I realized that the IRL project in which that idea was developed could be described by Fig. 5 and 8. This suggested that I apply Fig. 5 to other IRL projects, yielding analogies with curriculum development, software reuse, and corporate memory. After presenting an earlier, transparency version of this figure at an AI conference, I decided to drop an inner caption ("A knowledge base is not objective"), because it seemed to be a side issue for that audience. This reminding, adaptation, and theorization process subsumes diverse ideas and projects under a common way of seeing (Fig. 5), thus generalizing my understanding of the competing forces and processes involved in many settings. I came to see each project I reconsidered as involving multidisciplinary participants with diverse goals, but nevertheless constituting a community of practice. Thus the idea of a community of practice was changed. A similar experience occurred in the creation of Fig. 10.

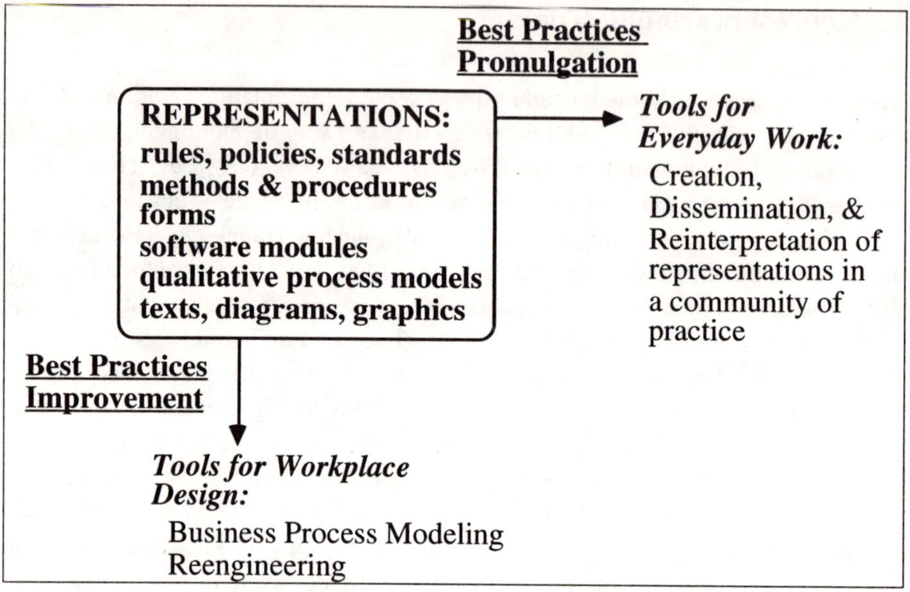

Fig. 10. Representations as tools for inquiry

This figure was conceived at a whiteboard during another IRL presentation for a corporate client. I wanted to convey our interest in how representations were created and used, showing that we were focusing on two domains of concern: tools for workers and tools for work redesign. The list of representations was generated on the fly, to illustrate how broadly we view the idea. The labels "best practices improvement" and "promulgation" actually came from my listener, as he described what he heard me saying. Schön (1987) calls this process of reconceptualization "backtalk" when it occurs within an individual's experience in reperceiving a previously created representation; it can obviously occur as well between individuals.

This example illustrates again how a picture can be used to "hold in place" different views and different activities. The "best practices" titles and the figure caption were added the next time I presented the figure, at a research conference. In effect, multiple interpretations and purposes are "held in place" by the representation itself as it is moved between and used in diverse settings. The cumulative, improvisatory and serendipitous process by which this representation developed are typical, illustrating how concepts develop in conversations that have different purposes and occur in different places over time. Representation "reuse" is therefore a process of adding elements and "reseeing" other activities, so the evolving model subsumes different sources and purposes -precisely what Fig. 5 aims to show.

In all of these diagram modifications and reconceptualizations, I was aware that I was improving prevalent views about the concept of "community of practice", multidisciplinary research groups, and how knowledge acquisition proceeds. That is, the driving force was not just a generalization which subsumed more and more territory, but a sense of being a deeper, more coherent descriptor of complex interactions occurring in any given domain. Each step is sensed as being constructive, in that something more generally useful is being developed. This theme is captured by the title of Engeström's (1987) dissertation, *Learning by Expanding*.

8. Conclusion

In *promoting* organizational learning, we are designing interactions - between people, machines, and their environment. We are developing a socio-technical system of social, physical, and information processing environments. Just as the architect, the writer, and the programmer cannot strictly predict or plan what they will produce, the organizational theorist and manager cannot strictly predict what workplace designs will evolve. Management and all participants must accept the fact that every intervention, every restructuring, every new tool for improved performance, is a naturally occurring experiment. The ongoing process of perceiving how we are doing, finding patterns, and justifying developments through new theories, is a *process of inquiry and discovery*. The manager, like the architect, the writer, and the programmer, must be open to serendipity and unashamed by unexpected interactions. The developing organization is like the emerging building design, the journal article, or the complex software system -there will always be surprises. What happens in practice is always one step ahead of theory.

Acknowledgments

I am indebted to my colleagues at the Institute for Research on Learning who have introduced and discussed many of these ideas: Kären Wieckert, Etienne Wenger, Susan Stucky, Jeremy Roschelle, Brigitte Jordan, Charlotte Linde, Ted Kahn, Peter Henschel, and Penny Eckert. Funding for this research is partially provided by grants from the Xerox Foundation and the NSF Knowledge Systems and Cognitive Models program, and project contracts with NYNEX and the Xerox Corporation.

References

Allen, C., Pea, R., de Vet, J., & de Vogel, J. (1991). Picasso Project Final Report. *IRL Technical Report*. November 1991

Bamberger, J. (1991). *The mind behind the musical ear*. Cambridge, MA: Harvard University Press

Bamberger, J., & Schön, D.A. (1983). Learning as reflective conversation with materials: Notes from work in progress. *Art Education*, March, 68-73

Bannon, L. (1991). From human factors to human actors. In: J. Greenbaum & M. Kyng (Eds.), *Design at Work: Cooperative Design of Computer Systems*. Hillsdale, NJ: Lawrence Erlbaum, 25-44

Barr, A. Bennett, J., & Clancey, W.J. (1979). *Transfer of expertise: A theme for AI research*. Working Paper HP-79-11, Stanford Univ. Dept. of Computer Science

Bartlett, F.C. (1977). *Remembering – A Study in Experimental and Social Psychology*. Cambridge: Cambridge University Press [1932]

Bateson, G. (1972). *Steps to an Ecology of Mind*. New York: Ballantine Books

Bickhard, M.H. & Terveen, L. (in preparation). *The Impasse of Artificial Intelligence and Cognitive Science*

Brown, J.S. (1991). Research that reinvents the corporation. *Harvard Business Review*, January-February, 102-111

Clancey, W.J. (1989). Viewing knowledge bases as qualitative models. *IEEE Expert*, (Summer 1989): 9-23

Clancey, W.J. (1991a). Situated cognition: Stepping out of representational flatland. *AI Communications*, 4 (2/3):107-112

Clancey, W.J. (1991b). Review of Rosenfield's "The Invention of Memory," *Artificial Intelligence*, 50(2):241-284

Clancey, W.J. (1992a). Overview of the Institute for Research on Learning. In proceedings of CHI, 1992 (New York: ACM). Monterey, CA. 571-572

Clancey, W.J. (1992b). Representations of knowing: In defense of cognitive apprenticeship. *Journal of Artificial Intelligence in Education*, 3(2),139-168

Clancey, W.J. (in press a). Situated action: A neuropsychological interpretation: Response to Vera and Simon. To appear in *Cognitive Science*

Clancey, W.J. (in press b). Notes on "Epistemology of a rule-Based expert system." To appear in *Artificial Intelligence*

Clancey, W.J. (in press c). Guidon-Manage revisited: A socio-technical systems approach. Submitted to the Journal of AI and Education

Dewey, J. [1896] (1981). The reflex arc concept in psychology. Psychological Review, III: 357-70, July. Reprinted in: J.J. McDermott (Ed.), The Philosophy of John Dewey, Chicago: University of Chicago Press, 136-148

Dewey, J. [1938] (1981). The criteria of experience. In: Experience and Education, New York: Macmillan Company, pp. 23-52. Reprinted in: J.J. McDermott (ed.), The Philosophy of John Dewey, Chicago: University of Chicago Press, 511-523

Dreyfus, H.L., & Dreyfus, S.E. (1986). *Mind Over Machine*. New York: The Free Press

Edelman, G.M. (1992). *Bright Air, Brilliant Fire: On the Matter of the Mind.* New York: Basic Books

Ehn, P. (1988) *Work-Oriented Design of Computer Artifacts,* Stockholm: Arbeslivscentrum

Engeström, Y. (1987). Learning by Expanding. Orienta-Konsultit, Helsinki. Unpublished dissertation

Gasser, L. (1991) Social conceptions of knowledge and action, *Artificial Intelligence,* 47(1-3)107-138

Greenbaum, J., & Kyng, M. (1991). Design at Work: Cooperative Design of Computer Systems. Hillsdale, NJ: Lawrence Erlbaum

Hammer, M. (1990). Reengineering work: Don't automate, obliterate. Harvard Business Review, 90(4): 104-112

Hirschorn, L. (1984). Beyond Mechanization: Work and Technology in the Postindustrial Age. Cambridge, MA: MIT Press

Jordan, B. (1992). New research methods for looking at productivity in knowledge-intensive organizations. In: H Van Dyke Parunak (Ed.), Productivity in Knowledge-Intensive Organizations: Integrating the Physical, Social, and Informational Environments. Working papers of the Grand Rapids Workshop, April 8-9, 1992. Industrial Technology Institute Technical Report 92-01. Ann Arbor, Michigan. pp. 194-216

Kling, R. (1991). Cooperation, coordination and control in computer-supported work. Communications of the ACM, 34(12)83-88

Kukla, C.D., Clemens, E.A., Morse, R.S., & Cash, D. (in press). An approach to designing effective manufacturing systems. To appear in Technology and the Future of Work

Lave, J. (1988). Cognition in Practice. Cambridge: Cambridge University Press

Lave, J., & Wenger, E. (1991). Situated Learning: Legitimate Peripheral Participation. Cambridge: Cambridge University Press

Nonaka, I. (1991). The knowledge-creating company. Harvard Business Review. November-December. 96-104

Roschelle, J, & Clancey, W.J. (in press). Learning as Social and Neural. To appear in Educational Psychologist

Rosenfield, I. (1988). The Invention of Memory: A New View of the Brain. New York: Basic Books

Ryle, G. (1949). The Concept of Mind. New York: Barnes & Noble

Schön, D.A. (1979). Generative metaphor: A perspective on problem-setting in social policy. In: A. Ortony (Ed.), Metaphor and Thought. Cambridge: Cambridge University Press. 254-283

Schön, D.A. (1987). Educating the Reflective Practitioner. San Francisco: Jossey-Bass Publishers

Scott, A.C., Clancey, W.J., Davis, R., & Shortliffe, E.H. (1984) Methods for generating explanations. In: B.G. Buchanan & E.H. Shortliffe (Eds.), Rule-Based Expert Systems: The MYCIN Experiments of the Heuristic Programming Project. Reading, MA: Addison-Wesley, pp. 338-362

Slezak, P. (1989). Scientific discovery by computer as empirical refutation of the Strong Programme. Social Studies of Science, Volume 19. London: Sage. pp. 563-600

Slezak, P. (1992). Situated cognition: Minds in machines or friendly photocopiers? Presented at the McDonnell Foundation Conference on The Science of Cognition, Sante Fe. Unpublished manuscript

Smith, B.C. (1991). The owl and the electric encyclopedia. Artificial Intelligence 47(1-3):251-288

Suchman, L.A. (1987). Plans and Situated Actions: The Problem of Human-Machine Communication. Cambridge: Cambridge Press

Tyler, S. (1978). The Said and the Unsaid: Mind, Meaning, and Culture. New York: Academic Press

Wallace, J.W., & Shortliffe, E.H. (1984) Customizing explanations using causal knowledge. In: B.G. Buchanan & E.H. Shortliffe (Eds.), Rule-Based Expert Systems: The MYCIN Experiments of the Heuristic Programming Project. Reading, MA: Addison-Wesley, pp. 371-388

Wenger, E. (in preparation). Toward a Theory of Cultural Transparency. Cambridge: Cambridge Press

Winograd, T., & Flores, F. (1986). Understanding Computers and Cognition: A New Foundation for Design. Norwood: Ablex

Wynn, E. (1991). Taking Practice Seriously. In: J. Greenbaum & M. Kyng (Eds.), Design at Work: Cooperative design of computer systems. Hillsdale, NJ: Lawrence Erlbaum Associates, pp. 45-64

Zuboff, S. (1988). In the Age of the Smart Machine: The future of work and power. New York: Basic Books

Ethnography and Organizational Learning: In Pursuit of Learning at Work

Julian E. Orr

Work Practice & Technology, Systems and Practices Laboratory,
Xerox Palo Alto Research Center, Palo Alto, CA 94304, USA

Abstract. This paper considers the question of "organizational learning" and finds the organization to be much more problematic than the learning. It reports on the successful experimental use of radios to support field service work, in particular to facilitate learning from experience on the job. This worked well, but the corporate response to the eperiment suggests that the organization enacted by management differs significantly from that enacted by those doing the work of the corporation.

Keywords. Applied anthropology, anthropology of work, organizational learning, work practice, organizational change.

1. Introduction

In this paper, I begin with a discussion of organizational learning in which I question the problematic nature of learning while finding 'organization' to be a very elusive entity. This discussion serves to introduce a description of an application of the ethnographic study of work to the changing of that work through the introduction of technology. The ethnographic perspective on this process raises some interesting questions about both learning and organization and suggests that a focus on organization removes some elements of reality from discourse on both work and learning. Our experiences with the field application would seem to confirm this suggestion, and yet the construct 'organizational learning' retains some interesting heuristic value. What follows will attempt to tease apart its strengths and weaknesses.

2. Organization

There is a tendency in writings about organizations to reify the organization, to grant it status as an actor comparable to persons, and yet it is clear that organizations do not exist in that way. Similarly, there is a tendency to see organization charts as portraying real structures, and while such charts are rarely complete fictions, the realities of what they portray seem invariably to be far more complex and problematic than their portrayal. Bittner tells us that the standard sociological conception of organizations is as follows:

> "... stable associations of persons engaged in concerted activities directed to the attainment of specific objectives. It is thought to be a decisive characteristic of such organizations that they are deliberately instituted relative to these objectives. Because organizations, in this sense, are implementing and implemented programs of action that involve a substantial dose of comprehensive and rational planning, they are identified as instances of formal or rational organization in order to differentiate them from other forms." (Bittner, 1965, p. 239)

However, the statement of an organization's design or structure originates with a member or members of the organization, and their intent in stating it is more often prescriptive than descriptive or analytic. To see the reality of an organization, one must see how the schema is used by competent members of the organization in their activities as members of the organization, and indeed, the criterion of 'competence' is that they be accepted as such by other members of the organization. Bittner then proposes "the concept of organization as a common-sense construct", in which:

> "... the meaning of the concept, and of all the terms and determinations that are subsumed under it, must be discovered by studying their use in real scenes of action by persons whose competence to use them is socially sanctioned." (p. 247)

If one conceives of organizations in this way, then the organization is enacted through a set of continuing relationships between some number of people whose activities with reference to the organization give meaning to the terms postulated to describe the organization. These relationships do evolve with time, although they are often portrayed as eternal and static, and include differences in status and role. Apart from Bittner's specification that members of organizations be "engaged in concerted activities directed to the attainment of specific objectives" and that the organizations be "deliberately instituted relative to these objectives" (p. 239), there would be little to distinguish a modern corporation from a kinship-based tribe as an organization. What differences there are emerge in the manner of recruitment, focus or purpose of the group's efforts, and even the percentage of time individuals devote to relationships within the organization. However, common usage of 'organization' connects it to business; in this sense, the common thread to the set of relationships constituting the organization is that they relate to the conduct of some enterprise.

The field work for the study discussed herein was done entirely within a major corporation which manufactures photocopiers and other office equipment and focussed on the work practice of service technicians, one of the lowest levels of the corporate hierarchy. The two levels of management in the district office were seen originally from the perspective of the service community; eventually they and higher level managers became participants in the project to change service work, shedding new light on the perspectives and concerns of the managers. However, even in this context of a study confined to a single corporation, one is struck by the elusive nature of the organization. The corporation clearly constitutes an organization, and yet membership in that organization seems problematic for some of the employees. Field service managers spoke of 'the corporation' as though it were distinct from service, their own organization, but field service is not a coherent entity either. There is 'this district' as distinct from 'that district', and then there is 'this team' as opposed to other teams or the district itself. Within teams there are workgroups, and within workgroups there are 'Customer Service Teams'. These are composed of two or three individual technicians. Thus, there is an array of organizations at play in the service organization, with substantially or wholly overlapping memberships, and yet there is a clear choosing of which membership is appropriate at a given time.

Selection of membership in a given organization within this array seems to be done on the basis of the alliances and oppositions of the moment, that is, one is doing these things with these people now, and the grouping is at least partially defined by *not* being those other people doing other things. For an anthropologist, this is oddly reminiscent of the oppositions within segmenting lineages of the Nuer (Evans-Pritchard, 1940) or the Afghans (Elphinstone, 1839), or of the *nisba*, an infinitely branching Moroccan system of personal identification which effectively can distinguish individuals from all those with whom they interact (Geertz, 1983). Participants in such systems can choose from a range of identities, and the selection is based upon with whom it is being used. Similar options seem to be available in discussing the corporation for which one works.

If the nature of the existence of an organization is problematic, surely its ability to act is equally so, and yet one speaks quite confidently of 'The organization doing this' and 'The Corporation doing that'. It is clear that such activity comes down to specific actions by specific members of the organization, and thus it seems a harmless Abstraction to aggregate many such activities as an action of the organization. The problem with this form of discourse is that the specific actions which accomplish the action of the organization remain invisible. With time and habit, they may be forgotten and those who want the organization to do things may not know how the desired activity must be translated into many specific actions. In the worst cases, it may not be possible for those actually performing the specific actions to do all that is necessary for the organizational action to occur. Ignorance of how the work is actually done has caused many problems in the organization observed in this study.

Yates (1989) suggests that the development of administrative techniques made possible the growth of the corporation by increasing the possible span of control. This growth also led to the development of layers of administration whose tasks are distinct from the work which provides the revenue and ostensible purpose for the organization, manufacturing, sales, or service, for example. Administrative tasks are, in fact, work *about* doing that work. It is possible to see the level of abstraction which permits or even requires one to think of organizations 'doing' as a concomitant of the development of this administrative superstructure. Neither administration nor abstraction need cause problems; however, the disappearance of the fundamental work of the organization from the discourse of the organization has caused many.

3. Organizational Learning

So, if the organization itself is the product of interactions by its members, and if that membership seems both optional and problematic, where does this leave organizational learning? Simon (1991) points out that learning is in fact all done by individuals, and organizations only acquire learning as their members learn or as they acquire new members with new knowledge. Learning, however, occurs in a social context, Simon continues, and any individual learning is influenced by the culture of the organization in which this is done. Both points seem unarguable, but, with reference to Bittner, there remains the question of how this learning affects the interactions by which the organization is enacted.

Having said that, one should note that discourse on organizational learning seems to find the learning more problematic than the organization, whereas ethnographic studies of people learning all seem to find that people learn easily, continuously, and virtually automatically. Learning only seems to become problematic when thought of in an institutional context, such as a school, or when there is a question of the relationship of that learning to an organization. This is, perhaps, related to the issue about the organization as putative actor; thus, organizational learning is the attribution to the organization of the activities of its members, with a consequent lack of precision about ownership or agency.

4. Community Memory

4.1 The Community of Practitioners

If one considers learning among service technicians not from the perspective of the corporation but through the study of their work practice, their learning is seen to be an aspect of the community memory (Orr, 1990a). This emphasizes both that the technicians constitute an occupational community and that the learning is the acquisition and circulation of information vital to practice as defined by the community itself. An occupational community, in van Maanen's and Barley's sense (1984), is a bounded work culture with little or no hierarchical structure and little possibility of career development. The only significant status in the community is that of a competent practitioner, which cannot be ascribed by the organization but can only be achieved by doing the job. The lack of hierarchy and career paths makes such a community remarkably independent of the organizations which employ them. Given that claims to status within the group are based on competent practice, learning and preserving the details of such practice becomes a major preoccupation of the group.

This becomes more important when one discovers that the nature of the work in practice differs from the way it is seen by the corporation. Service work is commonly thought of as the fixing of broken, identical machines. The corporation believes that it can provide documentation to direct the technicians through this work. However, the machines have individual differences, idiosyncrasies, and histories. Their behavior evolves, and they are not always broken when service is needed. A more accurate view of the work is as the maintenance of a fragile and partially understood triangular relationship between the user, the machine, and the technician. This work requires continuous improvisation on the basis of what the technician has learned through personal experience or from others in the community. This always changing and challenging work, then, requires the community to learn.

Learning happens primarily through diagnosis, the struggle of one or a few technicians to understand a particular problem. Diagnosis is observed to have a strong narrative component in the integration and assessment of known facts; the technicians tell themselves what they know about the machine. This narration prepares them to tell others of their experience, either in asking for help or telling of a new problem, and stories of interesting problems circulate quickly through the community. These stories inform the community; they also demonstrate and celebrate the competent practice in maintenance of the service situation which is the basis of the community.

4.2 The Community and the Corporation

It is clear that this community learns. However, it seems to do so quite independently of and at some distance from the corporation, the organization to which it belongs. Many members of the corporation recognize that there is a gulf between 'the field' and 'the corporation', and there is considerable debate as to exactly where that gulf lies. It is significant that the technician community also sees itself as estranged from the corporation.

One aspect of the distance between the technicians and the rest of the corporation may be seen when one compares the corporate approach to information with that of the technicians. Within the technician community, information is available to almost any practitioner, and indeed, stories are told whether the information is needed then or not. Given that the only status within the community is that of competent practitioner, fame can only be based on a reputation for extraordinarily competent practice, the ability to solve newer and harder problems. Since technicians normally work alone, achievements will only be known if the person responsible tells them. Moreover, technicians want the information to circulate, so that others can address similar problems. A team shares responsibility for its calls, so there is incentive to have all members competent for as many problems as possible. A final point is that since technicians are working with a partial understanding of the machines, there is no way of knowing definitively which details are significant and which are not. This encourages the preservation of stories with as much of the context as possible, to define both what was learned about the machine and the context in which it was known to be true.

The corporate attitude, in contrast, seems to be that facts are discernible and constant; this particularly appears in their assertion that the solutions to service problems can be provided in directive documentation. The provision of directive documentation makes the following statements: First, that the part of the organization providing the documentation has all the facts necessary to do the work, and second, that the documentation contains those facts necessary to do the work as they have defined it. There is an explicit rationing of information according to a formulation of what those in the field are thought to need and a withholding of information not deemed essential to approved activities. This is done to shape the activities of the technicians to the model of those activities held by those producing the documentation. (See Orr, 1991, for more discussion of this issue.) Despite the implicit and explicit claims of directive documentation, there is a recognition that field experience does yield some facts, problems, symptoms, or solutions which were not anticipated or not correctly anticipated. However, the corporation also seems to feel that what is learned in the field needs to be processed elsewhere in the organization before possibly, if approved, being returned to the field. New discoveries are to be reported through the appropriate channels; the technicians report that information so dispatched is rarely seen again. This requirement of reporting seems to be part of a widespread intuition that learning must be harvested, that left in place it will decay, and this intuition is not limited to the service organization.

The story-telling of the technicians is one of the most obvious aspects of their culture to any ethnographer. Reporting the importance of stories to many audiences, however, draws an apparently automatic response that those stories must be collected, preferably in a data-base. This view of information seems to require that information acquired by actually doing the work be hierarchically approved and then distributed in a formal, rationed manner. It seems remarkably close to scientific management's assertion that management must control information about work and dole it out piecemeal in order to control the workers. That information which is distributed to the technicians, however, is often so fragmentary that the technicians do not perceive it to be of any use at all, and information from the corporation is usually regarded as suspect until confirmed through community experience.

The corporate use of information for leverage and control contrasts with the technicians' attitude of sharing it among those who can use it. This contrast is strikingly similar to that reported in Eckert's (1989) ethnographic study of categories and class in American high school. The disaffected Burnouts, commonly with working-class background and futures, viewed information (and many other things) as most appropriately shared, while the corporate and managerial Jocks, middle-class in origin and destination, saw information as something to use, to barter with, to achieve control. Perhaps this similarity is not surprising, as most technicians have working-class or lower-middle-class backgrounds.

Learning then takes place within the technician community. The technicians and the rest of the corporation have different views of what should be done with what is learned in daily practice. As has been suggested, there are also differing views of what constitutes the job. The corporation asserts its view; the technicians act theirs out, doing what is necessary. It does not come to a contest (Orr, 1991), in that the corporation does not take steps to enforce its view, preferring that the work get done more than that their view be vindicated, while the technicians do not bother to contest the corporate view. In Bittner's terms, this failure of discourse means that the organization is never really defined as a single entity and that dramatically different versions of the organization exist for the managers and the technicians.

5. The Denver Project

5.1 History of the Project

There are, however, moments when managers appear to be talking about the same things as technicians when discussing field service work, and at those times it can appear that a more functional organization is in the process of being created and defined. The Denver Project was one such moment. It began with an invitation from the Denver District (which does sales and service in Colorado and parts of Wyoming for the corporation in question) to representatives of corporate research and a corporate expert systems development group to come discuss ways to enable technicians to enhance their own practice.

To sketch a brief history of the project, the initial meeting produced a suggestion that portable radios might permit the enhancements we sought to encourage. (See Orr and Crowfoot, 1992, for a fuller exposition of this process.) A preliminary ethnographic investigation sought to determine whether this tool made any sense to the technicians, while a parallel technical investigation sought to determine what radio technology would suit the technicians' needs. Radios were found which would do those things which made sense to the technicians; these radios were purchased and distributed to a sampling of both urban and rural teams. The technicians' use of the radios was then observed over the next six months; at the end of this time, technicians, ethnographer, and managers all believed that this was a valuable tool. The rest of the corporation then decided that this was a technology which should be investigated and began its own test.

5.2 Different Rationales

There was a perhaps surprising unanimity among those planning this project. The intention was to change the way the corporation does service, and we hoped to do this by making it easier for the technicians to do what they need to do. Implicit in these intentions are a set of uncommon recognitions: First, that service practice is far more than fixing the machines by following the directive service documentation; second, that the service technicians have more contact with the corporation's customers than any other member of the corporation and so should be valued by the corporation; and third, that some corporate policies and programs have not aided but rather interfered with necessary activities of the service force.

Given those recognitions, it would seem that from a manager's perspective the technicians must be seen as a very valuable if somewhat peripheral part of the corporation. However, it was not clear to any of those planning this project what technicians need, nor did we believe that any in the organization other than the technicians would know. In fact, it is not certain that the technicians would know, at least initially, since nothing in their environment has encouraged or even permitted reflection on what practice might be. It is also not clear how the technicians will respond to being more valued, as their experience suggests that being more valued is a prelude to being asked to do more with less, a condition they do not enjoy. Finally, for the manager's perspective, the technicians' somewhat marginal position in the company means that communications with the rest of the corporation are at best intermittent and frequently confused.

5.3 Goals for the Project

The goals we set for this project attempted to address these recognitions and concerns. The first goal toward changing service was to see if the present marginal position of the technicians with respect to the rest of the corporation is necessary or if some degree of rapprochement is possible. What would it take to make the technicians feel more a part of the corporation, and what would be gained by this? "Empowered" is a popular word in corporations today. What could it mean for technicians to be 'empowered', and under what circumstances would they be interested?

A second goal was to look for unnecessary hindrances to practice. If a corporate policy or program appears to interfere with the doing of service work, we wished to consider whether such interference could be eliminated without serious compromise to the corporation's goals. This would require negotiation with that portion of the organization where the policy or program originated, to determine what their original intent was and how it might best be accomplished.

The third goal was to see if there are ways in which technology can assist technicians' practice. This appears to be a deceptively easy goal, because numerous tools suggest themselves, but its execution requires persuading the corporation to commit funds to provide those tools to the technicians. In fact, the process of demonstrating to the corporation that a tool is a sufficiently valuable addition to the technicians' repertoire has rarely been successfully performed.

5.4 The Organizational Perspective

This set of intentions, recognitions, and goals contains some significant points if one attempts to see the corporation as a single organization, a common entity. Primarily, they reflect the recognition that learning does occur in the practice of field service, a part of the organization removed from the principal concentrations of power and resources. Somewhat more unusual was the recognition that learning functions where it occurs, that what the technicians learn is most useful to other technicians. The decision was then made to assist the circulation of that information at the level where it is most useful, among the technicians. Finally, there was a decision to keep the rest of the organization out of the loop, to make this circulation of information free from interference or even monitoring. It was a point of policy for the project that managers were not *allowed* to have radios, nor was there a radio in the district offices. Eventually one of the participating teams voted to let their manager have a radio, because they needed better access than could be managed with telephone and pager, but it is significant that it was their decision, not one from elsewhere in the corporation.

The decision to stay out of the loop seems counter to managerial tradition. In the first place, the tradition of scientific management mandates the control of knowledge by managers in order to distribute specific bits of information to those charged with performing specific tasks, all of which is intended to be planned by management. Secondly, there is a failure to use information for advantage, which is counter to the Jock mentality (in Eckert's terms) of those in the middle class who become managers. This relinquishment underscores a dilemma of modern managers: If they are not controlling, what then is their function? Given these pressures, the managerial decision to let the information remain available only to the technicians is a remarkable event and an achievement for those involved.

5.5 The Technicians' Perspective

As the technicians' use of the radios evolved, people in other parts of the corporation heard about the experiment and came to see if this could address their problems. There was a series of meetings in which the technicians came together to talk to visitors about what the radios had accomplished for them, and what follows summarizes their principal points. (All of these were also found in the ethnographic observation as well.)

The principal point for the technicians is that they are no longer alone. Although technicians have been organized into workgroups for several years, they still worked alone and saw their team–mates only at scheduled meetings, occasional lunches, or random encounters. Now they have the possibility of more or less continuous conversations with their team–mates and see this as an enormous advantage. Several reported that when working alone they had sometimes wondered whether they might be the only one of their team actually working; now, through over–hearing the conversations as their colleagues go about their business, they believe they know what everyone is doing.

The technicians say that the workgroups have become real through this tool, in that the groups now are based on real and continuing relationships which could not be achieved with weekly meetings (which were often even more infrequent). With this new ability for active cooperation and collaboration, the technicians found it possible to manage their call queue more efficiently, to arrange coverage of a territory for technicians who are ill, on vacation, or otherwise unavailable, or to coordinate dealings with sensitive customers. The radios may have reduced still further the frequency of meetings, in that much of the business can be done on the air in time available; they have also made it possible to have meetings on the spur of the moment when there is a slack period in the call rate.

Those planning the project had expected the possibility of technical consultation to be the most valuable attribute of the radio, and while it is probably not the most important attribute, it is one the technicians exploit. They like the fact that the accumulated expertise of the entire workgroup can be focused on any problem, and if circumstances are conspiring to intimidate the technician on the scene, those on the the radio cannot see the crowds waiting to use the broken copier. New technicians and teams with new technicians report that such technicians learn to be effective in the field much more rapidly when the radio permits more experienced technicians to support the less experienced. Finally, the corporation has provided consultation services over the telephone to help stuck technicians in the field; the technicians with the radio report that they use this 'call escalation' procedure much less frequently because collectively they can answer most of those problems.

Another important benefit from using the radios is related to the point about not being alone: The technicians can now provide moral support to each other. These technicians mostly work alone in other people's offices and rarely spoke to any colleagues; now they spend part of their time on the air supporting each other and often converse while driving home at the end of the day, unwinding and getting the day's events in perspective. In fact, when asked if data communications with a message system could substitute for the radios, their first response was that the stress in someone's voice cannot be conveyed in a typed message.

There are other benefits to using the radios: The technicians find it easier to share parts or to coordinate trips to the warehouse when necessary. They exchange information about weather or traffic. Through the telephone connection capability, they can call emergency services for accidents, and they are also much more available to their families. They also find that the customers like the radios. Since technicians do not always service the machines that are nominally 'theirs', they have found that when going into a new customer office, if the customers have heard them on the radio, they are much friendlier and more accepting. New technicians also report that their customers seem to have more confidence in them with the radio, since the customers know the technicians have the option of consulting with their colleagues if problems arise.

These are all benefits to using the radios, and they are quite real and important. In deploying the radios, we clearly changed the way the work gets done. It seems increasingly clear that despite the promising premises with which we began, relationships with the corporation did not change, and it is instructive to consider why this is so.

5.6 The Corporate Perspective

At higher levels in the corporation than the district, this experiment appears as a violation of the paradigm of technology development for service. To begin with, it did not originate at the center of the corporation in the part of the bureaucracy charged with developing technology for service but rather originated with a local initiative in an organization charged with doing service. Moreover, the planning was done with the technicians who would actually use the technology, something which does not happen with developments at the center. Finally, the radio technology was selected because it fit conditions in the Denver District in both metropolitan and rural areas; we were not sure whether it would work in other districts, although it seemed likely to be useful for at least some of them. Programs that originate in the national organization, on the other hand, are designed for all districts, which may mean that they fit no districts well.

Under pressure from the rest of the corporation, the original desire to help the technicians enhance their own practice disappeared from managerial discourse, and the talk shifted to demonstrating increased productivity. We were directly informed that true increases in productivity resulted in reduced headcount in service, at which point it is quite clear that the goal is no longer to help the technicians. In fact, the discourse has shifted away from work practice to organizational goals, and those organizational goals are in conflict with the way the work is actually done. It is not clear whether those driving this shift in discourse understand service work or not.

A final aspect to this transformation is that the focus shifted to the technology itself, the radios, from what we wanted the radios to do, which was to assist the technicians' practice and provide an opportunity for the technicians to reflect on their practice. It was hoped that this in turn would lead to new efforts to enhance practice. As the radios become a goal in themselves, it becomes reasonable for those with that goal to contemplate buying radios by reducing the number of technicians. It also becomes reasonable to save money on the purchase price of the radios by giving up some of the functionality of the radios, and this has been easier to do because those making the exchange have not known the usefulness of that functionality in the technicians' daily practice.

6. Conclusion

In this instance, the heuristic value of the concept 'organizational learning' would seem to be that it forces one to consider the relationship between the learning which is seen to occur in work practice and the corporation at large. This in turn calls into question the relationship between different parts of the organization, and one ends by contemplating the nature of the corporation's status as an organization. One reason the project was attractive was that there was a temporary matching of conceptions of the organization which allowed the alliance to form to do this experiment. The use of the radios, our experiment, did indeed take better advantage of those things learned through work practice than the previous organizational provisions for innovation had done, and in that we were successful.

In the larger discourse of the corporation, however, the radios were seen to be a good thing in that they helped the technicians at work, but the lessons about learning in the field or about how the organization might approach those things which occur in the field were lost. Harking back to Bittner's definition of the concept of organization, if the organization is defined through the actions of its members in their roles as members of the organization, then the organization we were discussing in Denver is not the same as that envisioned by those at the center of the corporation. A problem with the version enacted at the center of the corporation is precisely its lack of connection to those doing the fundamental work of the organization, which includes the service technicians, and this disjunction is a serious failure of organizational discourse. The coexistence of incongruent versions of the organization in what is nominally one corporation and the absence of any dialogue through which those groups holding these incongruent versions might negotiate their differences together constitute the greatest barrier to any real sense of organizational learning.

References

Bittner, E. (1965). The Concept of Organization. *Social Research* 232:230–255

Eckert, P. (1989). *Jocks and Burnouts: Social Categories and Identity in the High School.* New York and London: Teachers College Press

Elphinstone, M. (1839). *An Account of the Kingdom of Caubul.* London: Richard Bentley

Evans-Pritchard, E.E. (1940). *The Nuer*. New York and Oxford: Oxford University Press

Geertz, C. (1983) *"From the Native's Point of View": On the Nature of Anthropological Understanding*. *In* Local Knowledge. New York: Basic Books

Orr, J.E. (1990a). Sharing Knowledge, Celebrating Identity: War Stories and Community Memory in a Service Culture. In: D.S. Middleton & D. Edwards (Eds.), *Collective Remembering: Memory in Society*, pp. 169-189. London: Sage Publications

Orr, J.E. (1990b). *Talking about Machines: An ethnography of a modern job*. PhD Thesis, Cornell University. PARC Technical Report SSL 91-07 [P91-00132], Xerox PARC, Palo Alto, CA

Orr, J.E. (1991). Contested Knowledge. A paper presented at the 50th Annual Meetings of the Society for Applied Anthropology, Charleston, South Carolina. Published in *Anthropology of Work Review*, 12(3). pp. 12-17.

Orr, J.E., & Crowfoot, N. C. (1992). Design by Anecdote: The use of ethnography to guide the application of technology to practice. Proceedings of PDC '92: The Participatory Design Conference, Cambridge, MA

Simon, H.A. (1991). Bounded Rationality and Organization Learning. In Organization Science, 2(1), 1991

Van Maanen, J., & Barley, S.R. (1984). Occupational Communities: Culture and Control in Organizations. In: B.M. Staw & L.L. Cummings (Eds.), Research in Organizational Behavior, Vol. 6. Greenwich, CT, and London, UK: JAI Press

Yates, J. (1989). Control through Communication: The Rise of System in American Management. Baltimore and London: The Johns Hopkins University Press

Organizational and Cognitive Design of (Technological) Learning Environment

Cristina Zucchermaglio

Department of Social and Developmental Psychology
University of Rome "La Sapienza", Rome, Italy

Abstract. In order to survive in a period of rapid changes organizations must exploit their abilities to learn and innovate.The paper will explore how and to what extent the design of "educational systems" within organizations must consider the organizational context as a learning context in itself. In this perspective the design of any "educational system" must consider both the perspective of organizational design and the the specificities of the learning process highlighted by the recent researches in cognitive sciences. The design of any "educational system" must be strictly and coherently related to the design of the organizational context (for example, in terms of technology, reward and incentive systems, human resource practices, belief systems, communicative flows, access to information, etc.); it will be suggested the need for the search for a organizational coherence between the educational features of the organizational context (and of the learning that goes on in that context) and those of the formal educational context, through the definition of a common design project (in tems of design principles, methodology, professionals involved, etc.) of the organizational learning environment. The design of a formal and non formal educational context should be a part of the entire organizational design and that also reify a theory of organizational learning. The work context should be designed to take advantage of the learning potential embedded in "normal" productive activities, through the growth of a cognitive and educational point of view to manage and reunite learning and working in organizational settings. The design of such rich learning contexts, in which a model of real cognitive apprenticeship could be applied, would be a very promising research area for the cognitive science of instruction also contributing to the creation of a new learning culture in organizational contexts.

Keywords. Organizational, design, learning environment

1. Introduction

In analyzing how learning is fostered and implemented in organizational settings, two main characteristics which are strictly related can be identified:

1) The majority of educational contexts in organizations are designed as a *traditional school setting* (Resnick, 1987): a teacher talks to a group of learners about a specific topic for a definite number of hours. The learners listen and/or work individually with books or exercises, all of them do the same things at the same rate. These contexts presupposes that learning is mainly a process of *transferring "information"* from an expert (book, computer, teacher) to the heads of the learners, in different ways which are practically all predefined.The learning model is the "*pouring model*": information exists objectively and it is transferred into the head of the learner. The mayority of training operators and designers are still basically immured in a behavioristic paradigm: they treat "knowledge as a fluid, which pours into the student-vessels" (Kay, 1989), thus using a learning theory, often implicit, that is still not yet supported by recent developments in cognitive psychology of instruction, as we shall analyze further. For this reason, what gets learned is rarely what people are expected to learn or thought to have learned and moreover, knowledge acquired in these contexts is difficult to transfer and to use in the workplace. These educational practices could be termed "apparent education". As Seely Brown & Duguid (1991) said: *"The sort of learning that does go on in organizations does not fall within current paradigms for learning.....we suggest organizations need to change their conception of what learning is and how and where it takes place* (p. 45).

2) The mayority of *educational contexts in organizations are clearly separated from working contexts.* Employees working on production lines or in offices are sent to a Training Department, inside or outside of the organization itself, only when the time comes to learn something. There they find all the traditional school-like apparatus: teachers, exercises books, classrooms, time tables, specific courses, etc. And this is mostly the case for all employees, from managers to secretaries. The organizational structure and architecture assumes and implements a clear separation of knowing from doing (and of learning from working). In particular organizations do not recognize that *learning processes are continually present and active also in the workplaces,* so they are in danger of cutting themselves off from a particularly essential source of innovation.

   This traditional approach cannot work nowadays: organizations should manage and foster innovation processes as "normal" processes of organizational functioning. Learning should not be an activity to be carried out a few times a year, but should be recognized and used constantly: organizations must not lose the contribution of these distributed innovation processes. Learning practices should drastically change, a diffuse learning culture must become an essential part of the entire organizational culture. Learning should be considered as a strategic factor, essential for the survival of the organization[1]. In order to survive, organizations must exploit their abilities to learn and innovate; changing markets, technological innovations and therefore new jobs require enacting organizations, as Daft and Weick (1984) describe it: *"These organizations contructs their own environments. They gather information by trying new behaviors and seeing what happens. They experiment, test and simulate, and they ignore precedent, rules and traditional expectations"* (p. 288). Enacting organizations do not simply respond to a changing environment, they involve themselves in generating it.

In this view training and educational activities in an organization should comprehend both the design of organizational architecture, considering the learning carried out during work activities, and the design of corporate classrooms, always considering the specifics of the learning process highlighted by recent research in cognitive sciences. The growth of a cognitive and educational point of view to manage and reunite learning and working in organizational settings should be fostered.

2. The Design of Learning Contexts for Organizational Development

If learning is to be considered a strategic factor for organizational development, the way in which organizational training practices could be changed should first of all be analyzed. In particular an use of the working context as a learning context should be fostered and simultaneously the educational features and structure of the corporate classrooms should be substantially modified.

This new perspective, both organizational and cognitive, changes the approach to the design of the learning context drastically. In particular we are now aware that the learning needs of an organization cannot be solved simply by putting the personnel into a classroom once or twice a year. It is necessary instead to take into account the different contraints for an efficient design of organizational learning contexts and to find a coherent and balanced relationship between organizational choices and architecture and formal and non formal educational activities.

[1] At present organizations spend prodigiously in training and retraing all of their employees. Thus there are also economic reasons, not only theoretical ones, to create work contexts that are able to act as efficient learning contexts.

2.1 The Work Context as a Learning Context

How and to what extent should the design of any "educational system" within an organization be considered in the larger organizational context? The characteristics and features of the organization in which the "educational system" is to be "installed" have to be analyzed, not as an intervening variable, but as a learning context in itself (Scribner, 1984). Instead most organizational strucutures (in terms of working procedures, communication patterns, etc.) are based on a classical view of organizational development and of learning process, that assume, explicitly or implicitly, a strong separation of learning and innovation practices from working practices. As Seely Brown & Duguid (1991) said: "*We need a reconceptualization of the currently accepted epistemology of learning and of the ontology of work and as a consequence of this of the architecture of organizations*".

Thus one strong implication of this analysis is that the design of the organizational learning context should go beyond the design of formal and intentional educational programs (whether school or workplace based) and take into consideration also the design of the social organization of work itself (Scribner & Sachs, 1990). There is a growing body of research (Seely Brown & Duguid, 1991; Sachs & Scribner, 1990; Lei & Goldhar, 1992) that supports the conclusion that work environments differently organized promote, as results, different kinds of learning. So the *patterns of organizational learning can vary widely across firms*, depending upon both the existing organizational structure and functioning. This pattern of organizational learning is already in use in organizations, and as educational designers, we should, first of all, analyze it.

But we do not consider organizational analysis as a means for the identification of individual learning needs for the design of formal educational programs to be carried out in traditional classrooms; instead this analysis is needed to explore how learning is managed as an organizational factor and to decide what type of "educational" intervention is more suitable for the organization's development. To this end what is needed is an organizational model as a social system and a community of practice, not only as a set of information flows and execution procedures (Winograd & Flores, 1976).

In this view the intervention required for the development of the organization could be structural, cultural or technological, concerning, for example, a new structure or new instruments of office automation, new reward systems, new tools for distributing information, new processes of decision making, etc.. This organizational design should act as an educational theory in which learning and doing must be reunited, sustaining various forms of *incidental learning* and the dissemination of the *tacit forms of knowledge* that are embedded within and distributed across the people of the organization (Suchman, 1987).

This informal or tacit knowledge is generally ignored in the design of educational programs both in schools and in organizations (Scardamalia & Bereiter, 1992). Nevertheless the implicit part of work competence could be learned more quickly and effectively in practice (Martin, 1982), if the work contexts were specifically designed to sustain patterns of knowledge-building discourse (Orr, 1990; Jordan, 1987). Well-designed work contexts are intended as a means of sustaining these discourse-supporting functions under the peculiar conditions imposed by the organization itself.

Organizations should consider learning as an explicit "productive" objective and must create mechanisms (in terms of technology, reward and incentive systems, human resource practices, belief systems, access to information, communications and mobility patterns, performance appraisal systems, organizational practices and structures) for such learning to take place.

In the absence of explicit intentions and appropriate mechanisms and choices the learning potential of an organization may be lost (Ghoshal, 1987). Enacting organizations are constantly searching for new social "architecture" in which to better unleash the tacit (or implicit) and distributed forms of knowledge of their members, allowing and using all the nonformal educational activities embedded in work practices.

Through a specific organizational design and coherent communicative practices it may be possibile to "inculcate" dynamic and effective learning practices upon the members of an organization, such as also many studies of educational innovation in school contexts have demostrated (Scardamalia & Bereiter, 1992a).

In this perspective we should also revise the problem of the insertion of technology into an organization. If technology is to be considered one of the most powerful change factors in an organization, "the hardware side of the organization's investment in strategic flexibility" (Lei & Goldhar, 1992) should be accompanied by a corresponding investment in a new design of work flows and practices. Technology could be used, directly or indirectly, as a means sustaining organizational learning. If we consider technology as a tool that could deeply affect the communicative structure of the organization, we could use it as an effective tool for knowledge building.

The organizational design of technological tools should be explicitly focussed on social interactions and communication and they must be intended as cultural tools that mediate and support the social, distributed and situated construction of new knowledge. For example a technological system should not merely reveal itself to a user, but must be conceived as an instrument that helps to reflect on the working/learning process and check it consciously. It should also support communication and negotiation processes between members of a social community. If it is to become the intellectual technology of organizational culture, efforts at designing artificial systems will benefit from an analysis of the complexity of negotiation in the process of communication and of the nonformal educational activities embedded in "normal" work practices. Successful implementation and utilization of technology requires a commensurate organizational commitment to "generative" learning (Lei & Goldhar, 1992). In this sense the focus should be on the relationship between the organizational, cognitive and social aspects implied in the use of technological tools as tools for learning. What emerges is the need to define a set of design principles for the construction of (technological) organizational learning contexts, in which it should be possibile to use and take advantage of the distributed and situated educational activities embedded in work practices, that is, such forms of non formal education as on-the-job training, guided practice with a mentor and learning through doing on the job as a participant in a cultural community.

2.2 Corporate Classrooms as Knowledge Building Communities

This perspective also changes the design of formal educational contexts inside organizations: these educational contexts, whose design requires the consideration of various types of constraints, may be needed for the acquistion of some specific work-skills, even when the work context acts as an efficient learning context. Moreover, in a short term perspective, the major chances to change the learning culture of an organization and to modify something in its organizational training practices, should begin to change at least the ways by which the corporate classrooms are designed and managed. As Seely Brown & Duguid (1991) said: "*By preparing oversimplified courses for an assumedly unskilled audience, the organization loses the cooperation of the skilled people who attend the courses*" (p. 42).

To this end we must consider the large epistemological change in the research on learning that is taking place in recent years; the new approach (which can be termed as learning theory of instruction) fosters the "*studies of the acquistion of complex knowledge and skill relevant to the design of instruction in educational settings*" (Glaser, 1992). In most cases the attempt to understand learning is through the design of the educational context that sustains and produces it. The learning context is used as a specific research context on cognitive processes of learning and not only as a context in which the results obtained out of it should be applied. Educational contexts are being considered as social settings, in which it is possible to observe and understand "situated" processes of cognitive growth that can be affected by social exchange and by cultural transmission. This shift is directly connected to the recent interest in cognitive science for the study of the so called "everyday cognition" (Rogoff and Lave, 1984) or "situated cognition" (Suchman, 1987), that is the study of the cognitive, linguistic and thinking processes carried out in real and "ecological" contexts of life and work (hospitals, companies, factories, supermarkets, etc.) (cf. Brown, Colins & Duguid, 1989; Gardner, 1985; Seely Brown & Duguid, 1991; Latour & Wogar, 1986) The study of the learning processes of children and adults embedded and involved in real educational contexts highlighted some important aspects of the learning process, that also fit the organizational needs for distributed learning perfectly:

1) *Learning is not a matter of transfer, but of construction.* The learning process cannot be described by the metaphor of pouring something into an empty vessel: it is not true that information exist objectively and it is sufficient to transmit it into the learner's head. People are always active knowledge constructors: this knowledge interacts with the environment and the physical and social context in which they live. This constructive activity is not completely free and creative, but depends strictly on the specific constrains and characteristics of the environment. At this end it is necessary to design significative, rich and constrained learning contexts, contexts in which the activities carried out are the most relevant for the process of knowledge construction and for the socialization to the *practices*, also the cognitive ones, and to the artifacts, also the symbolic ones, of the culture (even the organizational one) in which we live. The keyword here is *constrained*, that is not getting learners free and lost in the environment, but giving them clues and pieces of information by which and through which they can construct new knowledge.

2) *The importance of content and context of thinking and learning*. Learning must be considered as a cultural development, a process of enculturation, of entry into a culture of practitioners, learning is like a complex social action embedded in historically and culturally defined frames of activity (Vygotsky, 1966; Cole, 1989). In this view it is necessary to evaluate how any symbolic system - oral and written language, music, mathematics - allows the development of different types of intelligence, that could be defined as "skills in a medium" (Bruner et al., 1966). This development presents a great variability not only between individuals but also in the same individual (Gardner, 1983), this means that the content of thinking is an important variable and that we think and learn in ways that are directly connected to instruments and symbolic systems of the cultural context in which we live (that is, the cognitive artifacts of Norman (1989) and the cultural amplifiers of Bruner (1989)). On the contrary in traditional educational contexts (also the orgnaizazional ones) the learning process is mainly considered as a decontextualized and solitary process, that should be carried out without those supporting aspects (other persons, information, instruments, notes, books, etc.) typical of the informal contexts of learning (Lave, 1988). In this sense the distance between learning "in" the educational contexts (intended as a specific ability) and learning "out" should be diminished (Resnick, 1987). Indeed the analysis of

learning processes, occurring in everyday contexts or among traditional populations that have not "invented" a school system, highlights that learning could be mainly described as a process of apprenticeship, that is as a graded, contextually embedded practice, inserted into a relevant frame of activity (Lave, 1988; Childs & Greenfield, 1980; Collins, Brown & Newman, 1989). Learning about an activity depends on authentically participating in that activity - not in an ersatz approximation to it. In this sense learning does not always involve abstract, decontextualized subject matter, but regards also work practices, social rules and communicative behaviour of the organization. In an apprenticeship practice much crucial learning happens without direct teaching: although there is very little "traditional teaching" - only occasional instructions or pointing out of errors by the "scaffolding" master - there is much learning through this graded, contextually embedded practice. In this view learning is in part the process of socially constructing a communal understanding: learning is not individual and receptive at all, but a collective, constructive social process.

On the basis of this growing body of research on learning and instruction it is possible to present some general conditions for effective learning (on which some authors agree, Glaser, 1992; Scardamalia & Bereiter, 1992a; 1992; Gardner, 1991) that could sustain and guide the design of educational contexts:

1. Foster the individual's *engagement in holistic situations,* in which learners are allowed to understand the goals and meanings of the situation as they attain specific competency. Learners need to be able to establish an apprentice-like relationship with practitioners, who model for learners the practices of expertise, explicitly or implicitly. Learners need to be allowed to begin on the periphery of authentic activity and to move progressively through increasingly full participation toward expertise. From a design point of view this means creating "apprenticeship" learning contexts in which competence could be developed through an expert's scaffolding during the realization of an authentic task or problem. *"The scaffold is a metaphor, originated by Wood, Bruner & Ross (1976) to describe the ideal role of the teacher. This metaphor is the basis for a theoretical model of the teacher in informal education. The scaffold, as it is known in building construction, has five characteristics: it provides a support, it functions as a tool; it extends the range of the worker; it allows the worker to accomplish a task not otherwise possible; and it is used selectively to aid the worker where needed"* (Greenfield, 1984, p. 118). Teacher or expert operate in the learner or novice zone of proximal development (Vygotsky, 1966) and it this sense scaffolding is strictly related to cooperative and social learning.

2. Foster *the social display and social modeling of cognitive competence* through group partecipation. In this regard Scardamalia & Bereiter (1992a) speak about knowledge building communities, in which a significant mechanism for instruction is that the *thinking of any learner is made overt* so that it can be examined, questioned and realized as an object of construction by the others, realizing the supportive nature of partecipatory instruction. From a design point of view this means designing learning contexts in which the focus is on collective goals of understanding and judgment rather than on individual learning and performance. As Scardamalia & Bereiter (1992a) say: " *There is little schoolwork of the conventional kind, where the students are working individually but all doing the same thing. More typically, students work in small groups;each group has a different task related to the central problem and plans how to distribute the work among its members*" (pp. 11). Moreover, in these contexts, because the cognitive role are shared between learners, complex tasks could be carried out and attained.

3. Foster the *proceduralization of knowledge,* in which declarative knowledge structures are connected to conditions of use and applicability. In this view of situated learning, knowing a domain is essentially knowing how to move in an environment and *knowing how to use its resources:* the conceptual components of knowledge domains, the physical features of the environment, and the interaction with knowledgeable persons. As Glaser says (1992): *"In the situated view, the concepts and their relations that an individual learns are not only represented in cognitive structures that are activated as problem solving proceeds, but also exist in the knowledge environment that is presented by social, working and disciplinary communities"* (pp.12). From a design point of view this means allowing an instrumental use of knowledge, that is creating contexts for meaningful and indexical learning and not only for "passive" trasmission of knowledge.

4. Foster an educational use of *prior knowledge. This could provide opportunities or constraints* on learning (in this sense replacing the older notion of aptitude for learning) and may be domain specific or involve more general cognitive processing. As Gardner (1991) says: " *What is going on here? Why are students not mastering what they ought to be learning? It is my belief that, until recently, those of us involved in education have not appreciated the strength of the initial conceptions, stereotypes and "scripts" that students bring to their school learning nor the difficulty of refashioning or eradicating them"* (pp.5). What is needed is the design of a context in which some cognitive and educational trajectories between prior knowledge and a model of the final competence to be acquired, are permitted and sustained and where resources that

fit in comfortably with the disparate learning styles and cultural backgrounds that exist in any group of learners, and in particular in adult learners, are made available.

5. Foster the *increasing coherence and integration of a domain of knowledge*, through the design of a learning environment in which it is possible to explore more or less complex topics without a sharp distinction between disciplines. Examples of these are the project models such as those described by Gardner (1991). A project provides an opportunity for learners to reunite previously mastered concepts and skills in the service of new objectives, this means also an eucational use of the prior experience and knowledge of learners..

6. Foster the growth, development and use of *generative metacognitive abilities and strategies for self-regulation and control*, as a means to facilitate the ability to learn to learn. An important educational objective is to foster learners to become increasingly skilled at being active agents of their own learning and of the learning of others. As Scardamalia & Bereiter (1992b) say: *" The challenge is to design environments in which students can use such questions to guide their building of knowledge, thus assuming a higher level of agency in learning,"*. This means fostering the develoment of strategic *abilities to acquire knowledge and skills*, as an important outcome of learning. From a design point of view this means creating learning contexts in which a significant amount of control of learning goals are the immediate responsability of the learners themselves, learners that "are willing to take the risk for understanding" (Gardner, 1991, pp.243).

This is only a first attempt to seek an agreement (or otherwise) on some basic principles of a "design science" of educational contexts. More research is needed to define more precisely the parameters to use in specific settings and especially in different domains of knowledge, but it is hoped that this first list could act both as a research agenda and as a common basis for discussion among those interested, at various levels, in organizational learning and education.

However, what is important to note is that these (or more definite) principles could act as a set of design principles also for the definition of the work context as a learning context. Functional differences between nonformal and more formal educational contexts cannot be seen, what is different is only the way by which these principles could be implemented, that is, the way in which the two contexts should be structurally realized.

For example you could ask: "How can social interaction at work and in the classroom be fostered?". If for the realization of this condition in the classroom, research is already available (for a comprehensive review, see Pontecorvo, Ajello,

Zucchermaglio, 1991), for its realization in the work context what is to be required is the fostering of group work practices, the implementation of a reward system that boosts group work, the availability of both a shared knowledge base and a distributed communicative system, etc.

Another example regards scaffolding. If learners need to be able to establish an apprentice-like relationship with practitioners, who model, explicitly or implicitly the practices of expertise, scaffolding behaviour should be an essential characteristic not only of professional teachers but also of more expert workers. Organizations should allow for this type of social construction of expertise between novice and expert, recognizing it as an essential way of learning through doing and not as a waste of time during work activities.

To find out more about these conditions of realization - here just hypothesized - what is needed is precisely those studies, suggested by Glaser (1992) that try to to understand learning through the design of specific educational context, both formal and non formal. This type of studies, mostly conducted to date in school contexts, should also analyze the organizational contexts and the corporate classrooms. As Scribner & Sachs (1990) say: *"Educational processes in the workplace, though occasionally acknowledge as ubiquitous and significant, remain largely invisible to the research and educational communities "*.

In this regard a study aimed at exploring the relationship between the acquistion of professional expertise and learning features of the work context, developing methods for studying work from a learning perspective and carry out ethnographic studies of work contexts is being planned (Ajello & Zucchermaglio, 1993). Through design and ethnographic observations of both innovative work settings and innovative corporate classrooms, data aiming at exploring both how specific knowledge and competence are acquired in the workplaces and in the corporate classrooms and how organizational contexts should be structured and designed to act as a learning context sustaining organizational innovation will be collected. This is a type of research that moves between a *descriptive* point of view (how competence is acquired in those contexts) and a *prescriptive* (that is how to design educational contexts sustianing those acquisitions: "*Progress occurs not only by moving basic research finding to application, but also by designing applications that test research and theory, and that can raise new questions of basic science for investigation*" (Glaser, 1992, p. 13).

3. Conclusions

Organizations should learn to learn. To this end the best way seems to be for organizations to become very effective learning contexts in themselves, as well as through a better design of corporate classrooms. The work context could be designed to take advantage of the learning potential embedded in "normal" productive activities. The design of such rich learning contexts, in which a model of real cognitive apprenticeship could be applied, would be a very promising research area for the cognitive science of instruction also contributing to the creation of a new learning culture in organizational contexts.

In particular starting from this analysis some points are emerging as problematic:

* the search for a organizational coherence between the educational features of the organizational context (and of the learning that goes on in that context) and those of the formal educational context, through the definition of a common design project (in tems of design principles, methodology, professionals involved, etc.) of the organizational learning environment. The design of a formal educational context should be a part of the entire organizational design and that also reify a theory of organizational learning;
* the need to design and analyze innovative organizational contexts as essential research contexts in order to understand the processes and characteristics of organizational learning. Only through this prescriptive-descriptive methodology could it be possible to reveal and define the characteristics of efficient organizational contexts.

References

Ajello, A.M., & Zucchermaglio, C. (1993). The acquisition of expertise in the organizational learning environments. Paper presented in Symposium "Learning in Work Context" 5th European Conference for Reseach on Learning and Instruction, Aix-en-Provence (France), August-September

Brown, J.S., Collins, A., & Duguid P. (1989). Situated cognition and the culture of learning. Education Researcher, 18, 1, 32-42

Bruner, J.S., Olver, R.R., Greenfield, P.M., et al. (1966). Studies in cognitive growth. Wiley & Sons: New York

Bruner, J. (1989). Culture and Human Development: a new look. Invited Address to Annual Meeting. Society for Research in Child Development. Kansas City, MO, April

Childs, C.P., & Greenfield, P.M. (1980). Informal modes of learning and teaching: the case of Zinanteco Weaving. In: N. Warren (Ed.), Studies in Cross Cultural Psychology (vol.2). New York: Academic Press

Collins, A. Brown, J.S., & Newman, S.E. (1989). Cognitive apprentiship: teaching the crafts of reading, writing and mathematics. In L.B. Resnick (Ed.), Knowing, learning and instruction: essays in honor of Robert Glaser. Hillsdale, NJ: LEA

Cole, M. (1989). Cultural Psychology: a once and future discipline? In: J. Berman (Ed.) Nebraska Symposium on Motivation: Cross cultural perspectives. Lincoln: University of Nebraska Press

Daft, R.L., & Weik, K.E. (1984). Towards a model of organizations as interpretation system. Academy of Management Review, 9, 2, 284-295

Gardner, H. (1983). Frames of mind. The theory of multiple intelligences. New York: Basic Books

Gardner, H. (1985). The mind's new science. An history of the cognitive revolution. New York: Basic Books

Gardner, H. (1991) The unschooled mind. New York: Basic Books

Glaser, R. (1992). Knowledge and learning: Rip Van Winkle awakens from learning theory. Paper presented at NATO ASI. In: S. Vosniadou, E. De Corte, & H. Mandl (Eds.), Technology-based learning environments. NATO ASI Series F, Vol. 137. Berlin: Springer-Verlag 1994

Greenfield. P.M. (1984). A Theory of the teacher in the learning activities of everyday life. In: B. Rogoff & J. Lave (Eds.), Everyday Cognition. Cambridge MA: Harvard University Press 1984

Halpern, J.Y., & Moses, Y. (1990). Knowledge and common knowledge in a distributed environment. In: J. Allen, R. Fikes, & E. Sandewall (Eds.), Principles of knowledge representation and reasoning. Morgan Kaufmann Publishers

Kay, A. (1989). Computers make slow progress in class. Science, 244, 906-909

Latour, B., & Woolgar, S. (1986). Laboratory life: the construction of scientific facts. Princeton: Princeton University Press

Lave, J. (1988). Cognition in practice: Mind, mathematics and culture in everyday life. New York: Cambridge University Press

Martin. J. (1982). Stories and scripts in organizational settings. In: A.H. Hastorf & A.M. Isen (Eds.), Cognitive and social psychology. Amsterdam: Elsevier

Norman, D. (1989). Cognitive artifacts. Workshop on Cognitive Theory and Design in Human-Computer Interaction, Kittle House Inn, Chappaqua, NY, June

Orr, J. (1990). Sharing knowledge, celebrating identity: war stories and community memory among service technicians. In: D.S. Middleton & D. Edwards (Eds.), Collective remembering: memory in society. Beverly Hills, CA: Sage Publications

Resnick, L.B. (1987). Learning in School and Out, Educational Researcher, 6 (9), 13-20

Rogoff,B., & Lave, J. (1984). Everyday cognition. Cambridge, MA: Harvard University Press

Scardamalia, M., & Bereiter, C. (1992a) Computer Support for Knowledge-Building Communities. Paper presented at the Symposium "Instructional theories underlying the use of networked computers in the classroom. AERA, San Francisco, CA, April

Scardamalia, M. & Bereiter, C. (1992b). Understanding expertise. La Salle, IL: Open Court

Scribner, S. (1984). Studying working intelligence. In B. Rogoff & J. Lave (Eds.), Everyday cognition. Cambridge, MA:Harvard University Press

Scribner, S., & Sachs, P. (1990). A study of on the job training. Technical Report, February

Seely Brown, J., & Duguid, P. (1991). Organizational learning and communities of practice: toward a unified view of working, learning and innovation. Organizational Science 2/1 (February), 40-57

Suchman, L. (1987). Plans and situated actions: the problem of human-machine communication. Cambridge, UK: Cambridge University Press

Vygotsky, L.S. (1966). Pensiero e linguaggio. Firenze: Giunti-Barbera

Winograd, T., & Flores, F. (1986). Understanding computers and cognition: a new foundation for design. Norwood, NJ: Ablex

Wood, D., Bruner, J.S., & Ross, G. (1976). The role of tutoring in problem solving. Journal of Child Psychology and Psychiatry, 17, 89-100

Learning is an Active and Personal Process

Bente Elkjær

Department of Informatics and Management Accounting, Copenhagen Business School, Howitzvej 60, DK-2000 Frederiksberg C, Denmark

Abstract. All participation in in-service training is always part of a personal learning-project. For the individual participation in in-service training makes sense either in relation to her present job-situation or in relation to her private future career plans - or most frequently in relation to both! To include both these elements within a personal learning-project the individual adopts a double perspective on evaluation of in-service training. This is comprised partly of an employee perspective, where the present organization and work-situation is the context of meaning, partly of an individual perspective where the context of meaning is constituted by private work and future career plans. The notion of a double perspective on evaluation of in-service training may also benefit the development of theories on organizational learning.

Keywords. In-service training, making-sense, individual and organizational learning.

1. Introduction

"He had come to her as a teacher, but they had forgotten her lessons. Still, their first meeting, their first lesson was a moment of learning above any other, for when the girl happened to glance at her feet, before she pulled her stockings up over them, she noticed something she would never have paid attention to before: how filthy her feet were." (Hald¢r Laxness: Salka Valka. Gyldendals Tranebõger, Copenhagen 1966/1934: 74)

I have included this quotation to illustrate that we are quite unaware of what goes on in processes of learning. Learners learn unpredictable things from learning processes, e.g. having dirty feet, even if this is not scheduled! Only the individual learner is able to assess the benefits derived from a learning process. The assessment is made on the basis of relevant criteria in relation to her present situation, hopes and future wishes. This is my main argument for focusing on the importance of the personal learning-project in evaluation of participation in training and education with relation to work, i.e., in-service training.

In evaluation of in-service training there may at least be two criteria of relevance inherent in the personal learning-project. These criteria could be of a private and an organizational nature reflecting the context of meaning for assessing the relevance of the learning-project and the benefits for the learner. One may also say that on the one hand, there is an individual perspective where the context of meaning is constituted by her private assessment of her present job-situation and career plans for the future. On the other hand, there is an employee perspective where her (present) organization and work-situation is the context of meaning. The following quotation can illustrate this:

> *"I learned a few basic things, got a little knowledge, but it wasn't anything I could go home and really use, as it is something we hardly use. So that was kind of a shame."*
> (Ip-4, D1).[1]

As a private individual this interview-person achieved a benefit, but it was irrelevant for her as an employee in her present organization and work-situation. Such answers to the question of derived benefits from participation in computer-courses puzzled me in the beginning of my interpretation of data from a research project on evaluation of participation in basic computer training made by members of the Danish trade union HK (the Union of Commercial and Clerical Employees in Denmark). I was not able to state an either-or benefit since the data comprised both. In the above quotation the benefit was only evaluated as relevant, when the interview-person regarded herself as a private individual, and not as an employee. This was the background for creating the concepts of an employee perspective and an individual perspective in order to grasp "the whole story" about trainees' assessment of their participation in in-service training.

When it is an employee perspective the assessment of participation in education is primarily viewed in relation to the context of the organization. And when it is an individual perspective the assessment is primarily viewed in relation to the context of private work and career plans. The two perspectives do not always exclude one another, but can exist side by side. There may, however, be a significant difference as to the emphasis the individual trainee places on the perspectives.

---

1 Ip means interview-person. D1 indicates that the interview-person has participated in the course Data Processing 1.

Before I go more into the design and results of the study I wish to introduce the context of the Danish society with regard to educational levels on the labour market.

2. Facts about In-Service Training in Denmark

The composition of the Danish labour force can be divided into unskilled workers (approximately 50 per cent), skilled workers (33 per cent) and employees with a further education constituting the remaining part. The tendency shows that those who are already well educated have the highest rate of participation in in-service training (Hansen, Laursen and Aarkrog, 1991). In other words, participation in in-service training activities shows a social distortion to the advantage of well educated employees[2] . While approximately half of the wage earners in the group with no more than 7-10 years of schooling have participated in in-service training at least once in their occupational career, the corresponding figure is 80 per cent among wage earners with a further education[3].

However, it is not only the very participation in in-service training that reflects a social distortion. So does the *motivation* for in-service training (Anker and Andersen 1991). Thus, there are fewer unskilled workers who express a wish for receiving more in-service training. According to Anker and Andersen (1991) this aspect is *firstly* connected with these workers' former (negative) experiences in the educational system. *Secondly*, the work functions performed by wage earners, who have had no more than 7-10 years of schooling, often require less in-service training than, for example, the work functions performed by wage earners with a further education. Or more correctly, it is *seen* in this way by wage earners on lower educational levels. Both of the above factors are assumed to have a negative influence on the motivation of these wage earners in relation to in-service training.

However, social distortion with regard to participation in in-service training is not only due to social distortion of motivation, i.e. due to employees' wishes - or lack of same - concerning in-service training. Employees also meet *opposition* from enterprise quarters concerning participation in in-service training during

---

2 The study is based on interviews with approximately 3,000 wage earners between 20 and 66 years of age and on approximately 1,300 replies from questionaires mailed to private enterprises with no less than five employees. See Anker and Andersen 1991.

3 The distortion with regard to employees' participation in in-service training is also confirmed in the so-called Price Waterhouse Cranfield project, which is a study of personnel conditions in 10 European countries. See Larsen 1991 (ed.).

working hours. For instance, a study on employee training for state-employed personnel shows that *office personnel* encounters most obstacles to participation in employee training. Apart from reasons such as economy and work pressure many believe that management has a negative or restraining attitude to the idea that office personnel should receive any other training than the most necessary *user training* in connection with learning about data processing, etc. (Hansen, Laursen and Aarkrog, 1991) A market analysis carried out by *HK ModulData* confirms that the public sector gives priority to in-service training of managers[4] . In-service training of rank and file employees, e. g. HK union members, has a lower priority. Thus, the public sector prefers to hold internal courses (i.e. courses arranged within the framework of the organization itself) for non-managerial personnel (i.e. the majority of HK-members) owing to its emphasis on a close connection to the technique and the specific, daily working process. In other words, it is assumed that rank and file employees have no need for anything else or more than *goal-directed and application-oriented* in-service training.

3. Design of Study

I wanted to examine how HK-members (primarily, office employees, but also shop assistants, laboratory workers, etc.) themselves evaluate the benefits they have derived from participation in basic computer training courses. The courses were offered by *HK ModulData*. HK ModulData is an educational organization founded by the trade union HK. I did this by studying trainees' evaluation of the benefits they derived from a number of basic computer training courses offered by a local HK ModulData centre. With evaluation of benefits as the centre of focus in my study I have, moreover, examined trainees' motivation for in-service training as well as their previous learning and educational experiences. The interpretation of all questions in my study, including trainees' evaluation of their benefits, has been based on the view that trainees are both employees in organizations and private individuals. This is an attempt to comprehend the meaning of participation in in-service training as something else and more than a project related to a specific work function. Thus, my intention has also been to illuminate the fact that all participation in learning and education is part of a personal learning-project. However, I also felt that this approach towards evaluation could provide us with ideas as to the development of so-called learning organizations.

HK ModulData's courses in Data Processing 1 and 2 and in Electronic Word Processing 1 and 2 were the objects of my study. The courses took place in a local

---

[4] See report: Development and Efficiency Increase of HK ModulData (1991).

HK ModulData centre in the spring term of 1990. The trainees in the study were employed women at the age of 36-50. Thus, my study focused on both the courses and the trainees that constitute the majority within the framework of HK ModulData. The method of my study was conducting interviews with trainees, who wished to participate in the study. Out of 47 women 32 participated[5].

My choice of method (i.e., interviews with trainees) was determined by my basic view on how evaluation of benefits derived from participation in training can be assessed. On this subject I have been inspired by the late American philosopher and educator, *John Dewey* (1859-1952). According to his view, to learn (or learning) is *an active and personal process*, in which the individual is the subject of the process of making sense of participating in learning. The individual is the one to pose the relevant criteria (Dewey 1966/1916; 1974). But why the interview form? The reason is that this study is of an exploratory nature, and application of a questionnaire study is only viable, when it is a matter of correlations between clearly delimited and unequivocal phenomena. This was not the case.

Of course, the problem with conducting interviews is whether the results will apply to other groups than the interviewed one. In my view, it is first and foremost a question of whether it leads to a better understanding of the field under study. In other words, does an interview study of trainees' evaluations of benefits give rise to a new and better account of in-service training of employees compared to the prevailing one?[6] I claim that if we involve employees' personal learning-projects, we will not only get a better understanding of employees' assessment of in-service training, but also of what it takes to develop learning organizations. Whether I am right or not is for the reader to assess!

4. The Organization HK ModulData

HK ModulData is a young organization. It was set up by the trade union HK in 1986, as they found that the public educational system did not provide satisfactory computer training for their members. Besides, by means of HK ModulData HK aimed at gaining an influence in the development of information technology in public and private organizations by training their members to become *qualified participants* in all processes concerning introduction and application of information

---

[5] For practical reasons, among other things transportation problems, the actual number of interviewed persons was 28.

[6] For a more detailed presentation of qualitative methods see, e.g., Giorgi 1975, Kvale 1983 and 1987. For a brief presentation see Jerslev, Tufte and Elkjær 1987.

technology. In other words, HK aimed at qualifying their members beyond a user level.

Today, HK ModulData has 22 local training centers with a total of 27 classrooms located all over Denmark. The pedagogical idea behind HK ModulData is precisely its structure of training, which consists of different levels, and which provides centrally produced training materials, extensively based on case studies. In addition, HK ModulData believes that a broad and general understanding of information technology will provide the vocational basis for any further specialization. This is why the course Data Processing 1 in principle is the basis for all other courses.

HK ModulData has been a rapidly growing organization. In 1986 it offered 252 courses with participation of 3,091 trainees, and by 1990 there were 1,613 courses and 17,368 trainees. Most courses are elementary courses in information technology and application (word processing, spreadsheet, etc.). In 1990, approximately 2/3 of the courses were held in Data Processing 1 (437), Data Processing 2 (122), Electronic Word Processing 1 (341) and Electronic Word Processing 2 (84), respectively. There were 111 courses in Spreadsheet and 128 in Computerized Accounting. By way of comparison it may be mentioned that only 9 courses aimed at actual systems analysis and systems design.

The trainees on the HK ModulData courses reflect the composition of members in HK with regard to gender and age. Thus, the majority of trainees on the HK ModulData courses is women (86 per cent in 1990), and the largest age group is trainees between 36-50 years of age (41 per cent). In 1989 almost 3/4 of the trainees were office employees. 30 per cent of the trainees participated during working hours on full pay, 28 per cent participated in their spare time, while 42 per cent received unemployment benefits. 37 per cent worked full time, 16 per cent worked part-time, 44 per cent were unemployed and 2 per cent were in process of training.

Studies[7] conducted by HK ModulData show that by and large the trainees are *satisfied* with HK ModulData as an educational organization. However, there is a certain dissatisfaction with the course administration (waiting lists and application procedures) and the computer equipment - both as regards machinery and programmes. It is considered to be *obsolete*.

7 Among others, the previously mentioned Market Analysis, but also a so-called Customer Relations Study, which covered 502 users, 58 decision makers and 120 internal employees in HK ModulData.

5. The Trainees in the Study

All the trainees in my study have participated in at least one of the previously mentioned computer training courses (Data Processing 1 & 2; Electronic Word Processing 1 & 2) at a local HK ModulData centre. The following five categories of HK members participated in my study: secretaries and case workers (9), routine office workers (e.g., switchboard operators) (5), librarian assistants (6), book-keepers (3) and non-office employees (5)[8].

In HK ModulData's own surveys they distinguish between private and public employment. And so do I. In my study 16 out of 28 trainees were *employed in the public sector*, while 12 were *employed in the private sector*. The division expressed in percentage corresponds fairly well to the overall composition of HK ModulData trainees. However, compared with all the trainees in HK ModulData there were relatively more trainees in my study who had *paid leave*, i.e. they were allowed to participate in a course during working hours instead of in their spare time. In percentage, 68 per cent had paid leave in my study, while, on the whole, only half of the trainees in HK ModulData have such working conditions.

In principle, Data Processing 1 is the basis for all other courses. But out of the 28 trainees that I interviewed only 21 had chosen Data Processing 1 as their first course. 6 trainees had Electronic Word Processing as their first course, while a single participant had started with a course in Data Base. Thus, a fourth of the trainees had chosen another course as their first choice. 16 trainees had participated in 2 courses, while 4 had participated in 3.

In total, 19 trainees had participated in courses during working hours and 9 in their spare time. The average age of the participants was 43 years, the youngest being 36 years old and the oldest 49 years old. 17 participants had taken a lower secondary school leaving examination or an intermediate school examination. 10 participants had left school after the 7th, 8th or 9th form, while a single participant had passed the examination for the G.C.E., A level. All participants with the exception of 3 had received vocational training. Altogether, the group had very little training in data processing prior to the HK ModulData courses, and they had only had few experiences with data processing on their jobs. Approximately half of the trainees (the secretaries, the case workers, the book-keepers and the laboratory technicians) used data processing in various ways in their work functions at the time of the interviews.

---

[8] The figures in brackets indicate the number of participants in each group. The non-office employees were shop assistants (2), laboratory technicians (2) and dental nurses (1).

6. Learning in Organizations and Individual Learning

Within organizational learning theories there seems to be a certain agreement about viewing an organization as an entity that does not learn - only individuals learn[9]. However, there seems to be some dispute about how to perceive of organizations and individuals, respectively. In my view, these two concepts are mutually constitutive. They do not constitute two well-defined groups, but they define each other exactly, so to speak, both in images presented (in theory) and in the actual life of an organization (in practice). In my perceiption, individuals are not containers of thoughts and ideas. They are being formed and developed through interaction with the social world (the organization)[10].

However, when we want to comprehend employees as members of an organization, it is relevant to include the available theories on learning in organizations. There are three schools of thought that often appear in the literature. Their common denominator is that learning first and foremost is viewed from the perspective of the organization or the enterprise. The crucial criterium of evaluation is whether participation of employees in in-service training contributes to the life and development of the enterprise. Thus, the dominant factor is first and foremost an employee perspective. However, the three schools differ in degrees in their approach to employees as individuals.

One of the schools within the theories on *learning in organizations* is called *organizational learning*. In this school it is primarily the organization that has to learn something. Organizational learning focuses on how an enterprise can collect and apply information and knowledge derived from itself and its environment. Through these processes an enterprise will learn something about itself and its environment[11]. In another school within the theories on learning in organizations the important aspect is to develop and train employees in such a way that they can adapt to the present technological and economic level of an enterprise. Here the focus is on the so-called *human resource development*[12]. A third theoretical school

---

9 See, e.g., Argyris & Schön 1978.

10 This is of course in line with the ideas of John Dewey (op. cit.; 1974). An early account of this view can also be found in Bernstein's (1960) introduction to the philosophies of John Dewey. Newer accounts with slightly different concepts may be found in, e.g., Lave, 1988; Lave & Wenger, 1991.

11 See, e.g., Fiol & Lyles, 1985; Levitt & March 1988; March & Olsen 1979; Shrivastava 1983.

12 See, e.g., Goldstein 1989; Goldstein & Gessner, 1988; Latham, 1988; Ostroff & Ford, 1989.

focuses on the question of how employees are *socialized* into participating in the work of an enterprise[13]. This is the school that is closest to the individual perspective, as it approaches employees not only from an organizational angel, but also from an individual angel.

The theme that is primarily emphasized by theories on learning in organizations is personnel's identity as employees. The theories are important in order to understand evaluations of benefits from in-service training in relation to trainees' identity as employees. Although, theories on socialization in organizations (also) comprise the individual perspective, it is primarily a concept of adaptation. If an individual's ideas about development of her/his working life can be combined with being a member of an organization, it would allow personal interest in development of qualifications to meet with the need of an enterprise for qualified employees. But perhaps this is not the case. And perhaps it is merely productive to view individuals on the basis of their ability to adapt as employees? However, if the aim first and foremost is adaptation, how is it possible to secure development and integration of new qualifications and skills in an enterprise?

The theories on *individual learning* do not have answers to all these questions. They have focused more extensively on understanding how individual employees acquire new knowledge and skills. However, it is a weak point that the organization as a unit for developing qualifications by and large is absent within these theories.

The purpose of employing theories of individual learning is to understand employees in enterprises from an *individual perspective*. In other words, to view learning as part of a personal learning-project. In doing so the concept of *experience* seems to be crucial. Once more, *Dewey* is my source of inspiration. His concept of experience is based on *the actions and subsequent reflections* of an individual (Dewey, 1974). Experiences are gained when individuals act and *reflect* in relation to themselves and their environment. For Dewey experience is not just the sum of an individual's *practical* skills and knowledge. Nor is it merely a passive reception of sense impressions. Experiences are gained through a *process*, which has both *an active and a passive side*. On the one hand, experience is gained by acting (actively), on the other hand something will take place in an individual, when he/she acts (passively). Dewey rejects a so-called dualistic view of experience, i.e., a separation of body and intellect; action and knowledge. On the contrary, gaining experience is always directed towards something. It is in relation to action (activity) that knowledge and skills (qualifications) arise, for instance, when an individual reflects over the *meaning* that she/he ascribes to an action.

---

13 See, e.g., Driver, 1988; Feldman, 1989; Granrose & Portwood, 1987; Van Maanen, 1977.

The value of viewing the process of experience (or acquisition of qualifications) with Dewey at hand is first and foremost to become aware that we need to include the *context of meaning*, on the basis of which individuals consider their benefits. In other words, to comprehend that the same person may consider evaluations of benefits both on the basis of a personal learning-project that includes a private perspective and from the perspective of an employee. In the latter context of meaning it is most often (the notions of) the benefits that an enterprise will derive from allowing an employee to participate in a course that are evaluated - and not private benefits and relevance.

Phrased a little differently, we can say it points out that acquisition of qualifications always is determined by the situation, in which it acquires a meaning for the individual[14]. However, evaluations of benefits from in-service training are primarily related to the value that benefits will have for an enterprise. By definition (and tradition) in-service training is aimed at improving qualifications in relation to the work functions of employees. This forms the basis for the very legitimation of participating in in-service training, specifically (but not only) when the training takes place during working hours (equal to the hours of the enterprise). But it are individuals who participate in in-service training. In other words, evaluations by participants in in-service training will always be based on their own learning-projects, too.

7. Experiences Gained in School and from Education

Other studies (Anker & Andersen, 1991) have shown that negative motivation will follow negative experiences with school and education. In my research this is not the case. The negative experiences can be changed due to voluntariness, pedagogical principles and the "role" of trainees in the learning environment.

In my results I have found that it is a characteristic feature that several trainees had negative primary school experiences, but later gained *positive* experiences in education and/or in-service training activities. Several factors have been mentioned as decisive for this change in perception. In the first place, it is important whether school attendance and/or training is *voluntary or compulsory*. Attending compulsory primary school could result in a negative experience, while voluntary participation in training may create a positive experience. In the second place, *pedagogical development* of teaching is mentioned as an important factor in connection with the change in experience. During their primary school years many

---

14 It is so-called situated learning. Jean Lave, among others, has dealt with this subject. See Lave, 1988; Lave & Wenger, 1991

trainees experienced an authoritarian educational practice, while in later years they have been subjected to a more dialogue-oriented pedagogy. A dialogue-oriented pedagogy brings out trainees' resources to a far greater extent than the authoritarian pedagogy. In the third place, the experience of belonging to a socially lower class in school can have a negative influence in terms of school experiences. On the whole, a stigmatization of being poor, stupid or merely "wrong" may result in a negative school experience. A change of context, f.ex. attending a continuation school, a commercial school or taking another education can lift the yoke of stigmatization and release learning resources of a hitherto unknown extent for a student/trainee. Or merely a parallel choice of context unrelated to the school may result in a feeling of being a success.

"I believe I found other places (than the school) where I could sort of stand up for myself, as I have never gone in for games with any of the pupils in my class. (...) It was kind of more real being on my own ground. Also because others do not remind you that there is something you are not good at."
(Ip-11, D1).

Our experiences in school and of educational activities are formed in relation to the social context that the learning environment constitutes. If a social context is experienced as negative, as for example due to the above mentioned reasons, a learning blockage may arise, which may be difficult to break through. In the case of Ip-11 her learning blockage was actually never removed in relation to the learning environment at school, but in turn she manifested herself in another social context: on the sports ground. Stigmatization as "a weak pupil" need not be in relation to the school in general. It may only be in relation to a single subject.

"I really liked going to school. I did not study especially hard, but I had fairly good grades despite very little effort. (...) But I was not strong in arithmetic and mathematics, and today I am working as a book-keeper. Isn't that strange?"
(Ip-14, D1).

"I despised any kind of arithmetic, and today I am doing a lot of arithmetic. It is kind of funny. It is part of my job that I must do a lot of calculation, (...) so now I am using what I didn't use back then. Now I am using it."
(Ip-21, D1 & D2).

Today, both Ip-14 and Ip-21 in practice work with a subject area - arithmetic - that they either disliked or did not perform too well in at school. The content of the subject has achieved a practical application value for both of them. The learning environment in the school where they studied arithmetic and mathematics did not provide a successful context. But the social practice does, i.e., the job environment in which they use their knowledge today. They are both surprised by this development. Probably, the reason is that we have grown accustomed to think of knowledge and qualifications as something we acquire at a certain time in life. Subsequently, we store them in a container made for this purpose (our head) and release them according to need. However, many people probably have had similar experiences in terms of knowledge and qualifications as Ip-14 and Ip-21. In other words, in one social context you may be weak in something in which you are competent in another social practice.

Thus, negative experiences in school - both in general and in terms of specific subjects - are gained in a specific learning environment, in a specific social context with a specific stigmatization of participants. By disintegrating a social context stigmatization can be lifted. As a result learning resources in individual participants can be released. In this way, previous, negative school experiences can be turned into positive experiences of education and in-service training. However, it is essential that the educational practice in the "new" learning environment does not make the student/trainee relive the negative effects of stigmatization. For instance, in the event that participation in learning and education is experienced as compulsory, if the pedagogy is authoritarian, and/or if (negative) stigmatization patterns are formed in the learning environment.

8. Motivation for In-Service Training

Research has shown (Anker & Andersen, 1991) that lower qualified jobs result in negative motivation. But my study shows that motivation does not depend on qualifications. Some trainees hide their motivation for fear of being overqualified, and as a result they define their wish for further training into their private-life.

According to the previously mentioned Market Analysis conducted by HK ModulData it appears that all HK-members are very interested in in-service training, and moreover, that they are active participants in various course activities[15]. In other words, HK-members are *motivated* for in-service training. Two analogous factors play a role in connection with motivation for in-service training. It is a combination of *necessity* in relation to continued employment in the enterprise and individual *inclinations* for further occupational development on the job. It is necessary (or such are the experiences of trainees) to qualify oneself in order to guard against being fired. At the same time, trainees also wish, through

---

15 See report: Development and Efficiency Increase of HK ModulData (op. cit.).

participation in training, to acquire qualifications that will enable them to participate actively in the development of information technology at their work place.

"If they install such machines (computers) on my (work place), I would like to be one of those who knows how to operate them. That is why I want in-service training."
(Ip-4, D1).

"You have to (take in-service training), because you have to keep up to date, or you will be fired."
(Ip-10, D1).

Both Ip-4 and Ip-10 assess their motivation for in-service training in relation to their respective enterprises. It is in order to stay in their jobs that they wish to attend in-service training. They both believe that their chances of staying employed will improve, if they participate in in-service training. But what should they do, if their work functions do not require any further training? How can they ensure their continued employment in the enterprise?[16]

"If I am competent in the work function I have been placed in, they (management) might think it's a bit silly that I dream about going to a course. If they (management) only can see me as placed where I am now and not in any other function, whereas I myself can imagine and see myself in other functions. Then, there's a difference."
(Ip-7, D1).

"You see, I have enough training for the job I do today. It cannot interest them (my workplace) greatly that I improve my training. Then, I might feel inclined to look for something quite different."
(Ip-10, D1).

Both Ip-7 and Ip-10 have (or find that they have) sufficient qualifications in order to perform their present work functions. For them further training will only mean that they would become "*overqualified*" for their present job - and who would be interested in that? According to their own assessment their respective managers are by no means interested in such a development. In my interpretation, they would overstep an invisible barrier for themselves as employees, if they demanded to participate in further in-service training during working hours. It is an invisible barrier, beyond which they begin to think and act solely on the basis of their own

---

16 See study previously referred to (Anker & Andersen, 1991) for a similar focus on this complex of problems.

personal learning-project, i.e., their individual wishes in terms of qualifications, they do not decidedly need in their present work function. On account of their employers they do not wish to overstep this barrier. In other words, they do not wish to signal openly that they would like another work function - more challenging - perhaps even in another enterprise by improving their qualifications. The solution is to participate in courses and other adult education in their spare time - and to define it explicitly as belonging to an individual, personal context of meaning.

> *am studying to take a bachelor degree in commerce belongs to my private life. (...) Even if it is very connected with my work, and I can use some of the things I learn at my job. (...) But it (the Bachelor of Commerce study) is something I do for myself, and then it is useful for me on my job.*
> (Ip-5, D1 & E1)[17].

> *"That (participating in in-service training) is something I do in my spare time hoping that at one point I will be able to use it (what I am learning), if I can find a job. (...) They (management) do not know that I am learning data processing. (...) That (I am qualifying myself) is definitely a private matter."*
> (Ip-21, D1 & D2).

Both Ip-5 and Ip-21 define their participation in in-service training as an entirely personal learning-project. It does not concern the enterprise. Thus, their personal learning-projects are not within the context of meaning, of which they are a part as employees. They do not wish to justify or to evaluate the importance of their participation in in-service training in relation to their present work function. However, it does not mean that the acquired qualifications are useless. But they are used on the basis of the trainees' own personal evaluations and future aspirations.

To a large extent motivation to participate in in-service training is based on the employee perspective, but this perspective may also act as a disincentive. Especially, for trainees who do not need (or do not find that they need) any further qualifications in order to perform their present work function. In such a situation, the personal learning-project will gain the upper hand for trainees who, nevertheless, choose to qualify themselves. This can be done by defining participation in in-service training as a private matter, on which the workplace has no influence. As a result, trainees do not have to justify any educational wishes to the enterprise. Conversely, the enterprise cannot demand to get more qualified

[17] E1 indicates that the interview person has participated in an Electronic Word Processing course.

employees - nor to get employees who are more independent and enterprising. In this way, qualifications will remain as an individual's private property instead of being used for the development of the enterprise - and for the development of employees in the enterprise.

9. Evaluations of Benefits from Courses

In general, satisfaction or dissatisfaction with participation in in-service training activities depends on whether a trainee's expectations have been met or not. These expectations are directed towards at least two aspects. First, there is the vocational - pedagogical aspect. In other words, evaluations of benefits derived from the subject content imparted, its relevance to trainees, and especially the teacher's influence on this aspect of the course. Secondly, it is necessary to include the context of meaning in relation to which trainees evaluate benefits, if we wish to understand why evaluations can be positive and negative, simultaneously. I.e., we have to distinguish between the enterprise and the employee perspective on the one hand and the private sphere and the individual perspective on the other as contexts of meaning. It depends on the individual trainee's evaluation whether it is in relation to an employee perspective or in relation to an individual perspective that she/he participates in in-service training. The two contexts of meaning co-exist, unless trainees, as illustrated above, have their own explicitly personal learning-projects with the intention of leaving their present workplace or merely to look for new challenges in their work.

First and foremost, the vocational-pedagogical aspect of computer training focuses on how trainees experience the relationship between theory (instruction) and practice. It is important for trainees to be allowed to test their theoretical knowledge on the keyboard, to convince themselves of the application value of the theory, and to feel that they are getting to know the specific tool they set out to learn.

"At any rate, when we had theory I was fairly impatient to go and turn it into practice, to test it with my fingers."
(Ip-17, E1).

"So we have all that theory, and at the same time we have a practical man, who can show us examples from our day-to-day lives. I thought that was pretty good."
(Ip-10, D1).

Many trainees found that the courses (especially the basic course in Data Processing 1) contained too much theory or referred to too many technical subtleties. In my view, this is a problem in all teaching, where a specific tool, f.ex., a computer, is employed. The problem is how much *background knowledge* about information technology is necessary in order to be an independent and creative computer user? For instance, how much does a trainee has to know about operating systems? About algorithms? About the binary number systems? There is no agreement among the so-called computer experts. But professionals within the field of information technology often believe that it is important to have a thorough foundation in information technology. They know that it is essential for independent work. On the other hand, beginners are primarily interested in learning how to use the tool data processing, i.e., to "try" the machines and the keyboard as much as possible. Here expertise (not very established) in information technology based on f.ex. computer science is met by requirements from the practical working life employing data processing as a tool. Each within its own perspective - science and everyday life - can claim to be "right". The question is how they can reach an agreement. I have come across a number of suggestions, but never any concrete proposals for how to solve this problem.[18]

As in all teaching the teacher is of course an important factor too. But in computer training it is especially the teacher's way of handling *the demand for practice* that is crucial to trainees. In other words, how well did teachers understand that trainees wanted to work on the machines.

> *"We had a teacher who had structured his teaching in such a way that we had so much theory before we were allowed to work on the machines. It was so abstract. (...) My fingers itched to try how it worked. (...) I wanted to get rid of that fear of the keyboard you have, when you have never worked with a computer before. You are afraid because "what do I do now, and how do I get back to where I should be?"*
> (Ip-19, D1 & E1).

The trainees are disappointed, when they learn how exposition of theory and machine operation are distributed in the teaching. The reason is that when they enroll in a computer training course, they expect that the teaching will focus primarily on machine operation. This is what distinguishes computer training courses from other types of instruction. In computer training courses there is a specific tool which trainees must learn how to use. Here the expertise of (the teachers and) the organizers of the courses collides with the trainees' expectations. The organizers wish to provide more than a course focusing on the keyboard,

---

[18] For instance, Bjarne Herskin has studied how to make a information technology pedagogy, which is based more on users' everyday life than on expertise in information technology. See Herskin, 1991.

while trainees first of all wish to have their fear of the keyboard removed, and therefore want to work on the keyboard as much as possible.

When trainees define *the enterprise* as the context of meaning for their evaluation of benefits derived from in-service training, their most important concern is that the course has (or that they find it has) a concrete *application value*. In this respect, it is important whether there is a correspondence of systems between work and course. If not, or if trainees only use data processing to a minor extent in their present job, they will experience their participation in a course as a "waste of time". In other words, trainees find that it is "wrong", or they feel "sorry" for the enterprise, if they have used time on participation in a course that has no direct application value.

> *"Well, (we learned) how we should enter the machine, and then they told us what it was, what took place inside the machine, and all that was very well. But there was something I thought I didn't get enough out of (...). I wanted to learn how to enter the machine I have at my job. You know, so I could use it immediately, but it was other codes and other keys you had to press, and therefore I didn't think I had anything to show for my efforts. (...) I thought it was a waste of time, but perhaps it is because I already work with data processing, and what I needed to learn were quite specific things I just could apply at work."*
> (Ip-16, E1).

> *"I guess, I feel that I had to be gone for a week, and they, that is my workplace, didn't really get anything out of it, as such. For my part I like to learn something new and all that. So for my part, I would jump at all the courses I could possibly go to, if they would let me. But I do believe that it should be a little related to your work, when it is during working hours."*
> (Ip-28, D1).

Both Ip-16 and Ip-28 are motivated to participate in in-service training and to learn something. They also find that they do, but the positive benefits are overshadowed by the lack of application value. It is the lack of application-oriented qualifications that trainees can use directly in their work, which makes them evaluate benefits negatively. In other words, it is in relation to the context of meaning in which they are employees that the negative benefits and the lack of relevance are significant. It is not in relation to an individual perspective of learning. This is pushed aside and defined as less important. In my interpretation it is owing to their loyalty (or their understanding of loyalty) to the enterprise that they only consider the direct application-oriented qualifications as important. Thus, they enter into and consider things from the perspective of the enterprise, namely that participation in in-service training must be goal-directed and application-oriented in order to be legitimate. Mind you, it is a goal-directed view that focuses on attending to present work functions and not on application of qualifications in the enterprise in general.

However, in order to tell "the whole story" about the trainees' evaluations of benefits derived from in-service training, it is important to include benefits in relation to an individual perspective. All the interview persons that I have referred to above admit that they have learned something, acquired a knowledge. But it is defined as being beside the point in relation to the context of meaning, in which they attend to quite specific work functions. Benefit evaluations in relation to an individual perspective include, as mentioned, the feeling of having learned something. It may be an experience of data processing as something they are able to learn; perhaps an experience of having acquired a background knowledge that goes beyond user courses, or perhaps they have found that the personal barriers and creativity they have had up till now have been extended.

> *"I think I got something out of it, (...) at any rate that I actually could learn some of that (data processing. Before the course) I thought: "But perhaps I can't figure it out at all." But now I think I can. So, at any rate, I got that much out of it."*
> (Ip-8, D1).

> *"But mostly I learned to use the machine and not to regard it as something dangerous. You know, I was very nervous about what I pressed and what I tried. (...) If I didn't know what I should do, then I didn't dare take any chances myself. (...) And afterwards I discovered that when you just try to play with it, you'll figure it out."*
> (Ip-10, D1).

Ip-8 and Ip-10 describe learning experiences from the courses they have participated in, and they do not solely relate to the enterprise as a context of meaning. The emphasis is not on direct application-oriented qualifications - perhaps, because the trainees do not use data processing that much in their work or only use it in connection with certain functions. But their evaluations of benefits are, nevertheless, essential - also in relation to being an employee. It is evaluations of benefits that may cancel any negative learning experiences they might have had and any resulting negative motivation to participate in in-service training. It is also evaluations of benefits that go beyond a wish to be merely a user. Thus, they express wishes to learn more about the machine, which will enable them to become their own "emergency assistants". Furthermore, it is evaluations of benefits, which emphasize the value of feeling that it is a human being and not the machine which is in control. And finally, the evaluations underline the value of playing and the courage to try the functions of unfamiliar keys.

In my view, all these evaluations of benefits are very important for the creative and independent computer user. However, when it is a question of in-service training during working hours, not everyone finds that such evaluations are legitimate, as the acquired qualifications apparently have no direct application value.

10. Learning Organizations Acknowledge Individual Self-directedness

Employees are exceptionally loyal, when they participate in in-service training during working hours. They look at themselves as employees, i.e., the qualifications they acquire during in-service training should be application-oriented. In other words, they are very goal-directed in their evaluations of benefits. But they may find themselves in situations, where they themselves have to take the lead in relation to in-service training. For instance, when they feel they have sufficient qualifications for their present work functions, and when they have their own private educational plans. In such situations they do not need to justify their participation in in-service training. They are realizing a personal learning-project, where they as individuals are the important factor.

When employees evaluate benefits from in-service training, they think double, i.e., they think both of themselves as employees and as individuals - although the employee perspective will always be present in some measure. It is not until we go beyond the employee perspective that employees' independence and creativity can be understood. However, the problem is that it is not quite legitimate, and therefore the importance of an individual perspective has to be belittled or concealed.

In my view, the employee perspective of both enterprises and employees constitutes a very traditional way of thinking based on (Tayloristic) management and control systems. The essence of their reasoning seems to be that provided in-service training is as goal-directed as possible, development of employees in terms of knowledge and qualifications can be controlled. As a result of information technology's entry into enterprises I believe it is precisely other types of control systems that are needed. Control systems that focus more on self-directness and *self control* than control. In other words, allowing employees to control their own development of qualifications based on the belief that they will think both as employees (and consequently of the requirements of the enterprise in terms of qualifications) and as individuals (thus rendering the total content of their personal learning-projects visible). However, it will require that enterprises accept the pursuit of individual perspectives in learning in in-service training activities. In turn, enterprises will have the advantage of broadly qualified, independent and creative employees.

I believe that the perception of employees as both employees and individuals can add valuable input to theories on learning in organizations and thus to the design of learning organizations. In essence, it is important to include theories on individual learning and always to view employees as pursuing a personal learning-project, when they participate in training and educational activities. A personal learning-project that comprises the individual as well as the employee. It is important to note that enterprises cannot develop creative employees by adhering to narrow ideas and exercising control in the event that their employees participate in in-service training. By doing so they will fail to derive any benefits from all the individuals who are hiding that they are (over) qualified or have taken initiatives to qualify themselves further. This does neither benefit organizations nor individuals.

References

Anker, N., & Andersen, D. (1991). Efteruddannelse. (In-Service Training, Report 91:13.) Copenhagen: The Danish National Institute of Social Research

Argyris, C., & Schîn, D.A. (1978). Organizational Learning: A Theory of Action Perspective. Reading: Addison-Wesley

Bernstein, R.J. (1960). Introduction. In: John Dewey: On Experience, Nature, and Freedom. Representative Selections. Indianapolis: The Liberal Arts Press

Dewey, J. (1916). Democracy and Education. An Introduction to the Philosophy of Education. New York: The Free Press

Dewey, J. (1974). Erfaring og opdragelse. (Experience and Upbringing.) Copenhagen: Dreyers forlag, Chr. Ejlers forlag

Driver, M. (1988). Careers. A Review of Personal and Organizational Research. In: C.L. Cooper & I. Robertson (Eds.). International Review of Industrial and Organizational Psychology. London: John Wiley & Sons

Feldman, D.C. (1989). Socialization, Resocialization and Training: Reframing the Research Agenda. In: I.L. Goldstein & Associates. Training and Development in Organizations. San Francisco: Jossey-Bass Publishers

Fiol, M.C., & Lyles, M.A. (1985). Organizational Learning. Academy of Management Review, 10(4), 803 - 813

Giorgi, A. (1975). An Application of Phenomenological Method in Psychology. In: A. Giorgi; C. Fischer, & E. Murray (Eds.), Duquesne Studies in Phenomenological Psychology, II. Pittsburgh: Duquesne University

Goldstein, I.L. (1989). Critical Training Issues: Past, Present and Future. In: I.L. Goldstein & Associates. Training and Development in Organizations. San Francisco: Jossey-Bass Publishers

Goldstein, I.L., & Gessner, M.J. (1988). Training and Development in Work Organizations. In: C.L. Cooper & I. Robertson (Eds.). International Review of Industrial and Organizational Psychology. London: John Wiley & Sons

Granrose, C.S., & Portwood, J.D. (1987). Matching Individual Career Plans and Organizational Career Management. Academy of Management Journal, 30(4), 699 - 720

Hansen, L., Laursen, P.F., & Aarkrog, V. (1991). Edb-efteruddannelse - oplysning og kvalificering. (In-Service Training in Data Processing - Information and Qualifications.) Copenhagen: Akademisk Forlag

Jerslev, A., Tufte, B., & Elkjèr, B. (1987). Forskerkursus. (Research Seminar.) DLH-Avisen, no. 1. p. 3

Kvale, S. (1983). The Qualitative Research Interview - a Phenomenological and a Hermeneutical Mode of Understanding. Journal of Phenomenological Psychology, 14. pp. 171 - 196

Kvale, S. (1987). Validity in the Qualitative Research Interview. In: Jette Fog et al. (Eds.). Interviewet som forskningsmetode. (The Interview as Research Method.) Aarhus: Psykologisk Skriftserie Vol. 12, nr. 1

Larsen, H. (Ed.) (1991). Human Resource Management i danske virksomheder. En rapport om strategisk personaleledelse i private og offentlige virksomheder. (Human Resource Management in Danish Enterprises. A Report on Strategic Personnel Management in Private and Public Organizations.) Copenhagen: The Danish Institute for Personnel Consultancy

Latham, G.P. (1988). Human Resource Training and Development. Annual Review of Psychology, 39, 545 - 582

Lave, J. (1988). Cognition in Practice. Cambridge: Cambridge University Press

Lave, J. ,& Wenger, E. (1991). Situated Learning: Legitimate Peripheral Participation. Cambridge: Cambridge University Press

Levitt, B., & March, J.G. (1988). Organizational Learning. Annual Review of Sociology, 14, 319 - 340

March, J.G., & Olsen, J.P. (1979) Organizational Learning and the Ambiguity of the Past. In: J.G. March & J.P. Olsen (Eds.), Ambiguity and Choice in Organizations. Bergen: Universitetsforlaget

Ostroff, C., & Ford, K. (1989). Assessing Training Needs: Critical Levels of Analysis. In: I.L. Goldstein & Associates. Training and Development in Organizations. San Francisco: Jossey-Bass Publishers

Shrivastava, P. (1983). A Typology af Organizational Learning Systems. Journal of Management Studies, 20(1), 7 - 28

Van Maanen, J. (1977). Experiencing Organization: Notes on the Meaning of Careers and Socialization. In: J. Van Maanen (Ed.), Organizational Careers: Some New Perspectives. London: John Wiley & Sons

Part II
Learning and Organizations

Introduction

Sebastiano Bagnara

Dept. of Communication, University of Siena, Via del Giglio 14, 53100 Siena, Italy

Until a decade ago, the most of people at work had to learn how to use prostheses and technological and organizational artifacts. A prosthesis is a tool that multiplies the power of one physical human capacity. Technological artifacts, instead, integrate two or more human mostly physical capacities. They multiply the effects of actions previously conducted with or without a prosthesis, and also allow to manipulate objects at distance in space: humans have been removed from direct manipulation of dangerous material and processes.

The organizational artifacts distribute in time and space different actions executed by many people in order to carry out activities directed to mid-, or long-term goals. They apply the human cognitive capacities of planning and coordination to the social environment. When performed within an organizational artifact, an action rarely has a clear-cut meaning: it becomes meaningful when taken as a component of a complex of actions, performed by different people and directed toward a common objective.

All prostheses and artifacts should be compatible with human limits and capacities. For prostheses and technological artifacts, both because they are part of the physical environment and require human physical actions, the compatibility concerns mostly the anthropometric, physiological, biological characteristics of human body. Human actions would be severely impaired, and even inhibited or dangerous, if the technological artifacts render the physical environment hostile to the human safety and health. Furthermore, the interfaces should make easy to the humans to handle a prosthesis or to act upon or through a technological artifact.

Some millions of years of experience have allowed a continuous refinement of the prostheses and made them easy to use and their potentiality in terms of effects on the environment immediately evident: the prostheses display how to use them and for what. This experience has been utilized to develop the technological artifacts and their interfaces. But it was not enough: they required learning. The study of human biological machine (i.e., ergonomics) was then applied for both installing the technological artifacts in the physical environment, where also the humans have to healthily stay and move, and for designing their interfaces.

Ergonomics has greatly improved the environment and the interfaces, but has not yet reached what long-lasting experience gained for the prostheses: the self-evidence of how and for what to use the technological artifacts. Training has been reduced in time, but remains a must.

Since technological artifacts are part of the physical environment and take space, they are visible, while the organizational ones are not. The latter are embedded in the social environment, influence the temporal rhythms and the workload and regulate the social roles. They are social, communication, decision, and control rules that regulate the inter-relationships among the people in an organization providing for its internal coherence so it is seen and acts as a unity. Their interfaces are also invisible when one becomes a permanent member of an organization. They operate tacitly and appear only when social breakdowns, or disruption in the distribution of actions occur, or discrepant individual behaviors take place. Instead, the rules governing the organizational artifacts are to be taught and have to be learnt by the new-comers, who clearly perceive them. The feeling of uneasiness and confusion the newcomers perceive when entering an organization is largely due to their ignorance of these rules and their interfaces.

It takes time to learn these rules. In family organizations it takes some years, and often, even in the more "democratic" ones, they have to be every now and then enforced. Unless the organizations one enters later mimic a family, learning is needed to interact with the such artifact. In the work, the basic principles of the functioning of the organizational artifact are taught, newcomers are instructed on how to properly handle them. Usually, details are learnt later through imitation (Tomasello et al., 1993), if the organization is stable. When the organizations are under a strategic change, they have to be found, negotiated and shared through co-operative cultural learning. It is no more a matter of the individuals to learn, rather the organization itself has to become a learning organization.

About twenty years ago, a new artifact has been introduced in work settings: the cognitive artifact. Of course, cognitive components are present in both the previously described artifacts and even in the prostheses. In fact, the relations among the tool, the actions it allows and the effects of these actions on the environment are embodied in the prosthesis: one has to see and recognize them to execute the action, once evaluated if allowed and needed within a specific context. The same applies to the technological artifacts, where the cognitive component is much more complex and requires the user to possess a refined mental model of its functioning and effects. In both cases, however, the core is on the physical manipulation of objects and processes in the physical environment.

In the organizational artifacts, the cognitive component is even greater and regards how to manipulate social objects, but it operates tacitly and is controlled through procedures and norms and by "roles" dedicated to. The majority of individuals in the traditional organizational artifacts have just to carry out simple operations, whose timing and boundaries are clearly defined and enforced by the artifact itself.

At variance with the previous artifacts that are centered on how and when one has to act, the peculiarity of cognitive artifacts consists in performing mental work which may or may not end up in actions. They are about computation, reasoning, information gathering and evaluation to be performed before a real action takes place and affects the environment. They multiply the power and the speed of specific mental abilities and integrate them with reference to a model of the environment they possess.

The cognitive artifacts have been and are still used as cognitive support tools for individuals' work. In this sense, they may be seen and often used as cognitive prostheses, even though they are characterized by representational properties (i.e., the relation between the state of the system a cognitive artifact acts upon and the representation of the system in the artifact), which are typical of cognitive artifacts (Norman, 1991).

A big effort has been made by cognitive ergonomics (see, e.g., Card et al. 1983; Helander, 1988; van der Veer et al., 1992) to make easy to use these artifacts, and the time and effort individuals need to learn to use them and for what have been greatly diminished.

However, at variance with technological artifacts, whose use and function can be learnt once for ever, the flexibility and rapid evolution of cognitive artifacts require a continuous learning process. The learning burden has been greatly reduced because these artifacts can also support the learning process itself: the very same artifact can be used both for work and to understand and practice on how to use it. It can be also an education technology. However, it is still an open question which should be the most satisfying solution in designing them both for an easier and efficient use (on this issue, see in this chapter, the contribution by Climaco et al.), and for supporting the learning process of the user, since some contradictory phenomena appear when novices are compared to experts (see, in this part, the contribution of Mandl et al.).

The flexibility of cognitive artifacts has allowed their integration with other artifacts. The first integration was with technological artifacts: the automation. Since the role of human in the interaction with the technological artifacts is to carry out standardized procedures, where the same cognitive processes are repeatedly conducted throughout the time, the cognitive artifacts have been used to substitute the humans, that have been removed from the interaction with machinery to the supervision of the integrated artifact.

Such a change has brought about a reduction in the work-force needed (this one of the major causes of unemployment) and new job profiles, and some contradictions. The emergence of new job profiles and the need for re-deployment of personnel have brought about the issue of adult and continuing education at a scale and for levels of education never known before. The contradictions have been well-identified by Bainbridge (1983), when she noticed, for instance, that people with automation have less chance to practice with a process, but they are asked to intervene in the most critical situations, i.e., if the automation fails.

All this made apparent the need for new learning and training strategies (Bainbridge, 1993). It has become apparent that learning implies error handling, and simulation (see, in this part the contribution by Frese, and by Bagnara et al., but see also Patrick, 1992; Reason, 1990; and Rizzo et al., 1987). Furthermore, the notion of mental model has emerged (Gentner and Stevens, 1983) because of the possible misunderstandings in fault diagnosis. Often, discrepancies arise among the model of the process in the cognitive artifact that controls the technological artifact, the mental model of the process possessed by the operator and the real process. There may be also lack of coherence and compatibility between the mental model of the cognitive artifact run by the operator and the actual functioning of the cognitive artifact. The issue of training and learning complex skills and how to improve, restructure, actively use knowledge and expertise has become a priority (Bainbridge and Ruiz Quintanilla, 1989).

The integration of the technological and cognitive artifacts made apparent that learning such an object in the organization tackles complex but vital questions for the reliability and quality of the processes and products and for the competitiveness of the organization itself. To render the issue even more vital was the integration of the cognitive with the organizational artifacts, such as the management information systems, or the computer integrated manufacturing, and computer supported cooperative work systems. These integrated cognitive and organizational artifacts have become the nervous system of the organization, and tend to be the communication system of the organizations. Since an organization is essentially a communication network, and lives until the needed information reaches every subsystem and component, the integration of the cognitive and organizational (actually also the technological) artifacts, have forced (together with a change in the competitive strategies) a wide-spread rethinking of the organizational model and the role of integrated artifacts in their reshaping (Winograd, 1987/8; Winograd and Flores, 1986).

The new principles adopted, also because the new integrated artifacts make possible to implement them, have been those of collaboration and co-operation in work within a flat organization, instead of the traditional division of work. This has changed and is changing the job profiles, the work content and the social relations within the companies. Sometimes, learning model and strategies anticipate this change, some others follow the dynamics of the technological change, or the organizational transformations. The turbulence has been high, the starting conditions very different, and the strategies adopted have very often been peculiar and "ad hoc". The actual organizations still comprises such different objects to be learnt as prostheses, technological, organizational, cognitive, and various types of integrated artifacts. However, the picture is becoming clear, and trends in strategies, models and methods toward coherence among learning strategies and organizational dynamics toward competitiveness can be identified.

This part mirrors the state of the art in such a complex and evolutionary field. Though some papers seems to focus on specific artifacts (e.g., *Mandl et al.*, and *Climaco et al.* on the cognitive artifacts; *Rogalski* on organizational artifacts), all of them try to face and explain how learning of artifacts in the organizations and of the organization itself take place in a changing context, where the models to refer to have to be continually revisited and updated.

The papers are ordered bottom-up. The analysis of differences between experts and novices to tackle with cognitive artifacts (*Mandl et al.*), is followed by a study of errors in interacting with them (*Frese*), and by the presentation of a case of design of cognitive artifacts (*Climaco et al.*). Then, the papers by *van der Veer* and *Rogalski*, which consider both individual and collective activities, face the question of learning and work in settings were inter-relations are fundamental to stay in organizations which move toward flat, network models. The final paper by *Bagnara et al.* aims at tentative systematization of the relationships between learning and organization in a very dynamic and apparently contradictory change.

References

Bainbridge, L. (1983). Ironies of automation. Automatica, 19, 775-779

Bainbridge, L. (1993). Planning the training of a complex skill. Le Travail Humain, 56, 211-232

Bainbridge, L. & Ruiz Quintanilla, S.A. (eds.) (1989). Developing skill with new technology. Chichester, UK: Wiley

Card, S.K., Moran, T.P., & Newell, A. (1983) The psychology of human-computer interaction. Hillsdale, NJ: Erlbaum

Gentner, D. & Stevens, A. L. (Eds.) (1983). Mental model. Hillsdale, NJ: Erlbaum

Hutchins, E., Hollan, J., & Norman, D.A. (1986). Direct manipulation interfaces. In: D.A. Norman & S.W. Draper (eds.), User centered design: New perspectives in human-computer interaction. Hillsdale, NJ: Erlbaum

Norman, D.A. (1991). Artefatti cognitivi. Sistemi Intelligenti, 3, 453-476

Patrick, J. (1992). Training: Research and practice. London: Academic Press

Reason, J. (1990.) Human error. Cambridge: Cambridge University Press

Rizzo, A., Bagnara, S., & Visciola, M. (1987). Human error detection processes. International Journal of Man-Machine Studies, 27, 555-570

Tomasello, M., Kruger, A.C., & Ratner, H.H. (1993). Cultural learning. Behavioral and Brain Sciences, 16, 495-552 van der Veer, G.C., Bagnara, S., & Kempen, G.A.M. (eds.) (1992). Cognitive ergonomics: Contributions from experimental psychology. Amsterdam: North-Holland

Winograd, T. (1987/8). A language/action perspective on the design of co-operative work. Human-Computer Interaction, 3, 3-30

Winograd ,T. & Flores, F. (1986). Understanding computers and cognition. Norwood, NJ: Ablex

Mental Models of Complex Systems: When Veridicality Decreases Functionality

Heinz Mandl, Hans Gruber, and Alexander Renkl

University of Munich, Institute for Educational Psychology, Leopoldstr. 13
D-80802 Munich, Germany

Abstract. The concept of mental models is employed for the analysis of information processing and problem solving in a semantically rich domain, that is, business management.

Mental models can be evaluated with regard to their consistency, their veridicality, and their functionality. We provide empirical evidence for the argument that functionality of mental models may be dissociated from consistency or veridicality. Looking for consistency or veridicality of one's mental model about a domain may even have detrimental consequences for problem solving, so that, under certain circumstances, expert subjects may be outperformed by novices. Quantitative data as well as protocol analyses are provided to substantiate these arguments.

Keywords. Complex system, computer simulation, expertise, knowledge application, functionality, mental models, veridicality

1. Introduction

Mental models are conceptualized as dynamic forms of information representation and information processing (Gentner & Stevens, 1983; Johnson-Laird, 1983; Schumacher & Czerwinski, 1992). Mental models integrate declarative knowledge and its application conditions. In addition, they allow the problem solver to internally simulate real world processes (simulative reasoning (Greeno, 1987); running the mental model (Dekleer & Brown, 1983)), and to reason about results and consequences of possible actions. Theories of mental models, therefore, do not only deal with describing the nature of knowledge representations in order to form a computational model (cognitive science approach (Rasmussen, 1990)), but they also allow for the analysis of competence in complex domains (cognitive engineering approach (Rasmussen, 1990)). For these reasons, the mental model approach is suitable for the analysis of information processing and problem solving in semantically rich domains. Complex cognitive activities can be regarded as the construction, modification, and evaluation of mental models (Johnson-Laird, 1983). The more expert the problem solver the more elaborated is his/her mental model of the domain under consideration.

However, handling mental models is not an activity exclusively restricted to experts. Even novices in a certain domain have already constructed mental models, although their models may be rudimentary and may presumably contain wrong assumptions (Frese & Brodbeck, 1983). For subjects of any level of expertise the initial mental model provides the basis for acting and problem solving (Kluwe & Haider, 1990).

Mental models can be described using three criteria: Consistency, veridicality, and functionality (Tergan, 1986). To evaluate the quality and adequacy of mental models, these features have to be considered. Adequate mental models should be consistent and should not contain internal contradictions. Only then an errorfree running of the mental model is possible. Another important feature of adequate mental models is veridicality; there has to exist a correspondence between the mental model and the reality. Consistency and veridicality are not independent constructs, because a completely veridical model necessarily is consistent. Both consistency and veridicality are also widely used evaluation criteria with respect to other complex forms of knowledge representation (e.g., schemata or scripts). The third feature is more specific to the mental model theories, because this approach especially stresses the utility or predictive power of representations (Kluwe & Haider, 1990; Norman, 1983). Functionality of mental models is not necessarily associated with consistency or veridicality. Empirical evidence shows that subjects may gain functional control over complex systems even if their mental models are far from veridicality. Mental models need not be completely veridical descriptions of the system to be functional (Ringelband, Misiak, & Kluwe, 1990).

One feature of elaborated mental models is that knowledge units are organized by chunking and by constructing hierarchies (Kluwe & Haider, 1990). This leads to a functional simplification of mental models so that they are manageable and save information processing demands. This assumption is supported by research on expertise which demonstrates that top experts' performance is, among others, due to chunking and hierarchical knowledge organization (Ericsson & Staszewski, 1989).

However, the modification of mental models, which can be described as learning process, does not always lead to better performance. Modifying or extending a mental model may lead to a more adequate description with regard to reality and at the same time to a temporal decrease of functionality, because the veridicality increases complexity that may be difficult to manage.

Temporal performance decreases in the development of expertise were reported several times (Lesgold, 1984; Lesgold et al., 1988). It is an open question whether these effects can be explained by the fact that intermediate experts compared to novices construct relatively complex and veridical mental models, but still lack the hierarchical chunking of top experts. This lack of an organization that simplifies problem solving for top experts may cause difficulties in effective use of the complex knowledge structures that are integrated into the mental model. Thus, intermediates might sometimes even be outperformed by novices in a problem solving task.

We examined the problem solving performance of novices and intermediate experts (in the following short "experts") in a computer simulation in the domain of business management. Furthermore, the relation between performance and the underlying mental model the subjects constructed during problem solving is investigated.

2. Methods

2.1 Subjects

Expert subjects were eighteen graduate economy students at the University of Munich. Novice subjects were seventeen students of education and psychology which were comparable to the experts with respect to age (M = 26.5 years for both groups).

2.2 Apparatus

We employed a computer simulation program that shows external validity, because it was constructed and evaluated by domain experts. The program JEANSFABRIK (jeans fabrication (Achtenhagen & Preiß, 1990; Preiß, 1992)) is situated in the domain of business management. The most important economics concepts and principles subjects have to deal with in the JEANSFABRIK program are profit, production quantity, stock, and price-sales relation. However, a complex economics situation is provided to them with several further information.

2.3 Procedure

In order to provide basic concepts for the novices and to reactivate the experts' domain knowledge, information about important economical concepts (e.g., proceeds, production quantity) and about basic principles (e.g., price-sales relation) was presented to the subjects in an instructional part of the study. A test assured that subjects knew the relevant declarative knowledge correctly before entering the simulation situation. Then the subjects were given the opportunity to explore the system for thirty minutes. Subsequently, they had to work on it for four planning periods, that was about twenty minutes. Quantitative parameters of the system control during the simulation procedure were automatically recorded (e.g., entered production quantity, entered selling price, gain, stock). During the whole procedure subjects had to think aloud; literal transcriptions of the recorded protocols were typed.

In order to diagnose the quality of the subjects' mental models after exploration of the system, that is, at the beginning of the problem solving phase, a teaching back procedure was used. Subjects had to explain and teach a third person how to maximize the profit in the simulation game according to their present knowledge state.

2.4 Results and Discussion

First, experts and novices are compared with regard to quantitative simulation control measures. Second, we analyze contrastive single cases with regard to the mental models that were constructed at the beginning of the problem solving phase.

Table 1. Profit made by experts and novices (in thousand DM)

Group	*N*	*M*	*s*	*t*	2-tailed *p*
Intermediate experts	18	401.0	584.2	–2.10	.043
Novices	17	804.0	547.7		

As Table 1 shows, experts compared to novices were not very successful in taking advantage from their larger knowledge base. The opposite was true with regard to realized profit. The novices significantly outperformed the experts. It should be mentioned that this effect evolved despite that the subjects had the opportunity to explore the simulation game for thirty minutes before the problem solving phase began.

Figure 1 shows that the difference in problem solving success is primarily due to the profit made in the first three planning periods, that is the first three months. In the fourth period, the difference decreased. Unfortunately, we lack adequate data to see whether in the long run the experts would surpass the novices or, at least,

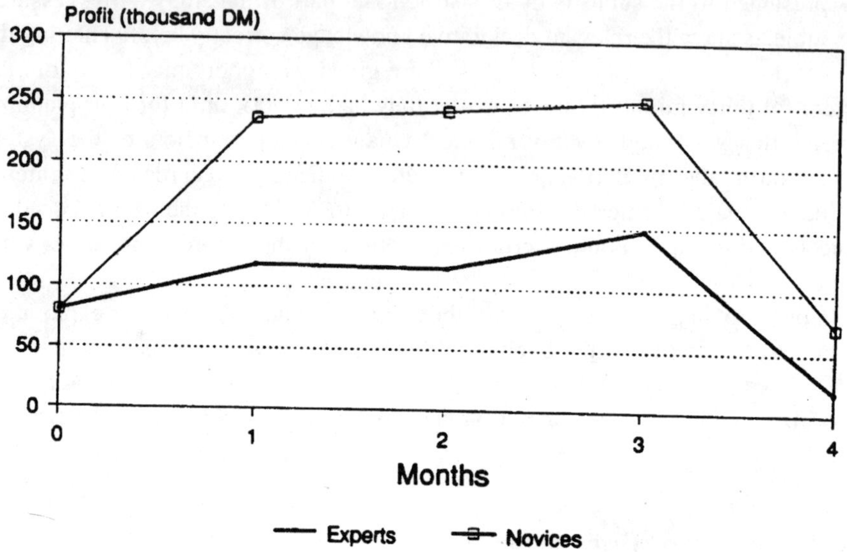

Fig. 1. Profit made by experts and novices (in thousand DM)

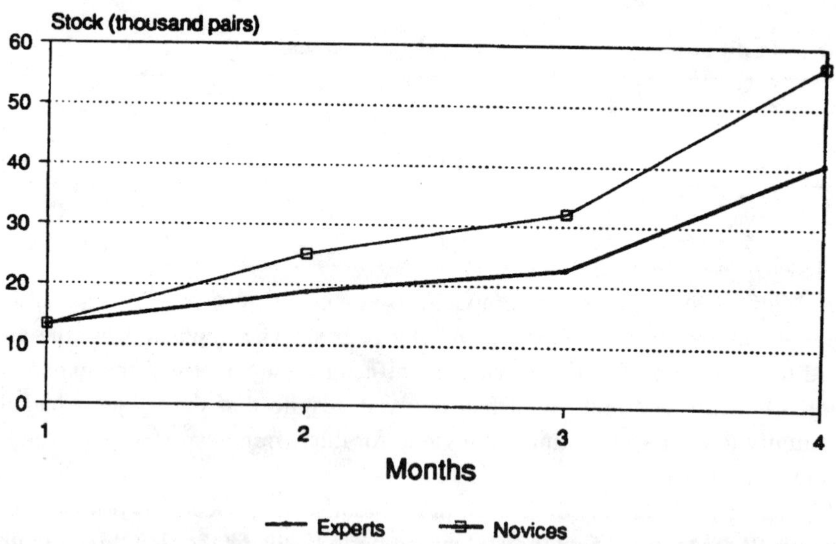

Fig. 2. Stock of experts and novices (in thousand pairs of jeans)

catch up with them. We have, however, some data that make this assumption plausible.

As Fig. 2 shows, the stock of the novices is rapidly increasing which indicates that the novices cannot successfully continue their business policy. There is a similar tendency for experts, but it is less dramatic. The experts' stock in the third and fourth months is significantly smaller.

At the first glance, the finding concerning the superiority of novices in problem solving performance is quite astonishing. To see how this could happen, we had a closer look at the teaching back protocols in order to get some ideas of the mental models that underlay the problem solving processes of experts and novices. We selected a typical novice to illustrate the types of mental model that these subjects constructed and used in a relatively effective way. In addition, two experts are compared with each other, one expert with poor problem solving performance, that is, low profit and the other one with high profit. This illustrates the differential hierarchical organization of experts with high and low profit.

The selected novice achieved an amount of profit that was nearly two standard deviations above the mean profit in our sample. Figure 3 shows the teaching back protocol.

> A moment please - one should keep constant the production quantity, the stock should equal about one third of the capacity, the price should not vary to much and it should be adapted to the opponents strategy and also adapted to the development of the customer, I mean, the development of the sales. Furthermore, it is useful to protocol the data. That's all.

Fig. 3. A typical novice's teaching back protocol

Although this novice was extremely successful with respect to the profit, his mental models comprises almost no correct facts and some prescriptions that do not make sense and that are not justified by economical theory (e.g., "One should keep constant the production quantity"). His explanation reflects only the "surface features" of his relatively successful decisions in the preceding exploration phase. The contrast between the low veridicality of the novice's mental model and his success with respect to profit asserts the assumption that there is often a dissociation between veridicality and functionality in complex problem solving.

Next, we present the teaching back protocol of an expert who was relatively successful, that is, the profit was nearly one standard deviation above the mean profit of the experts (Fig. 4). Due to the length of the protocol some less relevant parts are omitted.

First, the relation between production quantity and price is important. The program provides some information whether a certain combination of production quantity and price can result in profit (...) In addition to the price, the stock must be taken into account in planning the production quantity (...).
The second major point concerns the opponent and the customers. In fixing the price, one has to take into account the opponents price and the sensitivity of the customers with regard to price differences between our factory and the opponent (...).
Both aspects, the market and the situation concerning the production and production costs must be taken into account in defining the strategy for maximizing our profit (...).

Fig. 4. A successful expert's teaching back protocol

In comparison to the presented novice, the mental model of this expert was much more extensive. This result can be generalized to all experts and novices in our study. Furthermore, this expert's mental model comprises veridical information which is clustered into three areas: Production, market, and strategy. In addition, these areas were related to each other: "Both aspects, the market and the situation concerning the production and the production costs must be taken into account in defining the strategy for maximizing our profit." This means that there was a second organizing principle, namely the areas of market and production were integrated in the area of strategy planning.

Finally, we present the teaching back protocol of an expert whose profit was more than one standard deviation below the mean of experts (Fig. 5). Again, some parts of the protocol are omitted.

I think this is a duopoly market with tough competition. It is very difficult to sell the jeans by a price that is higher than the cost per jeans. It is important to note that the cost for materials per jeans are independent of the production quantity (...) Total costs are a linear function of production quantity (...). One strategy is to produce so many jeans as possible to reduce the cost per jeans (...).
It is important to note that the storekeeping capacity is limited. This must be consider in planning the price (...).
One major point is that the demand seems to be very sensitive to the price (...) Especially, if the opponent has also a high price, the total demand is dramatically reduced. This may lead to high stock (...).

Fig. 5. An unsuccessful expert's teaching back protocol

This unsuccessful expert constructed a relatively extensive mental model as well. The comprised information was veridical. However, this mental model resembles a list of different aspects more than an integrated and structured representation that provides a useful basis for problem solving.

Taking all examples together, two reasons can be identified, why the novices outperformed the experts.

(a) In the simulation game used, naive theories and simple mental models can lead to high profit, at least in the short run.

(b) Some experts put forward too complex hypotheses, that is, they constructed extensive mental models without simplifying structuring, such as hierarchical organization. They took into account more aspects than they could integrate for a sound decision.

Further informal analyses revealed two other reasons that were accountable for the relatively poor problem solving performance of experts:

(a) Experts were too cautious in manipulating price and production quantity.

(b) Experts held wrong assumptions which could be true in principle and did not hurt economical theory, but were not relevant in the present problem solving task.

3. Outlook

Do our results have any implication for higher education in economics? In the introductory part, we argued that performance decreases might be a feature of expertise development, because one can assume that the application of newly acquired knowledge and its integration into powerful mental model takes some time. On the other hand, part of the experts' difficulties might be due to the prevailing Abstract kind of university instruction that is only marginally concerned with application. Thus, it is seldom taught or practiced how to integrate domain knowledge into functional mental models. Recent instructional models, such as Cognitive Apprenticeship (Collins, Brown, & Newman, 1989) or Anchored Instruction (Bransford et al.,1989) might ameliorate the difficulties of applying newly acquired information. They try to teach knowledge in the context of application, that is, in the context of constructing mental models for coping with authentic problems. However, in how far the integration of instructional units in accord with these teaching models can prevent the difficulties in knowledge application and in the construction of powerful mental models is still an open question that should be attacked by further research.

References

Achtenhagen, F., & Preiß, P. (1990). Planspieleinsatz in der betriebswirtschaftlichen Erstausbildung [Using simulation games in primary economics education]. In: J. Biethahn, W. Hummeltenberg, & B. Schmidt (Eds.), Simulationsmodelle als betriebliche Entscheidungshilfen 2. pp. 23-34. Berlin: Springer

Bransford, J.D., Franks, J.J., Vye, N.J., & Sherwood, R.D. (1989): New approaches to learning and instruction: Because wisdom can't be told. In: S. Vosniadou & A. Ortony (Eds.), Similarity and analogical reasoning, pp. 470-497. Cambridge: Cambridge University Press

Collins, A., Brown, J.S., & Newman, S.E. (1989). Cognitive apprenticeship: Teaching the crafts of reading, writing, and mathematics. In: L.B. Resnick (Ed.), Knowing, learning, and instruction, pp. 453-494. Hillsdale, NJ: Erlbaum

eKleer, J., & Brown, J.S. (1983). Assumptions and ambiguities in mechanistic mental models. In: D. Gentner & A.L. Stevens (Eds.), Mental models, pp. 155-190. Hillsdale, NJ: Erlbaum

Ericsson, K.A., & Staszewski, J.J. (1989). Skilled memory and expertise: Mechanisms of exceptional performance. In: D. Klahr & K. Kotovsky (Eds.), Complex information processing: the impact of Herbert A. Simon, pp. 235-267. Hillsdale, NJ: Erlbaum

Frese, M., & Brodbeck, F.C. (1989). Computer in Büro und Verwaltung. Psychologisches Wissen für die Praxis [Computers in office and administration. Practical psychological knowledge]. Berlin: Springer

Gentner, D., & Stevens, A.L. (Eds.) (1987). Mental models. Hillsdale, NJ: Erlbaum

Greeno, J.G. (1987). Instructional representations based on research about understanding. In: A. Schoenfeld (Ed.), Cognitive science and math instruction, pp. 61-88. Hillsdale, NJ: Erlbaum

Johnson-Laird, P.N. (1983). Mental models. Towards a cognitive science of language, inference, and consciousness. Cambridge, MA: Harvard University Press

Kluwe, R.H., & Haider, H. (1990). Modelle zur internen Repräsentation komplexer technischer Systeme [Models for the internal representation of complex technical systems]. Sprache & Kognition 9, 173-192

Lesgold, A.M. (1984). Acquiring expertise. In: J.R. Anderson & M. Kosslyn (Eds.) Tutorials in learning and memory, pp. 31-60. San Francisco: Freeman

Lesgold, A., Rubinson, H., Feltovich, P., Glaser, R., Klopfer, D., & Wang, Y. (1988). Expertise in a complex skill: Diagnosing X-ray pictures. In: M.T.H. Chi, R. Glaser, & M.J. Farr (Eds.), The nature of expertise, pp. 311-342. Hillsdale, NJ: Erlbaum

Norman, D.A. (1983). Some observations on mental models. In: A.L. Stevens & D. Gentner (Eds.), Mental models, pp. 7-14. Hillsdale, NJ: Erlbaum

Preiß, P. (1992). Planspiel Jeansfabrik [Simulation game Jeans Manufacturing]. Göttingen: Seminar für Wirtschaftspädagogik der Georg-August-Universität

Rasmussen, J. (1990). Mental models and the control of action in complex environments. In: D. Ackermann & M.J. Tauber (Eds.), Mental models and human-computer interaction 1, pp. 41-69. Amsterdam: North-Holland

Ringelband, O.J., Misiak, C., & Kluwe, R.H. (1990). Mental models and strategies in the control of a complex system. In: D. Ackermann & M. J. Tauber (Eds.), Mental models and human-computer interaction 1, pp. 151-164. Amsterdam: North-Holland

Schumacher, R.M., & Czerwinski, M.P. (1992). Mental models and the acquisition of expert knowledge. In: R.R. Hoffman (Ed.), The psychology of expertise. Cognitive research and empirical AI, pp. 61-79. New York: Springer

Tergan, S.O. (1986). Modelle der Wissensrepräsentation als Grundlage qualitativer Wissensdiagnostik [Models of knowledge representation as a basis of qualitative knowledge diagnosis]. Opladen: Westdeutscher Verlag

Error Management in Training: Conceptual and Empirical Results

Michael Frese

Department of Psychology, University of Giessen, Otto-Behaghel-Str. 10F
D-35394 Giessen, Germany

Abstract. There are two strategies to deal with errors in training: Error prevention or error management. It is argued that the error management concept can be used to improve the quality of training. It has been shown repeatedly that error training - that is using errors as an enhancer of learning - leads to better performance. Various mechanisms for the superiority of error training can be distinguished: (a) Better emotional processing, (b) more effective error handling strategies, (c) higher motivation, and (d) errors as instigators of exploration. Mechanism (a) has partial empirical and (b) has no empirical support. Mechanism (d) has been shown to exist and to account for a large part albeit not all of the superiority of the error training concept.

Keywords. Error, training, management.

1. Introduction

There are two strategies that can be used to deal with the error problem in training: a) Error prevention, that is the reduction of errors as far as possible or b) error management.

Curiously, there have been more theoretical and empirical arguments for an error prevention approach. Skinner (1968) has explicitly argued for error prevention. Since errors were assumed to work like punishment, Skinner argued for an error free learning process. Erroneous behaviors disrupt good performance. Therefore, programmed learning, growing out of this tradition, made strenuous attempts to eliminate errors. Similarly, humanistic approaches have been critical of errors. While there does not exist a corpus of principles from humanistic psychology pertaining to errors, there is the concept that learning might get impeded by frustrations that go along with making errors.

2. The Concept of Error Management

The alternative concept is error management. Error management has to be distinguished from error handling. While error handling is a descriptive term and implies any type of response towards an error, error management is prescriptive. It means a useful approach to errors with the goal of reducing future errors, of avoiding negative error consequences and of dealing quickly with error consequences once they occur.

One prerequisite of the concept of error management is the differentiation between the error itself and the negative error consequences. Not every tripping leads to falling. The tripping is the error, the negative error consequence is to fall or to break one's arm. The problem is not the fact that one has made an error, the problem is that one has suffered from a negative error consequence. The concept of error management argues that the error per se does not have to be avoided or minimized - just the negative error consequences.

Figure 1 should explain the differences between error prevention and error avoidance. The strategy of error prevention attempts to reduce the number of faulty actions per se. Thus, a barrier is erected between the action and the error. The error management strategy is more non-cholant with regard to errors but it attempts to erect a barrier between the error and the potential negative error consequences. Learning how to trip without breaking one's arm or leg would be an error management training (as is routinely done in Judo).

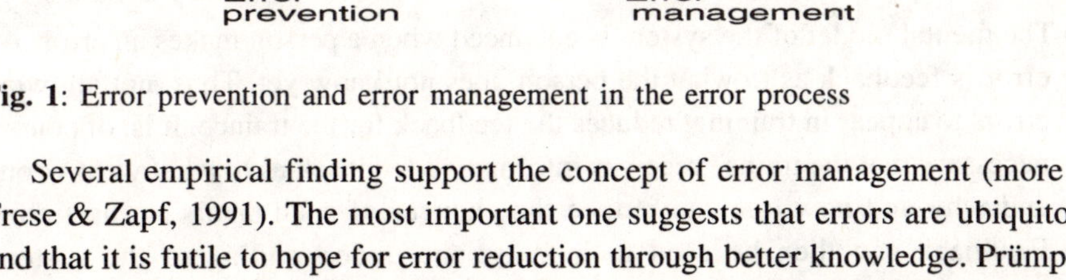

Fig. 1: Error prevention and error management in the error process

Several empirical finding support the concept of error management (more in Frese & Zapf, 1991). The most important one suggests that errors are ubiquitous and that it is futile to hope for error reduction through better knowledge. Prümper, Zapf, Brodbeck & Frese (1992) have shown that experts make as many errors as novices - as a matter of fact under certain conditions and with certain errors, experts err more often. For example, employees who know more than one program make on average 5.83 errors per hour working with the computer, while people, who just know one program, show only 3.87 errors on average - a significant difference. Thus, empirically, it is not necessarily a sign of higher expertise to be less erorr-prone. This is partly so because computers are systems that in principle support error management - it is easy to correct a mistake and the differentiation between the virtual world on the screen and the real world facilitates error handling[1]. From an error management point of view, it is more important whether or not people can quickly deal with an error because the concept of error management is more concerned with the negative consequences of errors than with the error per se. Loss of time is an important negative consequence. Indeed, empirically error handling is usually a bit faster by people who have more experience working with computers (Prümper et al., 1992).

For all of these reasons, it is useful to arrive at an explicit conceptualization of and some suggestions for how to deal with errors in training human-computer skills. In principle the concept of error management can be used to develop systems (cf. Frese, 1991; Zapf, Frese, Irmer & Brodbeck, 1991) *and* training methods. We shall concentrate on training methods in this article.

3. Error Training: The Concept

Most trainers want to reduce the number of errors because they are influenced either by Skinnerian or humanistic concepts that conceptualized errors to be frustrating punishing. We differ from these beliefs, because we suggest that (a) errors are a necessary part of the learning process; (b) errors may instigate exploration and creative solutions; (c) errors disrupt premature automatization of actions; (d) the transfer of what one learns in training into the work situation is enhanced when the training methods allows errors to appear; (e) errors will appear on the job frequently and error management strategies should therefore be taught; (f) additional frustrations because of errors should be reduced by learning how to deal with errors effectively.

(a) The mental model of the system is enhanced when a person makes an error. An error is feedback as to what the person does not know yet. Thus, not allowing errors to appear in training, reduces the feedback for the trainee. It is, of course, necessary that the trainee receives informative feedback as to what went wrong and why and how one can reduce the chances of such errors in the future. Feedback can then be used to increase the understanding of the system. Moreover, mental models are better when they also encompass potential pitfalls and error-prone problem areas.

---

1 Therefore, this finding of experts not being better in error making than non-experts does not translate into the use of other machines, for examples cars (Heinbokel, 1993).

(b) Sometimes errors can spur creative solutions to a problem and increase exploration. When an error occurs the person may be led to a part of the system that he or she did not know before and start exploring to find a better way of doing things. Thus, errors may instigate exploration. There is growing evidence that at least in adult learning of human-computer skills, trainees should have ample chance to explore the system (Carroll & Mack, 1984; Frese et al., 1988; Greif, 1986; Wendel & Frese, 1987). One particularly important reason for exploration is that only by learning to explore the system, the trainee is also able to transfer the knowledge learned in training to his or her actual work (v. Papstein & Frese, 1988). Errors are the bread and butter of exploration. There is no exploratory learning in which one does not make errors. Only by stifling exploration is it possible to eliminate errors. Thus, a good training program allows errors.

(c) People have a tendency to develop routine and automatic patterns of how to deal with the world. This automatization can occur quite quickly when the environment is redundant. Errors usually lead to a reintellectualization of action patterns. Thus, whenever one unthinkingly uses an action strategy in an automatic mode, an error helps to make the person think again (cf. Frese & Zapf, in press). Therefore, errors reduce the danger of premature automatization of action strategies.

(d) When an error-free training is aspired for, the trainer will restrict the kind of strategies used by the trainees, because unrestricted strategies increase the chances of errors. These restrictions in the training situation may lead to more rigid behavior patterns that may not be useful in the work situation. Thus, transfer of the training content to the work situation is made more difficult.

(e) Errors not only appear in training but also in the actual work situation. As a matter of fact, we think that this is the most important reason why error management strategies should be taught. At work one usually does not have the support that one receives in the training sessions. Moreover, people should be encouraged to explore the system in the work situation as well because otherwise they will not know enough about it to work flexibly and to adjust to new tasks. Even rather mundane tasks like text entry may turn out to include complicated elements that will sometimes lead to situations that are difficult to solve. For example, in one company we observed secretaries whose system set-up did not allow them to go into the MS-DOS system. Management thought, this was too error-prone. It turned out, however, that sometimes the only way to rescue a file after having made some other errors, was to use MS-DOS commands. Since the secretaries could not use it, the files were lost. Thus an error-reducing strategy led to higher costs once an error occurred. Furthermore, the work situation is always in flux - new versions of the software are introduced, new programs have to be learned. In each case, errors will be frequent and it is better for the worker to know how to deal with them.

(f) Errors do have an upsetting and frustrating quality, although this quality seems to dependent mainly on the error handling time in work situations (Brodbeck, Zapf, Prümper, & Frese, in press). Thus, the negative emotional feelings that accompany errors should be reduced. There are two ways by which this can be done: First, teaching effective strategies for error handling. Then the user knows what to do in case of an error and is, therefore, less upset when an error appears. Much of the aversiveness of errors in the human-computer field stems from the fact that errors produce states that are difficult to leave or that some part of one's work is lost. Error handling strategies help to overcome these problems.

Second, becoming more tolerant of one's error. Making an error always has two facets of first loosing time and of second being subjected to the additional frustration about oneself that one has "made such a stupid mistake." While time loss can only be reduced by good handling strategies, the additional emotional upset about the time lost and the fact of having made an error, can be reduced by developing a more positive attitude towards errors. But this positive attitude can only be done, if an integral part of the training process is to be positive about errors and perceive them as a chance to learn something from them. Such an attitude also decreases the tendency to give up in case of an errors. Thus, training should counter-act these tendencies and encourage a rational learning oriented approach to errors rather than becoming emotionally upset because of errors.

4. Empirical Studies on the Superiority of Error Training

The first experiment done in our lab on the function of errors in training (Frese, Brodbeck, Heinbokel, Mooser, Schleiffenbaum, & Thiemann, 1991) attempted quite a radical approach to show how errors could be used to learn a software: The group that was supposed to learn from errors received only a list of commands to be used for a set of tasks. Otherwise the trainees were on their own, a situation that produced many chances for errors. To be sure that they would be exposed to enough errors, they received an enforced error training in their 5th and 6th training hour. During this period two people were asked to work together on tasks that were clearly too difficult for the subjects (who were all complete novices): For example, they had to create a double column text with tables. In order to reduce the frustrations that result from errors, the subjects were given a set of heuristics that were repeated several times during the training. These heuristics specified that one should learn from errors and that one should take errors as an indication that one can learn something new. Examples are "I have made an error. Great!" or "There is always a way to leave the error situation."

In contrast, the error avoidant group received a training that mimicked commercial tutoring programs - a so-called sequential training. They had to solve the same tasks but they were told precisely how to go about it achieving this. Thus, they did not have the chance to makes errors. To keep the training time equal for both groups, this group was told that "practice makes perfect" and that they should go through the program again one more time if they finished before the allocated time.

Table 1: Performance effects of error training in different tasks

Study	Error Training	Error avoidant training	t or F
Frese et al (1991) (5-point rating)	2.47 (n=15)	1.81 (sequential group) (n=9)	2.57*
Thiemann (1990) (5-point rating)	3.37 (n=7)	2.39 (sequential group) (n=6)	2.78*
Irmer et al (1991)	54.1 (n=42)	31.1 (standard training) (n=12)	8.75*
Dormann & Frese (1994) (5-point rating)	3.93 (n=15)	2.60 (sequential group) (n=15)	3.03*

* = $p < 0.01$

Note: a higher score implies better performance

After the training, both groups received a set of tasks graded from easy to difficult and had to do other performance tests (e.g., free recall of commands learned). Our hypothesis was that error training would increase performance in difficult tasks (e.g., formatting tasks). The results were quite consistent with our hypothesis: The error training group was much better in the performance of difficult tasks after the training (cf. Table 1)[2].

We were quite concerned about replicating these results (replication results are also displayed in Table 2). Therefore, three types of replications were made. The first was a more or less exact replication of the Frese et al. (1991) study (Thiemann, 1990). The second one, replicated the results in a different setting

[2] It is interesting to note that independently, another research group (Greif & Janikowski, 1987; Greif, 1986) developed another type of error training at about the same time. While their approach was not as radical in terms of increasing the error experience of their subjects, the training the results point in the same direction: The power of using errors as an explicit training device.

(Irmer, Pfeffer & Frese, 1991); in this experiment a software training school participated. The standard training was the normal procedure that was used at this school to teach WORD. By and large errors were avoided in this training. Three different error training groups were contrasted with this standard training (more on the differences between these groups later - here the results are aggregated across these three groups). Another facet of this field experiment was that the subjects were not highly motivated volunteers for the experiment but normal employees or unemployed who were not particularly motivated to learn a text processing program. The last replication study Dormann & Frese (1994) was concerned with generalizing the results to another software area: a statistical software package (SPSS). This software package is not really well suited for error training because it does not provide good feedback. The error messages are partly obscure and one does not usually get feedback directly after an error has occured bad later after the job was sent off as a sort of batch job. Nevertheless, even with this system error training turned out to be superior again.

In all of these studies, several different indicators of performance were used. As hypothesized, easy tasks were usually not superior in error training. Other performance indicators, e.g. free recall was usually superior. In one of the studies, the normal test results from the school used for grading the students turned out to be superior as well (Irmer et al., 1991). Thus, we can be quite sure about the phenomenon that it is useful to use errors in training but we do not know yet which processes are important in the development of the superiority of error training. In the following, we shall concentrate on this issue.

5. The Processes of Error Training

Potentially, four processes may play a role in producing the superiority of the error training:

(1) The error training group learns emotional strategies to deal with errors more rationally and thus perform better even when errors occur.
(2) The error training group learns more effective strategies to deal with errors.
(3) The error training group may have been more motivated.
(4) Errors are instigators of exploration and, thus, lead to better performance.

These hypotheses do not have to contradict each other and they may all be true.

5.1 Learning Emotional Strategies

From our qualitative observations we think that the positive effects of errors are due to presenting and repeating the heuristics. Errors are stressful and they may even be more stressful in the German culture that highly values perfectionism. We could observe many stress responses in the trainees when they made an error (blushing, not wanting the experimenters to see the error, or becoming jittery). Moreover, a scale on "frustration with computer work" was used to tap emotional responses (Frese et al., 1991). The error avoidant group had a significant increase in frustrations after the performance test. Apparently, they were frustrated because they could not deal well with the performance task without receiving the detailed guidance provided during the training phase. It seems that our use of heuristics in the error training group had indeed helped to defray the negative frustrations due to errors.

Thus, there is some evidence that speaks for the emotional adjustment effect to deal with errors as a result of being exposed to the error training method. However, we were not able to replicate this result in other studies. We are not quite sure yet, whether this is due to differences in the design of these studies (we had used a different way of presenting the emotional scales) or whether the emotional processing is not so important. We are currently doing another study to look at this issue again. At this point in time, I have become more sceptical of the emotional hypothesis, however.

This does not mean that the emotionally toned heuristics on the positive notion of errors in the training process are unimportant ("You have made an error. Great!"). As a matter of fact, we once wanted to expose one group to error training and not give them these positive heuristics. But we decided that it was unethical to expose people to errors without giving a general notion that it was okay to make many errors. Thus, by all means, these heuristics are an important part of our error training. However, I have become sceptical about whether thinking of errors as a positive feature of learning is an important factor for the superiority of the error training method.

5.2 Learning Effective Strategies

Irmer et al. (1991) have looked at this issue in detail. Three different error training groups were contrasted with a three- day normal standard training of the school. The standard training was the normal procedure used by a very experienced software teacher at this school. As most training methods, it was little error oriented and attempted to reduce the number of errors. When an error happened, it would be quickly corrected by the trainer.

Three groups were exposed at least partly to an error training module: One group received the normal standard training of the school for just two days plus one day error training. This error training provided general emotionally toned heuristics as in the first study by Frese et al. (1991).

The second group also received the two days standard training plus one day error training. However, this time error training consisted of giving general emotionally oriented heuristics plus specific ideas on how to deal with potential errors. Thus, the subjects were trained to use better strategies to deal with errors, for example, they learned how to use the escape key, how to get back to the main menues, and how to use back-up files.

In the third error training group, there was no standard training but only specific error training (including general emotionally oriented heuristics). Thus, this group had the highest exposure to the error training method.

It turned out that there was no performance difference between these three error training groups. This suggests two conclusions: First, it is enough to receive an error training module of one day to produce the performance superiority typical of the training method. This has practical implications because there is no need to provide error training throughout - it is enough to expose people to an error training module during the normal training. Second, and most important for our discussion here, teaching specific strategies for error handling is *not* more effective than the more general error training method. Thus, the superiority of the error training is not due to being taught specific strategies to deal with errors. This implies that learning specific error handling strategies is not what counts in the superiority of error training. Rather, it seems something general that is taught in error training.

5.3 Higher Motivation in Error Training

We did not yet do a study on the issue of motivation. Therefore, there are no quantitative data to prove or disprove that higher motivation in the error training group leads to better performance. Qualitative observations suggested, however, that the error training group was quite motivated. While the participants in the error avoidant group went away after the training without waiting much longer, it was difficult to persuade the trainees of the error training group to leave the computer room. They "bombarded" us with further questions and seemed highly interested. Partly this may have been due to a sort of Zeigarnik effect (Ovsiankina, 1928) - the difficult problems had not been completely solved, thus, work on them was resumed again. Since all the difficult tasks of the enforced error training were not really solved, there were lots of tasks that had not been completed. This phenomenon also reminds of the literature on the achievement motive which showed that problems of moderate difficulty lead to higher motivation to solve them (Heckhausen, 1980) although in our case, the problems were of difficulty.

5.4 Errors as Instigators of Exploration

Dormann & Frese (1994) looked specifically at the issue that errors instigate exploration which in turn leads to better performance. This would mean, that a trainee in an error training group gets to know more of the system by having to explore it more actively (Greif & Janikowski, 1987; Greif, 1993).

The frequency and intensity of exploratory behavior in the training was measured by raters after the experiment (the rating was done on videos of the screen that were done during training). In the error avoidant group exploratory behavior was rated to exist if the subject deviated from the prescribed method. Since this was a sequential training in which the subjects were given a complete rundown on correct responses, in principle, they could not explore and not make errors. However, there was a sizable proportion of subjects who deviated from this approach. This meant that they started to explore *in spite of the training method.* In the error training group, exploratory behavior meant to go beyond the set of commands given to them.

Since exploratory behavior was not measured in the same way across the two groups, the correlations between exploratory behavior and performance were calculated separately for the two groups. In four out of six cases, there was a significant correlation between exploratory behavior and performance. The correlations between performance and exploratory behavior for the difficult tasks were between 0.44 (error training group) and 0.59 (error avoidant group). This finding is particularly interesting in the case of the error avoidant group. Because it was precisely the non-compliant subjects who resisted our sequential training procedure and just explored anyhow, who showed the highest performance. This implies that exploration is an effective mediator for good performance in training computer skills in general.

It is interesting to see that the same relationship holds in the error training group. Thus, exploratory behavior is an effective mediator of the error training procedure as well. This suggests that exploratory behavior leads to better performance when dealing with errors.

Thus, the correlations support the notion that exploratory behavior leads to better performance regardless of the specifics of the training method. There is no direct test in this experiment whether or not errors actually lead to exploration (we are currently doing a micro-analytic study on this issue). However, when the subjects were divided into those who explored a lot and those who explored little, the group with high exploratory behavior of the error avoidant training showed a similar performance to the group with little exploration in the error training condition (cf. Table 2). Thus, exploration reduces the differences between the two training groups. This suggests that there is some link between errors and exploration.

Table 2: Difficult Task Performance in Subgroups (Dormann & Frese, 1994)

Error avoidant training	
– Following instruction	2.2
– Exploration in spite of instruction	3.4
Error training	
– Medium exploration	3.7
– High exploration	4.4

However, there is still some superiority of the error training group. Although exploration may be an effective mediator learning the computer skills, there are additional reasons for the superiority of the error training group than just exploration.

The results should not be taken to mean that exploration is the only mediator - other mediators, for example, emotional or motivational factors may also be important (Frese et al., 1991, Irmer et al., 1991). However, these mechanisms need to be studied separately in depth.

Pfeffer (1992) has similarly suggested exploratory behavior to be an important explanatory concept for error training. He did a fine-grained key-stroke analysis of the Irmer et al. (1991) study. Pfeffer (1992) suggests that there is a difference in the way exploration is done: The standard training error avoidant group used exploration more in a trial and error sense while the error training group used exploration more specifically in an hypothesis-driven way. One indicator for this is the fact that the standard training group calls upon useless help much more frequently than the error training group (Pfeffer, 1992).

6. Conclusion

In summary, there is ample evidence that errors are an important part of the learning process and that allowing and encouraging errors to appear in training leads to better performance results. For practical purposes, it is enough to include an error training module in a general training. It may be useful to include the error training module in the middle of the training rather than in the beginning since in the beginning anxiety may play a more important role and in the end, error training may not fully develop its positive impact. However, practicioners are well advised to rely on error training as well (note, however, that the use of heuristics is very important here).

As for the specific process of why error training is superior, we have just started to scratch the surface. There is no evidence that learning specific strategies of dealing with errors is at the core of the superiority of error training. Thus, more general factors may be important. We think that one such factor seems to be that errors instigate exploration. We are not sure about the effects of dealing with the emotions that errors may bring about and there is no study that has looked at the motivational factors of uncompleted tasks (=enforced error training).

But the area of error training is an exciting new area that may change our concepts of what is important in training. It may not be so much the detailed training program but rather providing a general notion of dealing with computers with exploratory strategies, with little fears about making errors and with the use of errors as a way to learn.

References

Brodbeck, F.C., Zapf, D., Prümper, J., & Frese, M. (in press). Error handling in office work with computers: A field study. J. Occupational and Organizational Psychology

Carroll, J.M. & Mack, R.L. (1984). Actively learning to use a word processor. In: W.E. Cooper (Ed.), Cognitive aspects of skilled typewriting, pp. 259-281. New York: Springer

Carroll, J.M., Mack, R.L., Lewis, C.H., Grischkowski, N.L., & Robertson, S.R. (1985). Exploring a word processor. In: B. Shackel (Ed.), Human-computer interaction, pp. 283-307. Amsterdam: Elsevier

Ericsson, K.A. & Simon, H.A. (1980). Verbal reports as data. Psychological Review 87, 215-251

Frese, M., Albrecht, K., Altmann, A., Lang, J., Papstein, P.V., Peyerl, R., Prümper, J., Schulte-Göcking, H., Wankmüller, I., & Wendel, R. (1988). The effects of an active development of the mental model in the training process: Experimental results on a word processing system. Behavior and Information Technology 7(3), 295-304

Frese, M., Brodbeck, F., Heinbokel, T., Mooser, C., Schleiffenbaum, E., & Thiemann, P. (1991). Errors in training computer skills: On the positive function of errors. Human-Computer Interaction 6, 77-93

Frese, M. (1991). Error management or error prevention: Two strategies to deal with errors in software design. In: H.-J. Bullinger (Ed.), Human aspects in computing: Design and use of interactive systems and work with terminals. Amsterdam: Elsevier

Frese, M., & Zapf, D. (Eds.) (1991). Fehler bei der Arbeit mit dem Computer. Ergebnisse von Beobachtungen und Befragungen im Bürobereich, pp. 181-265. Bern: Huber

Greif, S., & Janikowski, A. (1987). Aktives Lernen durch systematische Fehlerexploration oder programmiertes Lernen durch Tutorials? Zeitschrift für Arbeits- und Organisationspsychologie 31, 94-99

Greif, S. (1986). Neue Kommunikationstechnologien - Entlastung oder mehr Stress? Beschreibung eines Computer Trainings zur "Stress-Immunisierung". In: K.H. Pullig, U. Schäkel, & J. Scholz (Eds.), Stress. Reihe. Betriebliche Weiterbildung, 8, Windmühle, Hamburg, pp. 178-201

Greif, S., & Keller, H. (1990). Innovation and the design of working and learning environments: The concept of explanation of exploration in human-computer interaction. In: M.A. West & J.L. Farr (Eds.), Innovation and creativity at work, pp. 231-249. New York: Wiley

Greif, S. (1993). Fehlertraining und Komplexität beim Software-Design. Osnabrück: Unpublished ms

Heckhausen, H. (1980). Motivation und Handeln. Berlin: Springer

Dormann, T., & Frese, M. (1994). Error Training: Replication and the Function of Exploratory Behavior. International Journal of Human-Computer Interaction 6(4), 365-372

Irmer, C., Pfeffer, S., & Frese, M. (1991). Konsequenzen von Fehleranalysen für das Training: Das Fehlertraining. In: M. Frese & D. Zapf (Eds.), Fehler bei der Arbeit mit dem Computer. Ergebnisse von Beobachtungen und Befragungen im Bürobereich. S. 181-265. Bern: Huber

Ovsiankina, M. (1928). Die Wiederaufnahme unterbrochener Handlungen. Psychologische Forschung 11, 302-379

Papstein, P.v., & Frese, M. (1988). Transferring skills from training to the actual work situation: The role of task application knowledge, action styles and job decision latitude. Human factors in computing systems. Special issue SIGCHI Bulletin.

Pfeffer, S. (1992). Fehlermanagement im Computertraining: Analyse eines Fehlertrainings mit Hilfe von Keystroke-Protokollen. Munich: unpubl. Dipl. thesis

Prümper, J., Zapf, D., Brodbeck, F.C., & Frese, M. (1992). Some surprising differences between novice and expert errors in computerized office work. Behavior & Information Technology 11(6), 319-328

Skinner, B.F. (1968). The technology of teaching. New York: Appleton-Century-Crofts

Thiemann, P. (1990). Aus Fehlern lernen? Fehlervermeidung vs. Fehlermanagement in der Mensch-Computer Interaktion [Learning from errors? Error prevention vs. error management in human-computer interaction]. Munich: unpubl. Dipl. thesis

Wendel, R., & Frese, M. (1987). Developing exploratory strategies in training: the general approach and a specific example for manual use. In: H.-J- Bullinger and B. Shacker (Eds.), Human-Computer Interaction-INTERACT '87. Elsevier North-Holland, pp. 943-948

Zapf, D., Frese, M., Irmer, C., & Brodbeck, F.C. (1991). Konsequenzen von Fehleranalysen für die Softwaregestaltung. In: M. Frese & D. Zapf (Eds.), Fehler und Schwierigkeiten bei der Arbeit mit dem Computer. Ergebnisse von Beobachtungen und Befragungen im Bürobereich. Bern: Huber

Decision Aiding in an Organizational Context

João Clímaco[1,3], C. Henggeler Antunes[2,3] and João P. Costa[1,3]

[1] Faculty of Economics, University of Coimbra, 3000 Coimbra, Portugal
[2] Dept. Electrical Engineering, University of Coimbra, 3000 Coimbra, Portugal
[3] INESC-Coimbra, R. Antero de Quental 231 cave, 3000 Coimbra, Portugal

Abstract. The increasing complexity of the social and economical environment, as well as the rapid pace of development in computer and communication sciences and technologies, have created situations which require not only the modernization of the organizations but also new types of planning and management. Under these changing conditions new forms of decision support must be developed, giving decision makers an active role throughout the decision making process. Decision support systems provide a technological support for learning in an organizational context. Our approach will be illustrated by means of operational decision support systems dedicated to rural telephone network planning and multicriteria problems.

Keywords. Decision support systems, learning, multiple criteria, interactivity, modelling, rural telephone network planning, energy planning.

1. Introduction

After the golden period of operational research during the 60's and the 70's, in which great part of the concepts and classic methods in this domain were developed, difficulties began to arise. From the 70's onwards the increasing complexity of the social and economical environment, and the rapid pace of development in computer and communication sciences and technologies, have created situations which require not only the modernization of the organizations but also new types of planning and management. The main question that arises is how shall the decision making process be approached under these changing conditions?

A lot has been written on this subject. In synthesis we believe that the essential ideas may be found in the principles enunciated by Ackoff (1979, 1981), the key ideas of which may be summarized as follows:

a) The decision process must be interactive and participative. It is essential that managers, operational researchers, computer engineers, etc., join their efforts into collective, mutually enriching work.

b) The plans must be continuously revised. This procedure is essential since the organizational environments where the plans are inserted change very rapidly, and this makes even more difficult to obtain reliable forecasts.

c) The units in the organizations must be planned and managed in an interdependent manner, that is, decisions must take into account all the interactions of the system.

It was in this context that Geoffrion (1983) wrote an article with the suggestive title: "Can Management Science/Operations Research Evolve Fast Enough?". We also believe that Decision Support Systems (DSS) must attempt to come closer to the principles formulated in this paper. We designate by DSS a computer-based system, generally interactive, aimed at supporting and improving the decision process particularly in complex and/or badly structured tasks which require critical appreciation by the decision makers. DSS do not intend to replace the decision maker (DM); instead they are aimed at effectively supporting to the DM in complex tasks. These systems usually integrate modern computer science techniques including sophisticated Human-computer interfaces, quantitative statistical methods, operational research methods, and more recently artificial intelligence techniques. At present we also observe the appearance of new technical bases for DSS, such as intelligent workstations and distributed systems or networks. It should be pointed out that the integration of operational research and artificial intelligence techniques opens new and very promising perspectives for the development of more effective DSS. However, it is also true that together with the advantages of integration, many difficulties subsist resulting from resistance which always exist when we look for synergies between different branches of knowledge which have been developed separately during decades.

In this paper we focus on our experience in the development and application of DSS, taking into account the following points:

a) We should not forget that human capabilities and technology are complementary tools in decision support.

b) Modelling is a key element, but we must have in mind that our conceptions of reality are "constructive and interpretative" (Pruznan, 1988). The classical assertion that we must not follow the temptation of considering ”tools first, tasks after”, by trying to force the models for which elegant analytical tools have been developed to adapt to reality, is obviously true in this context. This also entails that the choice and/or the combined use of the DSS models must have a high degree of flexibility. The new computing systems may also have an important role in the modelling stage of the DSS development by enabling, for example, the development of visual interactive modelling.

c) Interactivity is an essential feature of DSS to help DMs in the search for satisfactory solutions. We believe that the paradigm of searching for "satisfactory solutions" (as developed by Simon, 1981) is more adequate to the reality, than the one of "finding the optimal solution". In the implementation of interactive tools, in addition to the utilization of the most sophisticated available computational and algorithmic tools, attention must be paid to the cultural questions involved. In fact, the dialogues are generated by cognitive processes, and therefore the influence of the cultural and organizational environment on which the DSS is to be implemented can not be disregarded.

d) Concerning the scientific methodologies, it seems clear that domains such as multicriteria decision support, group decision support and negotiation, data collection techniques, treatment of uncertainty, validation of data and models, representation of knowledge, and integration of artificial intelligence and operational research techniques, are crucial ones. The requirements of DSS in the complex socio-economic environment in which we live and the prevailing idea (see, for instance, Keen, 1987) that the availability of information technologies is no more an essential limitation to the development, leads to the necessity of paying special attention to the methodological aspects. Therefore, the perspective of development must be the integration of appropriate methodological tools in order to obtain efficient but also flexible and easy to use DSS. As someone stated "because organizations are complex, we need sophisticated simplicity ".

e) As stressed by Pruznan (1988), it is not justifiable to separate the technical, political, social and ethical problems of management and planning. It is indispensable to consider the systemic character of the organizations, even if in the analysis phase the decomposition of the problem becomes necessary.

DSS provide thus technological support for learning in an organizational context. Benefits are not just the possibility of having decisions "automatically" produced based on complex models but mainly the aiding throughout the decision process, since this enables people involved to increase knowledge and skills concerning the object of the process. This learning aspect of DSS permits planners and DMs to adapt to changing environments, to develop new skills and to acquire expertise in those domains.

2. Experience in the Development and Application of DSS

The problems that arise in the context of organizations may be broadly classified as unstructured problems (such as the ones at top management level) and semistructured (such as the ones with technological basis, but with implications both at social and organization levels). We will illustrate our approach by means of case studies of this latter type, and distinct operational DSS we have developed will be referred.

One of the DSS is dedicated to help planners in rural telephone network planning tasks, using in an integrated manner operational research methods and artificial intelligence techniques. The other DSS are devoted to provide decision aid in multiple criteria problems, and a case study dealing with a power system expansion planning problem will be referred.

The problem addressed by the DSS dedicated to rural telephone network planning problems is of combinatorial nature. Models possessing this characteristic are often found when dealing with real problems. Due to its dimension this type of problems are of great computational complexity and it is often necessary to use heuristics to tackle them in an effective manner. Heuristics, unlike mathematical programming algorithms, do not guarantee the optimal solution, but produce "good" solutions in an acceptable computing time.

Most real-world problems are characterized by multiple, incommensurate and conflicting aspects of evaluation. Under these circumstances, single criteria models and methods fail to capture the complexity and conflicting nature of these problems. Mathematical models, as well as the perception of the problems by DMs become more realistic if multiple criteria are considered explicitly.

A model is a representation of reality for a particular purpose. As it can not represent all the complexity and interactions of real-world, a model is always an Abstraction in the sense that it can not encompass all aspects of reality. Essential attributes of a model are its reference (the object of the model), end (a purpose related to the reference) and practical cost (of using it instead of the reference). The model must then enable to deal with the real-world and cope with its complexity and irreversibility. Problems of combinatorial nature (such rural telephone network planning) and the ones characterized by the existence of multiple and conflicting criteria (such as energy planning) are particularly interesting to illustrate these issues.

The incorporation into the models of the possibility of intervention of the DM is a key issue not just to profit from his/her past experience and expertise, but also to amplify his/her capabilities of information processing and decision making. This must be accomplished by means of carefully designed Human-computer interfaces in a way that promotes effective user thinking (thus making the decision process more understandable) and learning during tasks (improve skill for suitable use of knowledge). The interaction environment must enhance the user's propensity and ability to benefit from the system (Sage, 1991). Special attention must be paid to the main limits of people (memory, accurate calculating, and patience) and their main strengths must be potentialized (visual scanning, learning highly integrated and complex material, pattern recognition). In this context, and taking into account the new technological means (such as powerful and affordable microcomputers, intelligent workstations and distributed computer systems). Human-computer interfacing has been gaining an outstanding importance in several domains of decision support, particularly in interactive multiple criteria DSS (Clímaco and Antunes, 1992).

3. Rural Telephone Network Planning

Rural telephone networks are a particular case of local telephone networks, characterized, among other factors, by their low telephone density, serving small population centres where most of the subscribers are located, the sparsity of the subscribers distribution and the existence of geographical difficulties to the expansion. At the economic level it is, in general, very difficult to amortize the necessary investments. This fact, as well as the social importance of this type of networks, implies that particular attention must be paid to the whole planning process.

The global objective is to minimize costs while satisfying demand (defined by a forecast model) and quality constraints (as established by the Telecom administration policy) throughout the planning period. The specific objectives pursued in the planning process supported by this DSS are: a) to specify the local exchanges in terms of number, location, year of installation and type of exchange (technical and capacity specification); b) to define the service areas of the local exchanges (which subscribers should be connected to each exchange, and when this connection should occur); c) to determine the secondary distribution network (which involves aggregating the subscribers in distribution points, and aggregating these ones in flexibility points; d) to define the primary distribution network (topology and cable capacity and when to connect each of the flexibility points of a service area to the corresponding exchange); e) to satisfy the demand forecast throughout the planning period; f) to satisfy the quality of service criteria; g) to minimize the total cost of network expansion.

Due to the complexity of the problem some approximations had to be made. The main source of imprecision derives from the fact that some heuristics require that the equipment cost is represented by means of a linear model, instead of piecewise linear schedules (as it is typical in modular equipment).

A key issue directly related to the acceptability and success of DSS is the possibility of profiting of the user's experience and expertise. This was accomplished in the development of this system both by means of the design of its functional structure and the combination of operational research and artificial intelligence techniques.

In the design of this system particular attention was paid to its flexibility and interface with the end-user to enable the DM to intervene in all planning phases and modify any parameters of the rural network under study. The main aim is to make the most of the DM's experience and expertise, by giving him/her the possibility of playing an active role in the planning process. The objective is twofold. On one hand, the human DM must not feel replaced by a computer system. On the other hand, since the system includes heuristics (which do not guarantee the "optimal" solution) the users experience can be used by means of suggestions which serve as input to the heuristics. In this manner it is possible to avoid *a-priori* the search of non-acceptable network configurations (for instance, by specifying non-acceptable locations for the exchanges).

Another way of attempting to support the DM in the consideration of factors of difficult representation (either not included explicitly in the algorithmic model or included in an approximate manner) was made by integrating artificial intelligence techniques into the DSS. This was implemented in the form of a rule-based system for evaluating the solutions produced by the algorithms of the DSS and minimize the effect of the above referred approximations of the mathematical model. In this manner it was possible to incorporate, up to a certain extent, qualitative factors which may be associated with the DM's experience and expertise since the knowledge used in this evaluating system is introduced and/or actualized by him/her. Note that when knowledge is being used it becomes procedural and when it is being improved it is declarative.

The possibility of intervention of the DM is important not just to attenuate the effects of the decomposition of the network in different levels (in order to have computationally manageable models), but also to deal with the various levels of interactions within the organizations where the planning process is being carried out. From the perspective of the characterization of its relation with the decision environment, this DSS was conceived in a perspective directed to the planning process. That is, by considering that the benefits from the planning process are not just the plan (and its implementation), but also the process itself in the sense that it enables the people involved to increase their knowledge about the network (the object of the planning process). This type of DSS enables a process of learning from observation by planners and DMs in telecommunication administrations or operating companies. According to the degree of interaction enabled by the DSS between decision agents in organizations and the external environments (rural telecommunication networks) this learning must be based on active experimentation in the sense that they may make changes in the environment and observe the results of those changes. This also contributes to a learning by analogy process, which involves transforming and extending (individual and organizational) knowledge and skill applicable in certain situations to perform similar tasks in other decision contexts.

The main components of the DSS environment are illustrated in Fig. 1.

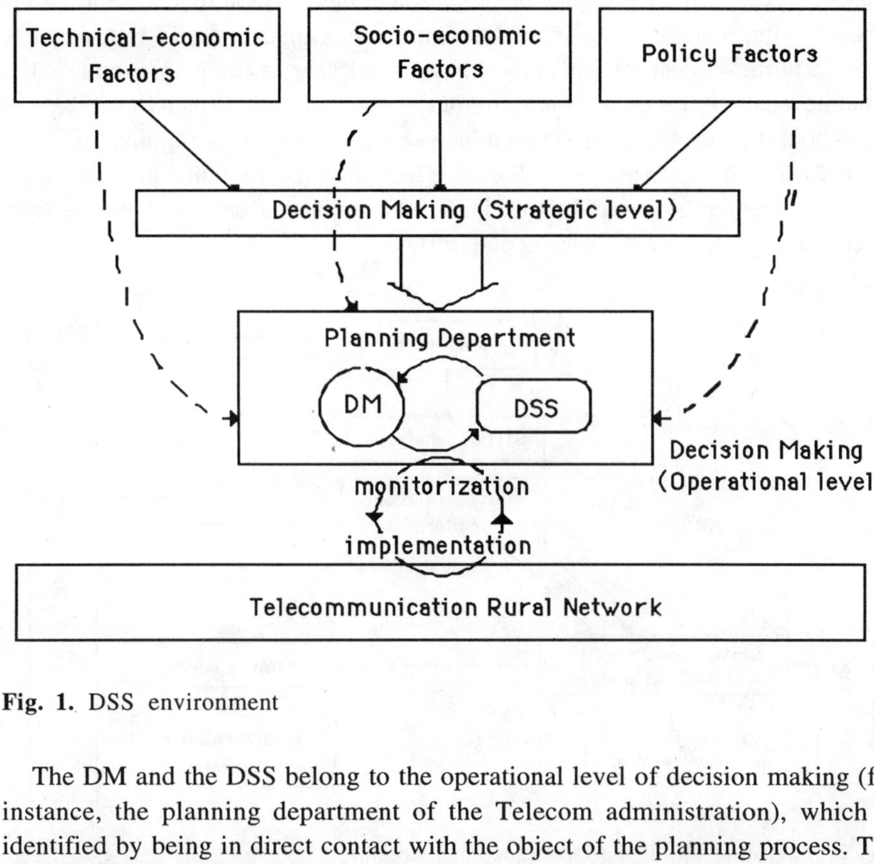

Fig. 1. DSS environment

The DM and the DSS belong to the operational level of decision making (for instance, the planning department of the Telecom administration), which is identified by being in direct contact with the object of the planning process. The DM can not be dissociated from some "environmental" factors. Typical influence sources over the DM derive from: technical-economic factors (such as equipment technical characteristics, its cost and availability); socio-economic factors (such as population growth, development of economical activities and their geographical distribution), and administration policy factors (such as budget constraints, network expansion and technological innovation diffusion policies). All these factors may influence the DM directly (via its internal representation of the factors), or indirectly (via the strategic decision making level, for instance expressed through administration directives).

The techniques generally classified as "visual interactive modelling" (Angehrn, 1992) can be very useful to integrate models with the DM's experience. This may be expressed directly on the screen on which the entity (network, etc.) under study is graphically presented, during the interactive process. The aim is to obtain with this new type of modelling a better approximation of reality both at the physical level and at the level of the cognitive process of the DM.

As far as the functional structure of this DSS is concerned, its main blocks are: a network editor which provides flexible capabilities for inputting and editing the network data; a static analysis module which determines a set of best network plans in terms of cost minimization for the final year of the planning period; a dynamic analysis module which computes a set of best expansion strategies throughout the planning period based on a static solution; a post-optimal analysis module which is supported on a rule-base where rules incorporating the knowledge about the problem are stored. See Fig. 2, and for further details about the DSS functional structure see Clímaco et al. (1991).

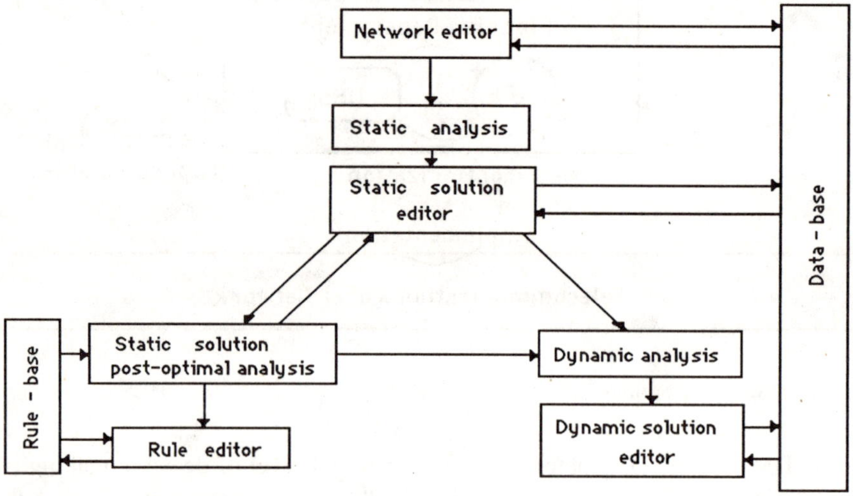

Fig. 2. Block diagram of the DSS functional structure

4. DSS for Multicriteria Problems

Since most real-world problems are characterized by multiple, incommensurate and conflicting criteria of evaluation, these must be explicitly incorporated into the models. In this manner not just the models become more realistic, but also the DM's comprehension of the complexity of the problems and trade-offs to be made is enhanced. Another important aspect related to the use of multicriteria analytical techniques is the active and participative role required from the DM in interactive procedures. These alternate computation phases and dialogue phases, in order to guide the search process to regions where solutions which more closely correspond to the DM's (evolutionary) preferences are located. Interactive procedures are nowadays recognized as the most effective way to deal with multicriteria problems (Steuer, 1986).

Besides its importance in stimulating the DM's capabilities in processing the information and in the search for satisfactory solutions, the development of an appropriate context for interaction potentializes the creation of new multicriteria methodologies (Clímaco and Antunes, 1992). In fact, some innovative features to be exploited in new methods would be impossible without the use of graphical means enabling a direct visual inspection.

The experience acquired with the computer implementation, application and evaluation of the conceptual characteristics of multiple objective linear programming interactive methods led to the development of a flexible tool in which different methods have been integrated in a computer package named TOMMIX (Antunes et al., 1992). TOMMIX aims at creating a flexible computational decision aid tool respecting the conceptual characteristics of the methods, the main concern being the possibilities of combining them while guaranteeing the transfer of usable information. The underlying principle is to support interactively the DM in the progressive narrowing of the scope of the search. TOMMIX integrates methods which are representative of distinct strategies for reducing the scope of the search, require different preference information from the DM and use different techniques to obtain nondominated solutions.

This work is now being extended in new directions:

a) Disaggregation of the methods in their elementary operations which can be invoked directly by the DM, or further combined in procedures which embody a sequence of those operations.
b) Construction of an interaction environment based on a control panel consisting of a set of distinct graphical displays and pictorial controls aimed at making the interface more flexible (see Fig. 3).

Since, in a multicriteria context, aggregation of preferences implies loss of information, it is very important to give the DM the possibility of using the elementary operations in a flexible way according to the underlying principle of progressively and interactively reduce the scope of the search.

Another DSS we developed integrates some representative discrete alternative multicriteria methods in the operational framework of a flexible and user-friendly computational tool (Antunes et al., 1992). This system allows the DM to experiment different preference information corresponding to distinct scenarios in terms of the context of the problem and observe interactively the changes in the results. This enables to evaluate in an interactive manner the robustness of the

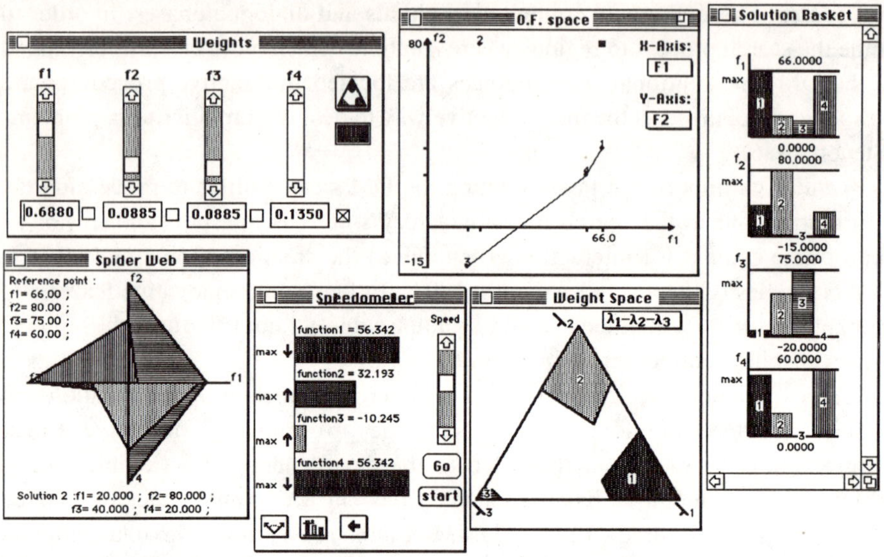

Fig. 3. The control panel displaying different type of graphical information

conclusions in relation to the lack of determination of the input parameters, which is an important issue in multicriteria studies.

A panel consisting of slide bars, check boxes and radio buttons which enables to control continuously the input parameters of one of the outranking family of methods is displayed in Fig. 4 (for an example problem of choosing a car using cost, comfort, consumption and speed criteria). The outranking graph shows automatically the influence of the parameters on the results, thus enabling to proceed by trial and error and grasp the parameters which reveal more influent in the conclusions of the study.

The DSS devoted to provide decision aid in power system expansion planning is based on multiple criteria models and methods, both in the mathematical programming and discrete alternative cases (Clímaco et al., 1990). In a first stage a multiple criteria linear programming model was developed which considered three objective functions: cost of the expansion plans (to be minimized), reliability of the supply system (to be maximized) and environmental impacts (to be minimized). Three categories of constraints expressed load requirements, operational capacity of the power units and budget restrictions. An interactive method named TRIMAP developed by some of the authors was used to deal with this problem (Clímaco and Antunes, 1989). TRIMAP is an interactive procedure aimed at enabling the DM to make a progressive and selective search for nondominated solutions (feasible solutions for which an improvement in an objective function value implies the degradation of at least other objective function).

A set of solutions selected with the aid of TRIMAP were in a second stage further evaluated against an expanded set of criteria related to issues which are difficult to be quantified (to be included in mathematical programming models) or for which the data available possess intrinsic uncertainty. Demand-side management concerns were considered to expand the set of criteria. Demand-side management is the planning and implementation of electric utility activities aimed at influencing customer use of electricity in ways that produce desired changes in the utility's load shape. In this manner it is possible to expand the environment and observe interactively (both by passive observation or by active experimentation) the changes in the results.

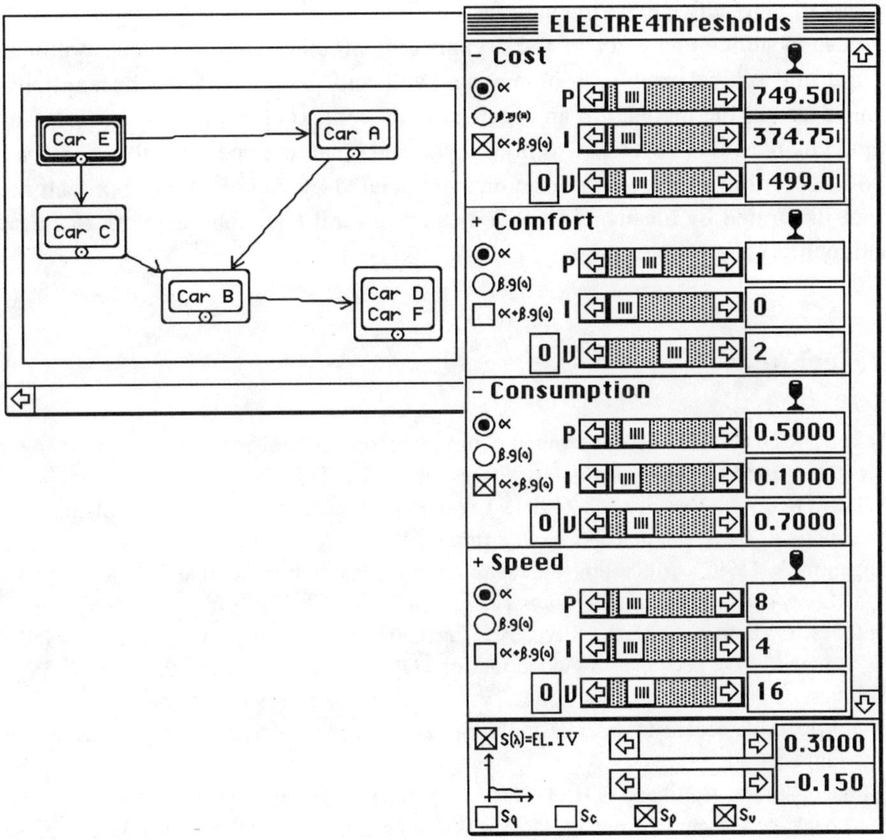

Fig. 4. A panel consisting of slide bars, check boxes and radio buttons for controlling the input parameters

5. Conclusions

In this paper we aimed at showing the importance of incorporating into the models the possibility of intervention of the DM, as a key issue to profit from his/her experience and expertise, and amplify his/her capabilities of information processing and decision making. Well designed Human-computer interfaces must promote the comprehension of the decision making process and improve skill for suitable use of knowledge. DSS provide a technological support for learning in an organizational context. Distinct types of learning may be supported: by analogy, involving transforming knowledge in one domain to perform a similar task in another domain; from examples, in an incremental concept acquisition process; from observations, involving various degrees of interaction, from passive observation to active experimentation.

The systemic character of the organizations (embodying technical, political, social and ethical problems of management and planning) must be explicitly considered in the models. In an organizational context characterized by deep and rapid changes, it is necessary to investigate and construct operational decision aid tools according to the main principles enunciated by Ackoff. Our approach has been illustrated by means of DSS dedicated to rural telephone network planning and multicriteria problems.

References

Ackoff, R.L. (1979) Ressurecting the future of operational research. *Journal of the Operational Research Society*, 30, 3, 189-199

Ackoff, R.L., & Vergare, E. (1981) Creativity in problem solving and planning: a review. *European Journal of Operational Research,* 7, 1-13

Angehrn, A. (1992). Modelling by example: new ideas for interactive decision support. *European Journal of Operational Research*

Antunes, C.H., Alves, M. J., Silva, A.L., & Clímaco, J. (1992). An integrated MOLP method base package - a guided tour of TOMMIX. C 19, 7, 609-625

Antunes, C.H., Almeida, L., Lopes, V., & Clímaco, J. (1993). A decision support system dedicated to discrete multicriteria problems. Decision Support Systems (to appear)

Clímaco, J., & Antunes, C.H. (1989). Implementation of a user friendly software package - a guided tour of TRIMAP. Mathematical and Computer Modelling, 12, 1299-1309

Clímaco, J., & Antunes, C.H. (1992). Man-machine interfacing in MCDA. Proc. 10th Int. Conf. on Multiple Criteria Decision Making, Taiwan, I, 231-244

Clímaco, J., Costa, J.P., Antunes, C.H., & Craveirinha, J. (1991). Decision support in rural telecommunication network planning. Engineering Optimisation, 18, 137-146

Clímaco, J., Martins, A.G., & Almeida, A.T. (1990) On the use of multicriteria optimisation for electric energy planning. Int. Journal of Global Energy Issues, 2, 3, 194-203

Geoffrion, A.M. (1983). Can management science/operations research evolve fast enough?. Interfaces, 13, 1, 10-25

Keen, P. (1987). Decision support systems: the next decade. Decision Support Systems, 3, 253- 265

Pruznan, P. (1988). Systemic OR and operational systems science. European Journal of Operational Research, 37, 34-41

Sage, A. (1991). Multiple perspective qualitative reasoning and dialog generation and management. In: M. Singh & L. Travé-Massuyès (Eds.), Decision Support Systems and Qualitative Reasoning, pp. 3-13. Elsevier

Simon, H. (1981) The Sciences of the Artificial. Cambridge, MA: MIT Press

User Interface Design from the Viewpoint of Human Learning

Gerrit C. van der Veer

University of Twente, Department of Ergonomics, Hengelo, The Netherlands
Vrije Universiteit, Department of Computer Science, Amsterdam, The Netherlands

Abstract. Information Technology is a special kind of artifact, in relation to its use by human users. IT systems and IT applications are characterised by a very complex structure, even if we restrict our focus to those parts of IT that an "end user" needs to understand in using it. At the same time the very nature of IT prohibits direct observation of the complexity of the processes and data structures that are relevant for the user. Consequently, users need to construct their own mental representations of the underlying complexity, in order to plan their use, to execute task delegation, and to interpret the actions of the system.

As far as the user is concerned, the relevant part of an IT system might be indicated by the concept "user interface", which in this case should be considered to indicate both the functionality provided to the user, the interaction means (language, media, representations), and all means to help the user maintain and improve interaction with the system (metacommunication).

In our contribution we will classify the different aspects of the design of the user interface (which might also be indicated as the user's "virtual machine") for which a focus on human learning is needed. This will include both the design of the functionality (object structure, events, task delegation), the interaction modes (including the lexicon, the graphic representations, and the interaction language), and the methods and facilities for metacommunication (learning, help facilities, metaphors and models).

We will illustrate these aspects with field studies and experimental results showing their relevance, and the relation to individual differences in human learning.

1. Introduction

Since prehistoric times humans have been using artifacts. From the moment artifacts were used, specialisms developed. For a very long period in history, the specialism was to a large extend related to skill. Hunting with bow and arrows, sewing clothes, playing the violin, could only be done successfully after considerable training.

Explicit learning, teaching, and apprenticeship were generally accepted activities in relation to mastering artifacts. At the same time it was recognized that not all people were equally gifted to start with. Both physical competence and psychological phenomena may be identified to contribute to the success of learning to use a tool, and often an interaction has been identified between individual characteristics and the process of teaching and learning.

It has only been in the last decades, however, that a type of artifacts was introduced into the world of general human tools, for which the general vision on learning and teaching needed to be reviewed.

Information technology and information systems show a general characteristic that is new in the world of artifacts and technology: In order to use these systems or to cope successfully with a situation in which these systems have a significant role, the human being has to understand the systems' complexity. Information technology applications are designed (and explicitly invented to do so) to behave in ways that can not be represented completely at the surface.

Computerized data-bases are intended to provide a richness in structure and detail that overloads human perception and attention if presented completely. Electronic data processing is chosen because of its speed, algorithmic power, and complexity, for situations that cannot be represented and elaborated by human beings in mechanical or formal ways.

Yet, the human user can only make apply information systems if the system can be "understood". The user needs enough conceptual understanding of the structure of information in the system and of the process of information processing inside the machine, to delegate tasks to the system with some basic "trust" in the relation between the user's own actions and the activities and resulting states of the system.

This is the case not only for the computer science experts, but for any person who applies information technology for "work"-situations, i.e., for situations where the machine is used to delegate tasks or to assist human activities like learning, reasoning, making decisions. Even a casual user of a relatively "simple" application of information technology needs basic understanding of the relation between his behaviour and the system's reactions. Otherwise the system will not be applied optimally, if at all, and the user will not be working in the best possible conditions.

The general problem in this case is that relevant processes and structures are invisible during actual interaction. The user will perceive only the surface of the machine, where only the basic elements of interaction are represented. Behind this perceptual part of the system, there is a conceptual system of which the user has to be aware. We will use the label "user's virtual machine" (UVM) to indicate the system, both perceptual and conceptual, as far as a certain (group of) users has to be aware of its internal complexity in relation to the task for which it is used.

Even if the surface of the system is designed well, reflecting the inside in a way that communicates the relevant structure and dynamics of the inside, the user still needs to translate this systems image in the underlying semantics of the system. In order to analyse this phenomenon, we will elaborate the concept of mental model of complex systems (Sect. 2). Section 3 will focus on the user interface and the design methods for the user interface, showing how to take care for the needs of human users. The analytical approach elaborated here enables to focus on learning and metacommunication early in the design process and to provide some generic ideas for taking care of the needs of the end-user.

Individual differences in competence will play a major role in the effect of measures and activities that aim on user learning. Some relevant human characteristics and some approaches towards optimal adaptation are reviewed in relation to design.

2. Mental Models

When people need to cope with situations in which the conceptual complexity surpasses the perceptual possibilities, they will normally rely on mental resources. Mental models of a variety of systems are relatively successfully used in order to understand the world. In culturally and socially complex situations like eating out, people apply a "scheme" that has been developed on the basis of past experience in analogous situations. The scheme provides slots and default "values" for the standard actors in this social situation, like guest (i.e., the person or group going to the restaurant), waiter, cook, cashier (in some cases identical with the waiter). The schema structures the situation into "scenes" like entering the restaurant, ordering, eating, setting the bill, each of which has a standard structure and related procedure. The procedure can be mentally "executed" in order to predict what activities can be expected from the different actors involved, and what effect each actor's behaviour will have on the behaviour of the others. Observed deviations between the actual situation and the scheme will result in an instant adjustment of the internal model and consequently in a different interpretation of the situation. E.g., if the menu announces "no credit cards accepted" the guest may check (physically or mentally) the content of his wallet before deciding on the order.

Interaction with an information system is a definite case where users need a mental model to cope with complexity. In general the situation is much more complex than the restaurant example. The only analogous situation is meaningful communication between human beings. In communicating between people, each partner has to presume that the other's utterances are based on a semantic basis that is generally equivalent to his own. As soon as one partner detects possible deviations in the underlying semantics, the communication will temporarily focus on this, allowing all partners to adjust their understanding of the discourse. This phenomenon may be labeled "metacommunication" - communication about the regular interaction (e.g., "do I understand you correctly", "what do you mean with...").

Metacommunication is indispensable for maintaining meaningfulnes in human to human interaction. In the same way it is an essential aspect in human-computer interaction. The mental model of the situation (including the human partner, respectively, the information system) are continuously adjusted on the basis of the interaction.

Metacommunication is a main source of explicit adjustment of mental models. In Sect. 2.1 we will elaborate the nature of mental models of information systems. Some scientific notions and methods for investigating mental models will be presented in Sect. 2.2, along with some empirical data on characteristics of mental representations.

The development of mental models in relation to metacommunication in interaction between users and information systems will be analysed in Sect. 2.3.

2.1 The Nature of Mental Models

Cognitive Psychology has contributed a vast amount of theory in relation to the notion of mental models. For an overview of approaches that are relevant to information technology, see Van der Veer (1990).

Norman's analysis of mental models of computer systems is of major relevance to our analysis (Norman, 1993). Unlike many cognitive psychological theoreticians, he approaches the concept from a practical point of view. The term is used to indicate any knowledge and representations applied by computer users in actually using a system. The main idea is that a mental model is to a certain extend "runnable" in the head of the user, in order to help him to cope with situations that are too complex to rely on observable aspects of the system only. In this respect Norman's concept is equivalent to Pask's notion of "entailment structure" (Pask, 1980), where understanding a complex system is described as the ability to mentally reconstruct relevant aspects of system knowledge.

Mental models play a role in all phases of the use of information technology. When a user is planning a task of which at least part might be delegated to a system, the possible course of action and the system's reaction are considered by mentally simulating task delegation. Once the decision has been taken to utilize the system, the mental model is used as an internal reference in guiding the detail steps of the interaction by feeding back the system's perceived behaviour to the mental model's equivalent performance. If a task delegation is supposed to be finished, the results are evaluated against the mental model's predicted outcome. Whenever the system shows behaviour that has not been expected (e.g., because of network events, actions of other users, or system failures) these actions are interpreted by comparing them to consequences of the mental model in reaction to possible outside events.

Norman (1983) points to a number of general characteristics of mental models of computer systems, which are inherent to the complexity of this types of artifacts and to the fact that information processing of the nature of computer processes can only be externally represented in an indirect and abstract way.

a. Mental models are incomplete. Systems are very complex, and the human information processing capacities are not fit for this complexity. Users of information systems are often well aware of the fragmentary character of their model. This characteristic, however, seriously limits the ability to mentally "run" the model.
b. Mental models are unstable. On the one hand, people tend to lose details of systems knowledge when not in use, and tend to suffer from (negative) transfer when other systems of a certain degree of similarity (either perceptive or functional) are being learned or used in between. On the other hand, each occurrence of system use where the mental model did not produce a complete and correct prediction or interpretation leads normally to a subsequent adjustment or extension of the mental model. As such, instability is not so much a negative characteristic but an indication of the evolutionary character of human knowledge in use.
c. The boundaries of mental models are vague. People tend to integrate knowledge of related parts of the outside world, sometimes resulting in mixing up different systems or losing the awareness of the borders between different systems or functions. E.g., facts of operating systems are incorrectly attributed to application systems if these are normally used only in combination with a certain operating system. Users may be aware of this in the sense of feeling uncomfortable about using e.g., the saving facilities of their favorite editor on somebody else's machine.
d. A certain flavor of superstition is part of mental models in relation to fragments of incomplete knowledge. In such cases, users tend to stick to ritual behaviour that gives them a feeling of safety or that induces the view that they are not to be blamed in case of a possible accident. Returning to the home directory before logging out or shutting down the system, or returning to the initial state before quitting an application are wide spread examples of this phenomenon.

   When asked, users will mostly agree this is not strictly necessary, but they know at least it will do no harm and intuitively it feels better.

e. Mental models are parsimonious. Users of information technology tend to restrict their knowledge to a subset of possibilities that enables them to apply the main functionality they need, although they are aware of the fact that for many functions there are other possible courses of action. They prefer to occasionally issueing a series of acts of delegation, although they know there may be shortcuts available. In this respect expertise consists partly in having collected a well chosen and relatively small subset of the system's functionality that allows to perform most of the tasks without having to learn the exhausting list of possibilities.

2.2 Empirical Analysis of Mental Representations

Mental models are mental. In fact, only certain aspects of them can be communicated. Especially as far as rules and skills are part of the mental model, this can only be investigated via observation of resulting behaviour and relating this to external models of the mental model (Moray, 1989; Rasmussen, 1990). As far as skill-based mental representations are concerned, performance is the only indicator. For rule-based aspects of mental models, representations can be indirectly investigated by using either performance or questionnaires. The current analysis focuses on knowledge-based mental representations. As far as the mental model consists of mental knowledge representation, we elaborated a teach-back technique (van der Veer, 1990) that aims at provoking users of information technology to externalize their mental representations. In comparison to the analysis of performance of questionnaires, this provides a relatively direct indication of the quality of mental representations, and, hence, of the cognitive aspects of mental models.

The main concern of our teach-back technique is to convince the users that we are interested in how they represent the system mentally.

After explaining this very briefly to them, we ask them to imagine a target audience that knows the relevant task domain but does not know the system under consideration (e.g., "explain to a colleague who never used system X"). Here we may use two variants. Either the target audience should have experience with other systems with equivalent functionality ("... who does not know your data-base system but who is a regular user of the system DATADUNGEON") or is explicitly indicated not to have used any information technology for this type of tasks.

Finally, a task sample is indicated that might be considered representative for the functionality of the system. Additionally the teach-back request should suggest that any form of representation is equally acceptable. Figure 1 provides an example of a teach-back request.

"We would like to know how you represent the UNIX mail system in your mind. Therefore we ask you the following question: using diagrams, drawings, and/or words, show how you would explain to someone who used "PC-mail" but who never used your UNIX mail system how to perform the following task: examine your mailbox, deleting any message from Tom of Dick, saving any message from Fred in a file called "fredmail" without their headings, forwarding a copy of any message from Michael to Peter, and identifying the date of the oldest message in your mailbox."

Fig. 1. Example of teach-back request

Based on a systematic analysis of teach-back protocols collected in studies of many different types of information technology systems, we developed indexes for analytic descriptions of mental representations.

The indexes were subsequently investigated for inter-rater reliability and precise scoring instructions have been published (van der Veer, 1990). The main dimensions on which cognitive mental representations of computer systems differ between individual users are style of representation, levels of representation, and completeness and correctness.

a. Style of Representation

Systems may be represented verbally or in the form of a spatial/graphical image. Sometimes both styles are combined. Figure 2 shows an example of a combined set of representations, where different system concepts are related both in a verbal way and in a graphical way.

Another aspect of representation style concerns the aspect of dynamic structure of the representation. Some mental representations show a procedural structure (see Fig. 3) whereas others are of a declarative nature, i.e., mainly consisting of a set of production rules without any indication of general dynamics or time dependency (Fig. 4).

A last aspect of style concerns the user of "icons", which sometimes take the form of representing an atomic element in a graphic representation (see the keyboard in Fig. 2), otherwise indicate concepts in a verbal representation for which the person is not able to or willing to give a verbal transcription (e.g., an angled arrow inside a square to indicate the enter key).

Generally, several different style elements are combined, for example, a verbal procedural representation with icons. Procedural and declarative representations, however, have not been found in the same protocols.

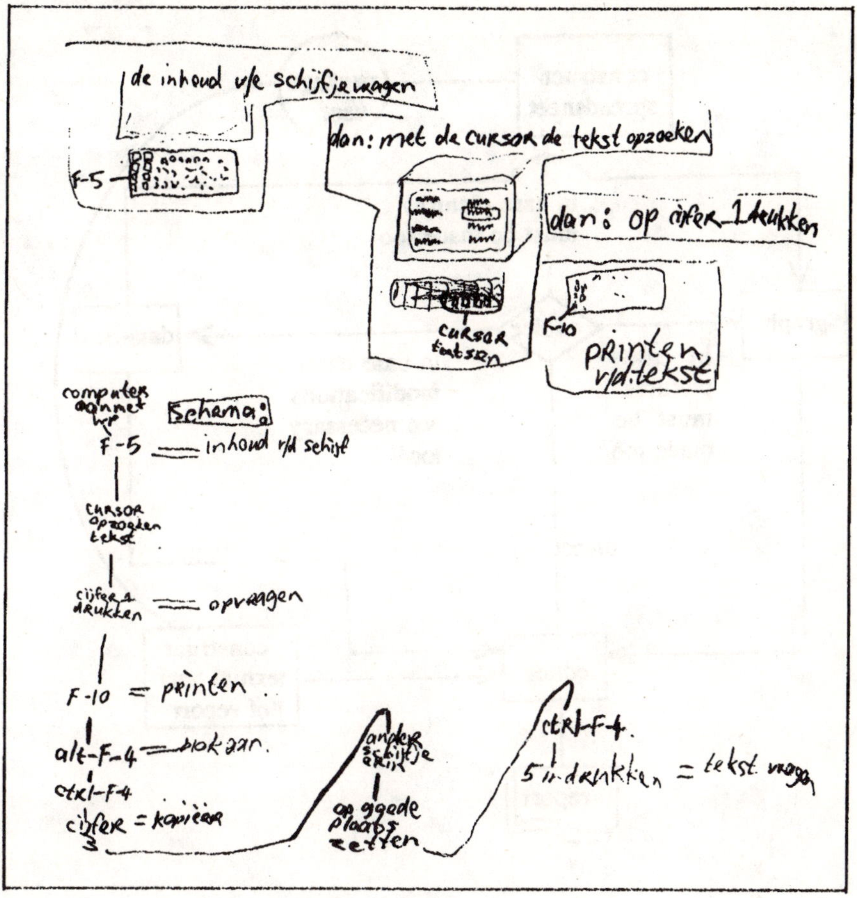

Fig. 2. Combination of verbal and graphical representation of a text processor

b. Levels of Representation

Levels of representation as found in the teach-back protocols closely follow the levels that Moran (1981) distinguishes in his CLG analysis.

Moran states that his approach not only can be used to analyse the user interface (the "design view") and the interaction language (the "linguistic view"), but also claims that it would have psychological relevance (the "psychologic view"), without, however, providing any support for this last claim. Our research showed that Moran's levels are a relevant analytic categorization for mental representations. Any teach-back protocol will be at one or more of the levels:

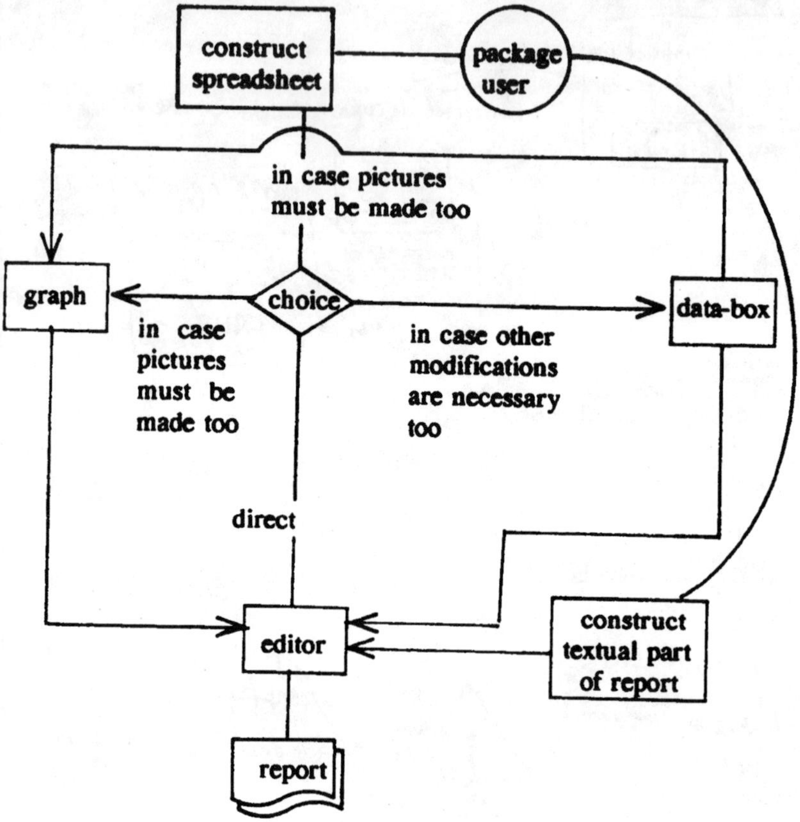

Fig. 3. Procedural representation of Golden Gate office system

examine your mail	:	type mail return
		efter that always type + for the consecutive letters
delete any messages from Tom or Dick	:	Type D at the top messages from tom or dick
Save any messages from Fred without their titles, in the file fredmail	:	type w (fredmail) at the top of all messages from Fred
forward any messages from Harry to Jhon	:	m(John) at the top of all messages from harry
delete the oldest message	:	via - you arrive at the aldest letter type D at the top of these

Fig. 4. Production rule representation of electronic mail system

- task level, referring to the system with respect to the delegation of certain subtasks and often at the same time indicating that other tasks have to be performed in other ways (e.g., by asking a person);
- semantic level, referring to the functionality of the system from the user's point of view (system actions, objects, attributes);
- syntax level, indicating lexicographic structure of the interaction language;
- key-stroke level, referring to physical actions related to input devices (mouse clicks, key presses) and output devices (blinking of the cursor).

If levels occur in a combined way, mostly adjacent levels are found together.

c. Correctness and Completeness

Teach-back protocols reveal the (relative) degree of correctness of the respondent's knowledge at those levels at which a representation is found. Moreover, the degree of completeness of the representation can be scored, for system knowledge related to the task indicated in the teach-back request.

2.3 Development of Mental Models

Mental models, at least as far as they may be considered knowledge based, are equivalent to structures in semantic memory.

Rumelhart and Norman (1977) discuss three ways in which a mental model or, in terms of Rumelhart and Norman, a schema, develops. This process may be categorised as either (a) accretion, (b) tuning, or (c) restructuring.

a. accretion indicates a process where new information is encoded in relation to an existing schema in semantic memory. The existing schema or model is extended with new knowledge that fills gaps in the previous version.
b. tuning occurs if, during interaction with the outside world, new information turns out to be inconsistent with some details of existing knowledge. This development can be characterised as evolution, where the general structure and meaning of the mental model is preserved.
c. restructuring, or schema creation, occurs if a new mental model is initiated. In this case the new schema will be based on analogies with existing knowledge and metaphoric resemblance with available schema, as far as these provide a structural or functional equivalence at some level of abstraction.

The general lesson from this developmental view is that adequate metaphors, as well as representations at different levels, are needed in order to promote the development of mental models in new circumstances. Once an initially adequate core mental model is formed, accretion and tuning will be the mechanisms that help users of information systems to improve their understanding.

3. Designing an Interface to the User

In analysing the user interface from the point of view of human learning, a distinction has to be made between the aspects of functionality, interaction, and metacommunication.

From the viewpoint of the user the system is a seen as a total machine, that can be used in a certain task domain to delegate some of the tasks or parts of the task (the functionality), in a way that is defined by the design of the system (the interaction), and with certain possibilities to get some help about how to interact (metacommunication). The total machine as it appears to the user may be referred to as a "virtual machine", of which the precise meaning may differ depending on the user group and task domain for which the system is used. Hence, we will refer to this as the "user virtual machine" (Van der Veer, Tauber, Waern, & van Muylwijk, 1985). This user virtual machine includes the "system image" as defined by Norman (1983) but also all conceptual knowledge the user needs in order to understand what is going on inside the system between perceptible actions of user and system. The user virtual machine, in other words, refers to all aspects of the system the user needs to know about in using the system for the task domain, both perceptually and conceptually.

When we refer to user interface design we aim at the design of the whole user virtual machine, including both the functionality, the interaction, and the metacommunication. This concept combines the concepts of application interface and user interface (in the restricted sense) as defined in the Seeheim model (Pfaff, 1985).

One should keep in mind that this meaning of "user interface" is broader that what is often found in literature, where the concept is restricted to interaction, or even to the perceptual aspects of interaction. We prefer to use the concept in the broad sense, since for the user the virtual machine is conceived as a whole, whatever the architecture behind the system's design.

3.1 Functionality

The most important aspect of the user interface, from the point of view of the user, is WHAT the system can do for the user or what the user can do with the help of the machine. This is what is called the functionality of the system or application functionality. The functionality refers to the tasks the system is designed for as well as the system entities designed for enabling the user to delegate these tasks to the machine. These tasks can be referred to as "primary tasks", indicating that these are the functions of the system that are or intrinsic interest to the user.

A side-effect of using any artifact for performing primary tasks is the need to utilize the tool or artifact. If the primary task is to connect wooden boards to the wooden framework of a house, the side-effect of using nails and a hammer create tasks like finding or choosing appropriate nails and manipulating a hammer. These tasks we label secondary tasks. In using information technology, the delegation of primary tasks to an application system can only be accomplished by interacting with the system. The secondary tasks of interaction will be analysed in Sect. 3.2. Information technology as such is different from other artifacts that humans apply in performing their tasks, in that the virtual system has to be mastered far beyond perceptual aspects.

The user has to understand what is going on in the system at a conceptual level, from what can be perceived via the physical actions of the user and the system. In order to learn this, as well as in order to solve problems during interaction, a third set of tasks has to be incorporated in user interface design, aimed at assisting the user in using the system. This type of tasks, referred to as metacommunication (communication between user and system on the subject of functionality and interaction) will be analysed in Sect. 3.3.

3.1.1 Primary Tasks for the System

Although the functionality of a system regards the primary tasks, this does not mean that the external task domain is completely and equivalently reflected in the system. The external task structure in the situation where no system is present, or where an "old" system is used, may well be different from the task structure where the new system is applied. Design of a system or introduction of a system in a task environment will lead to changes in task structure, to re-allocation of tasks to different people and to machines, and to combination or decomposition of tasks that have hitherto be considered "unit tasks". Information technology may be designed just to assist people to perform well established tasks, improving speed, reliability and accuracy or relieving people from boring routine activities. More often than not, however, information technology will aim at adding value to the task situation in enabling new functions, providing new possibilities, or allowing combinations or variants of tasks that were not even thinkable in previous situations.

The functionality of a well designed system will enable the user flexibility in delegating tasks where the system can be of maximum service, at the same time allowing the user to concentrate on tasks that people like to perform themselves or that can better be performed by human beings, like making decisions, comparing "hard" information with intuitive feelings, judging ethical and aesthetical values.

3.1.2 Specifying Functionality

In order to specify which tasks a user can delegate to the system, an analysis has to be made of the user's task domain. To this end, task analysis methods should depart from the potential users' task knowledge. Looking at the different design decisions at the level of functionality, it becomes clear that we need at least a clear view of the decomposition of tasks into subtasks, and a description of each task in terms of the goal, the conditions under which the task can be performed, and its structure. At the same time we need a description of objects involved in task performance, including a taxonomy of objects and semantic relations between them, as well as knowledge of attributes, and the relation between tasks and objects. In tasks of some complexity, moreover, different users will have different roles, and tasks may be split up into parts that will be considered by different individuals.

Professionals in any task domain may well have knowledge only of part of the total domain. Knowledge elicitation techniques and systematic observations that focus on different aspects may have to be combined with formal analytical methods.

Several approaches have been developed in this field, none of which can up to now be regarded as complete. Sebillotte (1988) and Scapin et al. (1990) developed MAD, an approach of structured interviews that allows to create a hierarchical task structure that may be the base for delegation of subtasks to a system. Johnson and Johnson (1990) focus more on relevant objects and on the concept of roles, applying a broad range of observation methods and knowledge elicitation techniques.

Once the task knowledge and the task domain has been systematically analysed, the specification of the functionality of the system to be designed (or redesigned) can proceed. ETAG ("extended task action grammar", Tauber, 1990) is a recent approach for this phase of design. It provides the designer with a formal framework to specify the object structure of the system, including type hierarchy and attribute relations, and it requests the designer to design a dictionary of "basic tasks", being the system's functional units of delegation which are counterparts to the original "unit tasks" as have been detected in previous task analysis. Based on the knowledge of the users' goals for the unit tasks, the designer can now define the basic tasks in terms of system events. subsequently, ETAG requests the designer to further specify task delegation in terms of interaction style and communication to the system (these concepts will be dealt with in the next sections).

3.2 Interaction

This section is about *how* the user can interact with a system.

Interaction can be analysed along several dimensions. In actual design each of these dimensions represents a separable set of design decisions. The design decisions related to the different dimensions should be considered separately and explicitly, but, at the same time, they effects are strongly interacting, so the decisions should be related and design methods should be used in order to explicitly attend to the complexity. The most important dimensions of interaction are the interaction style (3.2.1), the ways the user can communicate to the system (3.2.2) and the way the system represents itself to the user (3.2.3). Some manufacturers developed a typical combination of design choices at all three dimensions and promote this as their "look and feel". Many authors take certain combinations as standards and have invented labels like "WIMPS" interface, which might lead to the invalid assumption that certain design decisions have to be automatically coupled.

3.2.1 Interaction Style

There are many styles of human-computer interaction. New technological developments are likely to give rise to new possibilities. Hence, the list of styles presented here is in no way final or complete. Even the way of structuring styles is only tentative.

a. **Commanding** The oldest interactive paradigm can be described by the user issuing commands to the system. In most cases a special formal language has to be used, of which the syntax is strict and of which the semantics are very restricted. Command languages most of the time focus on a single application domain, like text processing, searching or querying a data structure. Often the user has the possibility to create complex commands by combining commands into a structure like a program, script, or macro, that can be issued with parameters or of which the execution time meaning may be programmed to depend on the actual state of the system. With this interaction style users feel completely in control.
   Users need to learn the special language and to remember the lexicon and the syntax, and they have to be precise in issuing a command.
b. **Negotiation** Certain interaction styles have a flavor of negotiation between the user and the system. A system may be designed to ask questions in order to be able to continue the dialogue (in quasi natural language dialogues in order to solve semantic interpretation problems, in command language dialogues for completing the set of arguments), for reasons of safety (to make sure the user really wants to have an action performed that has consequences beyond return) or authorization (e.g., asking for a password). In some cases the system action takes the form of a suggestion that may be ignored ("the batteries are running out of power, please insert a power plug soon", or "an edit buffer has been

saved, in order to retrieve it, use the command..."). The user will have the feeling the system is in control. Negotiation may be designed in such a way that possible user actions or answers are indicated.

c. **Choosing** If there is a limited set of possible next user actions, the interaction style can take the form of a choice. Users may be provided with a menu (on the screen, by labeled keys, by voice output from the system), or have the possibility to get a menu by "pulling it out" at a certain location. In case of a choice from a large set, menus can be constructed as hierarchies or as scrolling lists. Chosing can also occur if points or areas in a graphical space allow the user to zoom in at details by selecting them. In the case of this interaction style, the user feels in control. The possible choices can be percieved at the interface.

d. **Form filling** Another interaction style where the user will need not much effort in retrieving what possibilities are available, consists in providing a template in which slots are available for all elements of the user's action that may vary. Slots may be provided with a default value, which, if well chosen, may speed up interaction and save user actions. and the range of possible user actions for each slot may be indicated. In most cases the user is free to decide in which order the slots will be filled. Users will often have the impression the system is in control.

e. **Exemplifying** An interaction style that seems to be specially attractive to inexperienced users, is exemplifying. In this case, the user himself performs the task to be delegated to the system, either in order to teach the system the meaning of subsequent equivalent task delegations (programming or query by example), or in order to communicate the intention of a single task delegation more "directly" than with the help of commands or menus. "Direct manipulation" indicates a paradigm where the user can show his intentions by manipulating the system objects directly on the screen (e.g., stretching a cell range in a spread-sheet), or manipulating a symbolic representation of the object (an icon, the name of an object, or a combination of the two). As the symbolic manipulation can only be done by the user in the space that is represented at the user interface, direct manipulation seems to prevent the user from making interaction errors (slips), although intention errors (mistakes) are in no way prohibited. In virtual reality environments, symbols or objects can be moved and changed in a 3-dimensional world. In all cases the user delegates the task to the system by providing the example of what the system should do, often in a metaphoric way. A different case of exemplifying consists of programming by example. In this case the user "teaches" the system to perform a structure of actions. Query by example is a similar way to teach the system the relevant characteristics or the intended set of items to be retrieved, without being obliged to issue commands in a formal query language. In most widely available direct manipulation systems, however, teaching the system is not possible, and especially experienced users will feel annoyed by this if they need to repeatedly perform the same sequence of symbolic manipulations. Apart from this, exemplifying interfaces will be experienced as direct, and user controlled.

Interaction styles will often be combined (e.g., certain MS-DOS commands like "delete *.*" will be followed by negotiation of the type "are you sure? y/n"), or users may be offered a choice. Sometimes the choice exists for the user even if the system image does not provide different representations (e.g., in LOTUS-123 users can either point the cursor to the intended menu choice, or type in the first character of the choice - for trained users the latter behaviour is experienced as issuing a command).

Expert and regular users of a system tend to prefer commands in most occasions and use this style even if the interface is not designed for this behaviour, e.g., by automatically pointing to the location of the menu options they need without looking at it, by answering questions before they are asked by the interface, and, if at all possible, by constructing their own scripts to take care of certain negotiations in a tailor made default way.

3.2.2 User Language

From the point of view of user interface design the user language can be analysed separate from the interaction style. Different decisions on language can be taken with the same style and vice versa, although certain combinations will be preferred for certain situations (dependent on characteristics of the task domain and of the intended user group, see Mayhew, 1992). Linguistic aspects of the user's actions have several aspects that are worth to be considered.

a. **Syntax** In interfaces where the user has the feeling to be in control, the general characteristic is that there is some syntax that is not trivial. Simple command languages often have a general syntactic rule where the action is specified first, followed by the arguments (objects and attributes). Alas, some well known systems are less consistent, disabling the user to quickly learn, and, worse, preventing educated guesses when knowledge is not present. Menu driven interaction often shows a standard syntax where the core object of a task delegation is indicated first, followed by a choice of action.
The same lexicographic structure is characteristic of most direct manipulation styles.

b. **Lexicon** The actual representations of the elements of interaction comprise the lexicon. Command names, abbreviations, or other verbal or abstract symbolic labels are generally used in command and query languages, and will sometimes be found in menu interactions.

In cases of choice and for indication of symbolic manipulation, icons may be used. Icons may be applied to indicate actions (in menu interfaces), attributes (character type, size, colour, texture), or object types (in the case of object classes like files, folders, multiple printers, the actual instance may be indicated by connecting a verbal label to the icon). Gestures may also be part of the lexicon, like 2-dimensional patterns indicating the crossing out of an item or a "check" to symbolize a choice.

The choice of lexical items is relevant in relation to the type of interface. Each type of representation can be selected or designed in relation to its mnemonic value, with the aim of both ease of recognition (in the case of choice paradigms and symbolic manipulation) or recall (for command languages and form filling).

c. **Specification** The specification of the individual elements of the interaction language can be done in several ways. Elements can be referred to by the name to be input by the user, by pointing to the system's representation of the element, or by labeling the element (e.g., by number, by first character). The style of specification has to be related both to the interaction style (i.e., pointing presupposes the element is represented on the screen or in the virtual world), and to the hardware and physical activities involved (whether a keyboard or a touch screen will be used. For users with handicaps and for special environments it is often the choice of specification and physical actions that enable adaptation of a general purpose system.

d. **Physical actions** The physical actions are designed in relation to the hardware available and characteristics of the user. Generally available commercial systems often are flexible in respect to whether pointing may be done by mouse movement, arrow keys, or touch screen. Relatively new interaction styles in the domain of virtual reality often allow physical actions like hand and body movement and head movement to be applied to specify objects, actions, or arguments of commands. In the more traditional situations the keyboard still general being used, although in recent systems function keys and chords of keys (combinations of 2 or more keys to label a lexical element) are often used. Voice input is relatively rare, although generally thought to be an obvious way to offer the user several input channels in parallel. Voice recognition still produces a high error rate and is requires extensive learning by the system in order to recognize individual voice characteristics.

Formal specification methods like CLG (Moran, 1981), and, more recently, ETAG (Tauber, 1990) explicitly request the designer to specify the user language regarding each of the aspects: syntax, lexicon, specification, and physical actions. At the same time this type of specification methods help designing consistent user languages and provide analytic methods for keeping track of the complexity of the secondary task that concerns interacting with the system, separately from the primary task, for which the functionality reflects the user's possibilities.

3.2.3 System Language

As a counterpart to the user language, the system can be designed to represent for the user the relevant aspects of the current system state and of the history of interaction. In this respect the system can show the user where he is, what is possible in the current situation, and how the previous interaction lead to the current state.

If a designer wants to adjust the system to the user, the system language will offer a relatively broad spectrum of possibilities. The user interface can in this respect be considered to maintain aspects of representing the working environment. In addition the system's representation for the user regards aspects of the interaction language, the lexicon, and the physics of communicating towards the user.

a. **Environments** The user interface may represent for the user the status of the working environment. If several different processes are relevant, each can be represented in a different window (or, depending on hardware availability and spatial arrangements, on different screens). Often, from the point of view of the user, a single process can be related to different spaces and subspaces (e.g., searching in a hierarchical tree of directories or folders), each of which, again, may be represented by a separate window. Moreover, windows may be used to represent processes that belong to other users or user groups (a video look into a colleague's workplace or into a meeting room). If a process is shared by several users, all users may be represented in the window (by a marked cursor, by showing the moving hand or pen of each). If multiple windows are represented, there is a choice to allow them to overlap partly or wholy, to the limit of treating each separate process as a "virtual terminal" (each of which sharing the same screen and input devices, where only one at a time can be addressed by the user). Apart from the obvious visual images, environments may be represented by sound, in which case either complete overlap or careful separation of spatial location of the different sources will be needed.

b. **Interaction language** The different interaction styles that have been mentioned earlier will bring their characteristic representation styles for the system language. Alphanumeric screens can only manage some of them, and many situations of direct manipulation require a graphical representation.

   Graphical languages will have their own variety of syntax, enabling the reference to the semantics of the interaction and the system state.

   System prompts can be used to show the user the current state of interaction and of system processes, including indications of what has just been changed (feed back), what can currently be done by the user, or what is being in progress from the side of the system and how long this will take. Sound could be used for the same feed back and prompt functions. Sound and 3-dimensional visual images together can be applied to create virtual worlds to represent the working environment where the systems functionality can be applied.

c. **Lexicon** The system's part of the lexicon covers the actual names, graphics, colours, and sounds used to indicate the information units and the elements of the system language, equivalent to those specified by the user in his actions towards the system. Rohr (1987) studied the effect of the system's surface representation on users' understanding, mainly focusing on visual aspects of the interface. In interfaces where movements can be accepted as input, physical resistance (e.g., build in the data glove) can be added to the lexical expression power, enabling physical feed back to 3-dimensional manipulation.

d. **Physics** The physical aspects of the system's output at the user interface concern qualities that are related to the limits and differential sensitivity of human perception. Relevant factors are colour saturation, screen resolution, sound quality and loudness.
For this aspect of design "classical" ergonomic knowledge should be applied. The system language will have a strong effect on what the user will learn. Bad systems will misguide the user. But, on the other hand, systems that are good in representing the state of the interaction and of the system state will enlighten the work of the user, sometimes to the extend that the user does not need to keep a very elaborate mental model of the system since the system is showing its state and dynamics clearly. In van der Veer (1990) we reviewed a field study in secondary education where two large and comparable groups learned to use personal computers. The only systematic difference was that one group used a PC with a simple PC-DOS or MS-DOS type interface and the other group used Macintoshes or related systems. There were striking and statistically significant differences in the mental representations that the students turned out to have after 10 lessons. The DOS students represented more knowledge on the physical interaction with the system, and they also used more iconic elements in their representations. In fact they seem to include in their mental models precisely those features that were lacking in the system language: feedback and representation of actions, and iconic representation of relevant interface objects. Another significant difference was that the DOS group represented more program like knowledge structures.
Macintosh users (of this level) do not seem to need a program representation of the interaction since the interaction style seems to be guided by the system language.

3.3 Metacommunication

Metacommunication refers to HELPING the user to utilize the system.

This is an aspect of the user interface that can be analysed separately, focusing on how the user learns to use the system and how use of the system can be assisted if interaction problems arise. The concept in general denotes all communication between the user and the system as far as it concerns the interaction or communication, both regarding the application and functionality (the primary tasks) and the interaction language in all its aspects.

3.3.1 Implicit Metacommunication

Metacommunication can take place even when the user is not aware of it. This implicit type of metacommunication may be due to metaphoric aspects of the interface, both in respect to interaction style and in relation to the lexicon. A well chosen interaction style may communicate to the user what possibilities exist for continuation, both regarding functionality and regarding the way to perform the next action. The screen design for direct manipulation may represent all relevant objects, a menu may illustrate the relevant attributes or the possible values of a single attribute, a form may indicate for what arguments default values are not available and what type of values have to be provided by the user. Implicit metacommunication, if not designed properly of if not designed at all, will still take place, and, hence, might convey incorrect ideas to the user.

Moving the symbol of a disk into a folder in a direct manipulation interface might be interpreted as making the disk available in that folder, although actually it results in creating a copy of the content of the disk and placing the copy in the folder. In the same type of interaction style moving the disk to the trash can might implicitly communicate to user the incorrect information that disk could be retrieved by opening the trash can.

Another source of implicit metacommunication is the lexicon.

The system prompt may indicate interaction mode as well as reveal the "current" environment (the name and location of the directory). Verbal elements of the interaction language will suggest a meaning to the user. Command labels like "d" in UNIX mail systems might suggest that the related system action results in deletion of messages. In fact the command labels the messages for possible deletion if later on a certain exit procedure is chosen, but the messages are still available for certain commands, and may be referred to by their original numbers. Icons can be designed to transfer semantic knowledge like the type of object they symbolize (text files, folders) or the semantic structure related to the command they represent. Figure 5 (from Tauber, 1987) illustrating commands from a menu

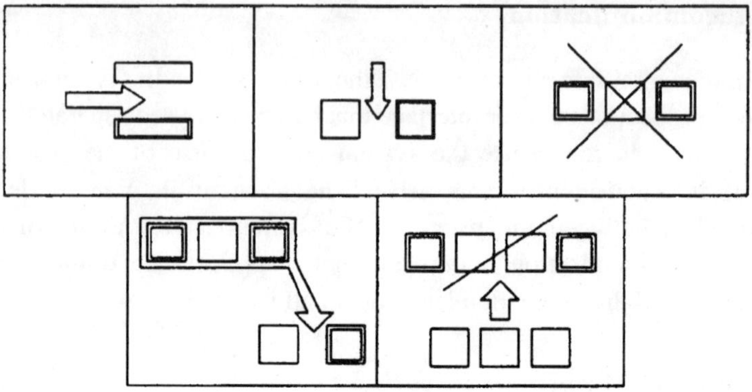

Fig. 5. Heidelberg icon set (Tauber, 1987)

interface that clearly indicate which objects have to be specified for commands like "insert before character" or "insert before line".

All aspects that have been mentioned when we analysed the "system language (3.2.3) are in fact potential sources of implicit metacommunication, and, hence, designers should be aware of this aspect whenever they decide on specifying the system's representation of the task environment, the screen layout, and the system's lexicon.

3.3.2 Explicit Metacommunication

Explicit metacommunication occurs if an action taken that is dedicated to helping the user to understand and learn about the system.

Generally speaking, the goal of this action is either to improve the mental model of the user (in order to establish relatively stable knowledge about the task, the system and the interaction) or to provide information of restricted temporal importance (about the current state of interaction or the current state of the task environment and system). Especially in the latter case the metacommunication will mostly be on line, either in the form of help that is available on request of the user, or in the form of system initiated messages ("spontaneous" help from the system or "error messages").

The forms of metacommunication that aim at learning, or, in other words, improving the mental model of the user, can be either on line ("guided tours" with simulated interaction fragments, and on line tutorials can be added to the user interface, but also help facilities can have the function of a learning aid). Metacommunication outside the system can take various forms, from incidental help from colleagues and local gurus to explicitly designed documentation and training. The latter forms may even be considered to be part of the user interface design.

The same approach that has been mentioned for specification of the functionality and the user language can be of use in the design of explicit metacommunication. Formal languages like ETAG (Tauber, 1990) describe the virtual machine in many respects, including the system's functionality, and the user language. This type of description is aimed at system aspects as far as they are relevant to the end user, and, hence, can be conceived as specifying completely what knowledge of the system is needed for the user. In other words, such a specification is equivalent to the system knowledge that would be needed for a competent user. At the same time, such methods do not describe the system's side of the coin (i.e., what is visible on the screen, etc.), but this is not relevant, precisely since this can readily be perceived by the user. Using ETAG-like description methods for design specifications does automatically provide a knowledge source for manuals (as far as the system's functionality and the user language is concerned) and for on-line question answering systems, and even for "intelligent" diagnostic modules in the user interface that are able to detect which basic task a user is trying to delegate. In this way the method used for designing the user interface provides at the same time a source for metacommunication. See van der Veer et al (1990) for examples of automatic generated metacommunication based on ETAG formalisms, applied to manuals, task diagnosis, and user initiated help requests.

3.3.3 Metacommunication Outside of the System Proper

Documentation and teaching are major sources of explicit metacommunication. Although often not located in the physical system, the design of this facilities is strongly related to designing the user interface. As far as the goal is to enable the user to learn the system and to help easy use, the design of this form of metacommunication should either be done in parallel with the development of the system or at least be based on a thorough analysis of the virtual machine, i.e., a complete description of the functionality and the interaction language. Subsequently this knowledge should be structured in such a way that it provides a user centered review of the system. This does not mean all should be completely described and certainly the user should not be forced to learn a considerable amount of details regarding the interaction before hands on experience allows his mental model to develop in the individual way best fit to each person.

Minimal information to help the user get started could be designed around the system's functionality and its relation to the user's task knowledge. The main guide for the user to allow fruitful exploration may often consist of a global view of the system semantics (the system's object world, the main operations that can be performed in this world and their effects as far as directly relevant to the user).

If the system is indeed well designed and on-line metacommunication is available when interaction gets stuck, the external information does not need to contain much more. In stead of having the user look up the meaning of an error message in a manual, this could and should be provided on line on request, without disturbing relevant information that is on display at the moment when the problem emerges.

Providing an overview of system semantics may be done by using metaphorical methods. In this way the relevant information can be represented in such a way that the concepts to be understood are shown in a relational structure that resembles the way human semantic structures tend to be mentally represented. Figure 6 shows a simple representation of an MS-DOS system. This simple item of external metacommunication was used in a field experiment reviewed in van der Veer (1990) where two comparable large groups of secondary education students learned to use an MS-DOS system. The only systematic difference between the two groups was the fact that in one case Figure 6 was presented on paper to the students and the teacher was asked to spend about 10 minutes to refer to the figure and relate it to the system structure, mainly focusing on the semantics of the operating system. Statistical analysis showed the group with this item of external metacommunication showed more mental representations of the system at the lower levels (syntax and keystroke), although the metacommunication did not represent this. Apparently it helped them to explore the system and urged them to put this self discovered knowledge in their mental models. The group with the extra metacommunication (which was of an image type) showed to represent significantly more in verbal information as well as more icons. Again the metacommunication was not copied in the mental model but it helped the students to collect and develop their own representation of other types than was made available. The last finding in this study was that the metacommunication increased the correctness and completeness of mental representations. The design of a relatively simple item of metacommunication, that illustrates for the user the relevant knowledge of system semantics that happen not to be available at the interface (another type of operating system like Macintosh does represent this knowledge in the system language) proves to increase the learning and development of relevant mental representations.

3.3.4 Metacommunication and Individual Differences

Individual computer users differ on several variables that may be of importance for learning to use the system. Metacommunication should be designed to cope with this variety. In van der Veer (1990) a systematic analysis is made of user

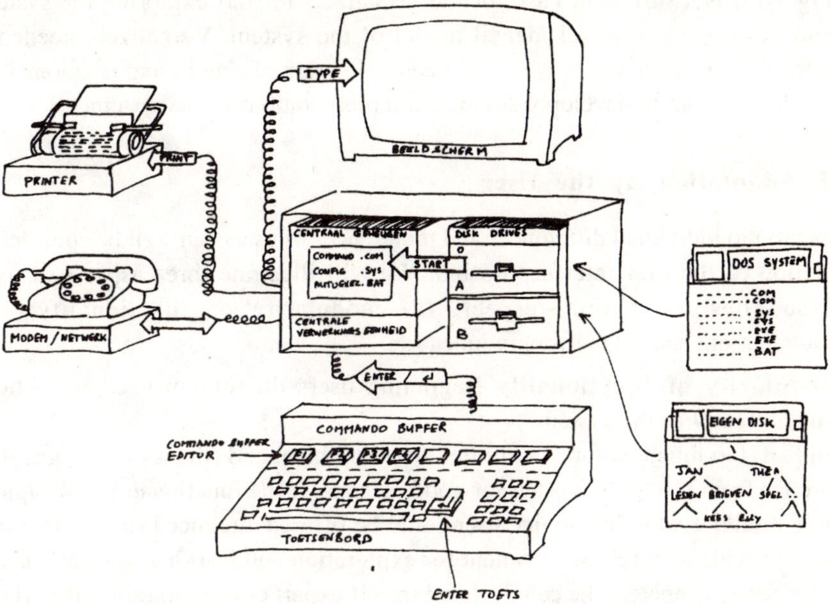

Fig. 6. Visual-spatial representation of information transport and information structure of MS-DOS type operating systems

variables in relation to learning to use a system. We will illustrate here only two examples that regard the design of metacommunication.

a. **Spatial ability** Spatial ability can be regarded an aspect of intelligence, in which computer users tend to differ to such an extend that this is known to influence the use of systems. Especially when the application concerns large object structures that are too big to be represented on the screen (spread-sheets, large text files, electronic mail systems), users with low spatial ability are reported to have problems. A way to overcome the problems seems to be to provide facilities for orientation in the "space" of the object structure (structural overviews) and to provide search facilities to help people locate items that they feel they have lost.
b. **Visualizers - verbalizers** The style of mentally representing information is a personal preference that is often shown to be quite stable. Some people show a preference for verbal representations, others prefer visual-spatial images. It turns out people are either reluctant or unable to cope with situations where their preferred kind of representation is not available. In the case of a course in using an integrated office information system for managers, it turned out a graphical representation of the system semantics (in the style of

Fig. 6) was a sufficient base for the visualizers to start exploring the system and developing a useful mental model of the system. Verbalizers needed a different type of training, where guided exploration of actions and reactions was needed in order to develop valid mental representations of the system.

3.3.5 Adaptation by the User

Adaptation to individual differences and preferences of users can well be considered a function of the user interface design. We will illustrate three aspects of user interface design in this respect: (a) modularity of functionality; (b) communication; and (c) metacommunication.

a. **Modularity of functionality** Beginning users do seldom need the whole functionality of the system.
   In fact, too much features will often frighten them, and (thanks to designs that are far from fool proof) often for good reasons. If the functionality is designed in a modular way, the novice user could be offered a reduced set of functions that provide a "safe" environment for exploration and learning. As soon as the user feels competent, he can declare himself expert enough to utilise the whole system, although he can always be allowed to go back to the safer version if he feels a need for this. In fact this is the essence of the "training wheels" approach of Carroll (1985).
b. **Communication** The second example of user adaptation to improve learning related to the user language. In a field study on learning the spread-sheet system LOTUS 123 (van der Veer, 1990) it showed that beginning users are happy to make use of the menu interaction style provided by the interface. After an average of 60 hours of hands on experience, however, users of many types of systems seem to have learned the main system command structure to such an extend that they prefer to change to a command language style. Systems like LOTUS 123 provide users this opportunity by the way in which the user may select the menu items: either by pointing (as many novices prefer to do) or by typing in the first (and always unique) character of the next menu choice. relative experienced users turn out to type in sequences of characters without looking to the menu on the screen anymore. In fact they have transferred to a command style that the interface provides them, but they can still fall back to menu choices for situations of commands that have seldom or never been used. The design of the user interface provides the flexibility that allows both novices and experts to work in the style that best fits their knowledge and skills, and, hence, learning to improve system use seems to proceed automatic and unnoticed.
c. **Metacommunication** Users differ in the amount of help they need, and depending on their expertise and the amount of use they experienced of certain

system functions, they might need more or less metacommunication. Too much metacommunication quickly becomes a burden of its own and leads to irritation. In designing the interface adaptation to the user's momentary need of metacommunication may be done in different ways.

Some systems allow the user to indicate the preferred level of "verbosity", i.e., the length of explanations and error messages that will be provided by the system if something seems to go wrong in the interaction (i.e., when the system is in control). A different type of design measure has to be taken for cases where the user is in control.

If the user feels he needs some explanation, this can be provided on request, like in the case of the "balloons" in the newer releases of the Macintosh operating system. The user can switch this facility on and off at will. Another user controlled type of metacommunication is related to the machine language, or, more specific, the system's representation of user objects. It is again the design of the Macintosh that allows the user to change the representation of the content of a directory in different ways ("views"): iconic representation stresses the type of the objects, verbal representations can be chosen to represent different attributes of the objects like date of last change, size. Moreover, both types of representation can be combined. Designs of this kind allow the user to adopt the interface to current needs.

4. Conclusions

We advocate a systematic analysis of the design problem. The different aspects of the user interface show us where design decisions can be taken that will influence and enable learning by the user. The main result of systematic design will be to provoke systematic decisions that will provide ways to cope with the problem of users of complex systems. Different aspects of human coping and the need for an adequate mental model can be taken care of. If the design is adequate, human learning will proceed during actual use and exploration of the system. If needed, the system shall have to adopt to the user, and in well designed systems the user may be allowed to control adaptation to the current needs for knowledge, representation, and metacommunication.

References

Carroll, J.M. (1985). Minimalist design for active users. In: B. Shackel (Ed.), Human-computer interaction; Interact'84. Amsterdam: North-Holland

Johnson, H., & Johnson, P. (1990) Designer-identified requirements for tools to support task analysis. In: D. Diaper, G. Cockton, & B. Shackel (Eds.), Human-computer interaction: Interact'90. Amsterdam: North-Holland

Mayhew, D.J. (1992). Principles and guidelines in software user interface design. Englewood Cliffs, NJ: Prentice Hall

Moran, T.P. (1981). The command language grammar: a representation for the user interface of interactive computer systems. International Journal of Man-Machine Studies, 15, 3-50

Moray, N. (1988). A lattice theory of mental models of complex systems. Report EPRL-88-08, Engineering Psychology Research Laboratory, University of Illinois at Urbana-Champaign, Urbana, Il

Norman, D.A. (1983). Some observations on mental models. In: D. Gentner & A.L. Stevens (Eds.), Mental models. Hillsdale, NJ: Lawrence Erlbaum

Pask, G. (1980), Developments in conversation theory - part I. International Journal of Man-Machine Studies, 13, 357-411

Pfaff, G.E. (1985). User interface management systems, proceedings of the Seeheim workshop. Berlin: Springer

Rasmussen, J. (1990). Mental models and the control of action in complex environments. In: D. Ackermann & M. Tauber (eds.) Mental models and human-computer interaction 1. Amsterdam: North-Holland

Rohr, G. (1987). How people comprehend unknown system structures: conceptual primitives in system's surface representations. In: P. Gorny and M.J. Tauber (eds) Visualization in programming. Berlin: Springer-Verlag

Rumelhart, D.E., & Norman ,D.A. (1977). Analogical processes in learning. In: J.R. Anderson (ed.) Cognitive skills and their acquisition. Hillsdale, NJ: Lawrence Erlbaum

Scapin ,D. & Golbreich, C.P. (1990). Towards a method for task description: MAD. In: L. Berlinguet & D. Berthelette (Eds.), Work with display units 89. Amsterdam: North-Holland

Sebillotte, S. (1988). Hierarchical planning as method for task analysis: the example of office task analysis. Behaviour and Information Technology, 7 (3), 275-293

Tauber, M.J. (1987). On visual languages and their conceptual analysis. In: P. Gorny & M.J. Tauber (Eds.), Visualization in programming. Berlin: Springer

Tauber, M.J. (1990). ETAG: Extended task action grammar – a language for the description of the user's task language. In: D. Diaper, G. Cockton, & B. Shackel (Eds.), Human-computer interaction: Interact'90. Amsterdam: North-Holland

van der Veer, G.C., Tauber, M.J., Waern, Y., & van Muylwijk, B. (1985). On the interaction between system and user characteristics. Behaviour and Information Technology, 4, 284-308

van der Veer, G.C. (1990). Human-computer interaction: learning, individual differences, and design recommendations. Alblasserdam: Offsetdrukkerij Haveka

Organizational Setting and Task Complexity: Interactions with Collective and Individual Cognitive Activities

Elements for Elaborating a Framework

Janine Rogalski

Senior Researcher, CNRS-University Paris VIII.2, rue de la Liberté, F-93256 Saint-Denis Cedex 2, France

Abstract. A comprehensive approach is proposed for analysing interactions between organizational setting, task complexity, and individual cognitive activities. A triadic relationship "actors-task-operational system" enables to take into account the "vertical" and the "hoizontal" dimension involved in collective work, through considering a task as a node in a network. Actors are defined as (system of) prescribers and (system of) operators linked by a "contract" about the task. A task may be analyzed along three components: a target-state to be reached (a goal related to higher level purposes), a set of criteria for evaluation (related to more general values), and a set of resources and constraints. The notion of "virtual operator" allows to use this framework for analysing individual or collective activities as well as for designing operational devices (including human-machine interaction). Examples are given for open dynamic environment management.

Keywords. Task complexity, organization, cognitive activities.

1. Introduction

Our purpose is to build a comprehensive approach for analyzing interactions between organizational settings, task complexity and collective and individual cognitive activities, inside a framework allowing to link individual, collective and organizational points of view. The adopted perspective is centered on a triadic relationship "actors/tasks/operational system" between *actors* performing a *task* in a given *operational system*. The actors may act at any organizational level (from the highest decision level of a firm to the physical executive level of operators on the field); in the same way, the task may be defined at various levels (from elaborating a global strategy to modifying a physical component); the operational system is an entity defined by respect to a given network of tasks and actors: it is not an organization of social groups, nor limited to a work team.

The chosen approach does not aim at capturing all issues related to "organizational learning": we do not claim to tackle the whole set of questions about how an organization can "learn" and evolve, and under which conditions; or how individual operators or teams can learn - or at the opposite may loss their competences - through working in a given organizational setting. We want to articulate in a comprehensive framework a cognitive perspective on the psychological subject (individual and inter-individual cognitive activities) and a "systemic" perspective which integrates in a common frame subjects cognitive activities and functioning of an operational system. (Questions about sociological, juridical and economical constraints on operational system are outside the scope of our framework).

This framework is intended to be a basis for analysing cognitive activities related to any task performing, with common concepts and methodological tools whatever the level at which these activities are considered: individual or collective activities, as well as "distributed" task performing through a large operational system with numerous hierarchical levels. The task oriented approach allows also to take into account "activities" of automatized systems in human-machine joint cognitive systems (Hollnagel & Woods, 1983).

Before developing the main ideas of this comprehensive framework, we will briefly present:

a) issues concerning properties of organization as variables affecting actors' cognitive activities and

b) methodologies using activity analysis as revealing properties of operational systems or organizations.

We will then specify a general framework for task analysis (whatever the domain): a task is considered as a node in a network, with prescriber(s) and operator(s) linked by a "contract" about the task; three components define a task: its goal - a target state to be reached; the criteria for evaluation; the set of resources and constraints. This frame expanses Leplat's definition of task as "a goal in a given context". Two notions are presented for the case of open dynamic environment management:

1) a virtual operator in charge of the task, with an a priori task analysis based on a method for tactical reasoning, parallel to Rasmussen's well-known model for process control (Rasmussen, 1986);

2) a virtual operational system related to this task analysis.

"Distributed decision making" (Rasmussen et al, 1991) is defined at the intersection of task analysis for a virtual operator and design of the related operational system. Temporalities and organizational levels are briefly sketched for risk management considered as an operational unit. Finally, possible extensions and needs for further developments are discussed.

2. Work Activities and Organization

Interactions between organizational setting and work activities are manyfold. In litterature concerning psychological and ergonomical approaches, they mainly appear in studies dealing with collective activities. Relationships between organization and activities as "objects in the world" may be considered form two symmetrical points of view: organization as a factor affecting activities and actors' activities as modifying organization. Another point of view is a methodological one, related to analyst's purposes: activity may be seen as revealing properties of the organization, or – less frequently – organization reveals crucial conditions for activity.

2.1 Organization and Work Activities

Organizational settings may be considered as components of task complexity. For instance, in open dynamic environment management - and particularly in crisis management - task is more or less complex depending on deepness of embedding of decision and execution levels and heterogeneity versus homogeneity of systems of actors interacting in "distributed decision making" (Baron et al., 1989). The dynamics in implementing an appropriate operational system form previous permanent organizational settings is more or less difficult to manage, depending of intra and inter-organizational cohesion (Sorensen et al., 1985). Organizational settings are also considered as general factors on efficiency (Bagnara et al., 1991; Bagnara & Failla, 1994). For instance, collective activities and decision processes are depending on the characteristics of the sociotechnical organization. A "mechanistic" model of organization with hierarchical control and functional dependency develops strong tasks specification, and strictly constrains collaborative codecision. By contrast, an "organistic" model, characterized by decision autonomy and functions interdependance defined with "fuzzy" boundaries is favorable to strong distributed decision making and collaborative problem-solving. These relationships were underlined at the beginning of automation concerning the impact of automation on work teams autonomy and on task specialization (Susman, 1972).

The role of organizational factors in safety measures and risk control is also emphasized. Marek et al. developed an empirical study on risk control and safety measures on an oil-platform, which provides general results for evaluating the safety climate of an enterprise, including operators mental representations on risk and on existing safety measures as well as effects of organizational decisions on risk management (Marek et al., 1987). Training all actors involved in a global task such as nuclear power plant control was also considered as a way for increasing reliability through communicaiton and collective activities (Jansens et al., 1989). This training aims at "counterbalancing" negative effects of tasks specialization on risk management by providing occasions for the various actors to identify the others points of view and "styles" of action, depending on their organizational position.

Handling task complexity may lead to adapt previous organizational settings. For instance hierarchical organization can be seen as a response to complexity by limiting load and time pressure through "distributed decision making" (Brehmer, 1991). Collective activities themselves may introduce changes in effective work organization: teams may construct new procedures or new tasks allocation for handling new situations or face unforeseen disturbances; these procedures or local work organization are then object of mean or long-term negotiation inside the organization (de Terssac & Chabaud, 1990; Norros, 1989)

Cognitive activities related to risk modified may be changed through collective activities, as was shown in a study comparing mental representations of operators just engaged in a firm, and experienced operators used to team work (Favaro et al., 1991). As a consequence of these type of data increasing human-systems reliability may lead to propose socio-organizational modifications. For instance, one conclusion of a study on flight safety was that revise careers, systems and (when possible) cultures were appropriate means to improve flight safety (Amalberti, 1992).

From actors' point of view, cooperative work produces complex effects on individual cognitive activities and competencies, depending on organizational flexibility (de Terssac & Chabaud, 1990), task complexity (Navarro, 1987). Internal properties (size, structuration, cohesiveness) of a group, as well as position of the group in the whole organizational setting intervene in task cognitive requirements (Curtis, 1987) as well as for group efficiency (Dailey, 1978; Rogalski & Samurçay, 1993).

An interresting point was underlined by Hutchins, concerning the role played by cognitive tools used as tasks supports: the "openness" (perceptual and cognitive visibility) to others actors on ship desk of the use of theses tools by an actor appears as a means for improving robusteness of collective activity and for ensuring group training and group "conservation" against operators' mobility (Hutchins, 1990). So that properties of resources for action (here, operative tools) interact with organizational properties (spatial organization and mobility of operators) in group functionning and evolution when considered as a self-active system (von Cranach et al., 1986).

Individual mental representations on task or actors may intervene on tasks allocation, that is on the functioning of an operational system. For instance, the so-called “machist” conceptions about hierarchy and task allocation in the cockpit may lead to serious human errors on the flight deck (Green, 1990). Implicit mental representations about task “tractability” and other actors’ competences may interfere with cognitive dimensions in task performing, producing specific biases on decision making, such as direct control taking instead of "mediatizing" (delegating) decision to other actors, as was shown for financial decisions by US managers during the Mexico crisis (Kuklan, 1988). For mediatization as well for collaboration, confidence in other actors appears of peculiar importance for collective work, in hierarchized or “horizontal” operational systems (Bagnara & al., 1992; Cicourel, 1990; Petermann et al., 1992); the same is true for mediatization through automatized systems (Lee & Moray, 1992).

2.2 Collective Cognitive Activities as Revealing Organizational Properties

Collective activities may be used as "observables" in order to analyse organizational properties. Studying the flow of activities between actors or sub-systems in an organization may reveal hidden properties of the studied organization, or "decalages" between the prescribed organization and the effective operational system implemented by actors for a given set of tasks.

A main stream of empirical research in cognitive ergonomics is concerned with communication studies. Their functions in work activities were analysed by Savoyant and Leplat (1983), while Navarro has considered interaction in work situations from a cognitive point of view as special problem-solving (Navarro, 1991). They may reveal differences in expertise for individuals or teams (Cellier & Mariné, 1984; de Keyser et al., 1985). Analysing specific points such as conflits episodes can be used as a way for studying teams, as was done by Walz and al. for software design (Walz et al., 1987).

Analysing communication may aim at analysing effects of task complexity about collective activities and not organizational properties for themselves. They can also reveal interactions between task complexity and organization. As a typical example, we analysed the communication flow between actors in a group (Operational Center Staff) acting as a support system for a decision maker in Rescue Operations, for a phase of preparation, without strong time pressure, and for a crisis (a risky forest fire) just following the preparation phase. Figure 1 presents the graph of interactions between actors for these two phases.

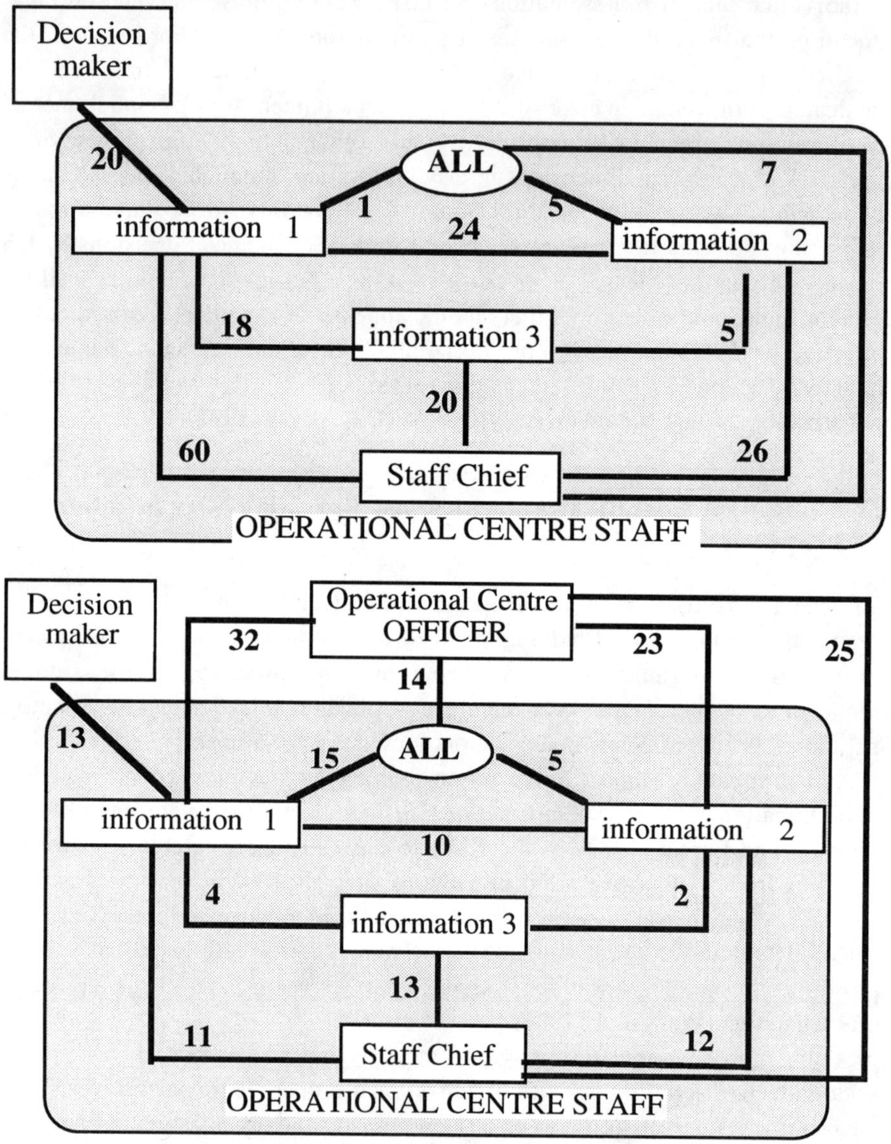

Fig. 1. The two schemas exemplify how differences in task complexity are expressed in the communication flow (number of communication units exchanged by the actors) and organizational functioning of an Operational Center (sub-system acting at the "strategical" level) in Rescue Services. The upper schema represents the communication flow in preparation phase; the above schema represents the communication flow during crisis management. Time spans were equal (45 minutes) "All" denotes no specified receiver. Actors were the same in the two situations.

From an organizational point of view, a first result is the presence of the Operational Center Officer, acting at the upper level, only for the crisis phase (similar observations could be made in control rooms for process control). A second point is the change in distribution of exchanges between actors: emergency results in a lot of direct exchanges between Officer and Operational Center Staff team (in current formal organization outside crisis Officer's decisions are mediatized by Staff Chief). Communications also reveals differences in the roles of the "information" actors: "Information1" appears as a center in information exchanges (from the decision maker and inside the Staff) in the preparation phase; during crisis this role is partly shared by "Information2". (Content analysis would also show changes in the set of staff tasks depending on the situation.)

3. Elements for a Comprehensive Framework

3.1 Reasons, Choices and Principles in Modelling Complex Tasks in a Comprehensive Framework

We chose modelling open dynamic environment management in the specific case of Rescue Operations as they provide externalization of organizational processes, and time acceleration. The definition of a task as a complex: <a node in a tasks network; a contract between prescriber and operator; and three components -goals, values, resources & constraints> shows how to integrate organizational dimensions as "objects of the world" that individual or collective operator have to tackle for task performing. This integration allows to apply concepts and methods of cognitive psychology and ergonomics, and aims at articulating this psychological level of subject's activity analysis to approaches dealing with concepts defined at a higher level (as in work sociology, management). The notion of virtual operator for a task allows to use common methods for analysing individual activities and group activities (or even activitites of human-automatized systems) from the point of view of task performing. Task analysis and virtual operational system are then coordinated to analyse distributed decision making. Issues related to relationships between temporalities, organizational levels and tasks are sketched for the case of risk management.

3.1.1 Principles and General Concepts

The proposed elements for a theoretical framework are based on a systemic and task oriented approach of relationships between tasks, organization and actors' activities. We will refer to the following theoretical approaches, as a general background.

First we use the differentiation developed by Leplat between task (point of view of what is expected to be performed between initial situation and final expected state) and activity (point of view of individual or collective activities developed by subjects when achieving a task) (Leplat 1990).

Secondly, we follow the orientation taken by Savoyant and Leplat to use the existing individual models for analysing collective activity (Savoaynt, 1984; Leplat, 1991). This choice is coherent with the conception, developed by von Cranach (von Cranach et al., 1986) of a group as a self-active system, which can be considered as a whole striving towards achievement of goals. Thirdly, we adopt the categorisation proposed by Schmidt for analyzing purposes and types of cooperative work (Schmidt, 1991).

Finally, organizational settings are considered as ways for managing tasks complexity (Brehmer, 1991) and more generally to respond to requirements of tasks interactions and conflicts in a task field (Stoelwinder & Charns, 1981). For open dynamic environment management, we "derive" – from a cognitive task analysis – a virtual operational device in form of a bidimensional organization, with a strong hierarchization. Effective Rescue Services and Military organizations share this hierarchical organization. Nevertheless data in other domains, such as large projects management, show that vicariance appears not only for activity in performing a task, but also in operational organization: several operational systems, with different tasks specialization, may have the same efficiency.

3.1.2 Modelling Open Dynamic Environment Management: Reasons for a Choice

The elements, presented below, of a framework for analysing interactions between organization, task complexity and actors' cognitive activities were elaborated for field studies, about cognitive activities and competences required by the complexity of open dynamic environment management. Instances of dynamic environment management (DEM) are process control, aircraft or ship piloting, crisis management: pollution, chemical fire, forest fire, and military operations. "Openess" of DEM is defined by the lack or strong restriction of automatized systems for control (information taking) and command (action implementing), of predefined procedures, and limitation of computable models of the process itself. We consider ODEM as a paradigm allowing analysis of relationships between tasks and organization, and actors activities and competences (operational knowledge), and their dynamics.

Open dynamic environment management (ODEM) may require numerous physical and human resources specifically devoted to this purpose. An operational system has to be designed and implemented: it consists in a transitory organizational setting, adapted to operational task constraints. As procedures are not predefined and means for action are not directley linked to the process, relationships between the target state, the sub-goals and the actions to be implemented at a physical level with adapted means, are made visible. Moreover design and implementation of the operational system can be analyzed for themselves as specifically organizational tasks. The dramatic unities of place, time and action are respected in ODEM: this allows to analyse actors' activities in direct relationship with organizational tasks allocation and communication flow, using methods of cognitive ergonomics for real time activity analysis.

3.2 The Task as a Complex: <a Node in a Network; a Contract; three Components>

Analysing activities in relationship with organizational settings starts from inserting the task in a tasks field, expliciting tasks components, and taking into account existence of contractual relationships, which are the objects of cognitive and/or social regulation processes (Faverge, 1966; Reynaud, 1988).

3.2.1 The Task: A Node in a Network

A task is situated: inside a system, with generic and circumstancial properties. It can be analyzed from its functional position by respect to a global task. Figure 2 expresses a functional local view of a "target" task (for which individual or collective activities are studied) as a node in a tasks network. The target task T^0 is derived from the decomposition of an upper level task. It has itself derived sub-tasks. The vertical relationships represent hierarchical task levels. The target task may also be related to other tasks by "horizontal" functional relationships; these tasks are not necessarily defined at the same level. This functional decomposition may fit with an organizational structural decomposition, but it can also be considered a priori, for instance for designing an organizational setting.

With respect to Stoelwinder's approach this schema of task as a node in a network is more "local" and the purpose is to be directly related to subjects' activities (individual or collective), not to design appropriate organizational settings. From this network representation, it is possible to define types of collective work. Mediatization characterizes the collective activity of operators in charge of task T* and operators in charge of the target task. Operators in charge of the different tasks T1, T2 and T3 are in a distributive cooperation if there is a mutual dependance between their activities, from their point of view. If several operators are in charge of the same task Ti they are in a situation of collaboration. When operators share same resources without performing linked tasks they are in co-action. This categorization for collective situations is dependent on the

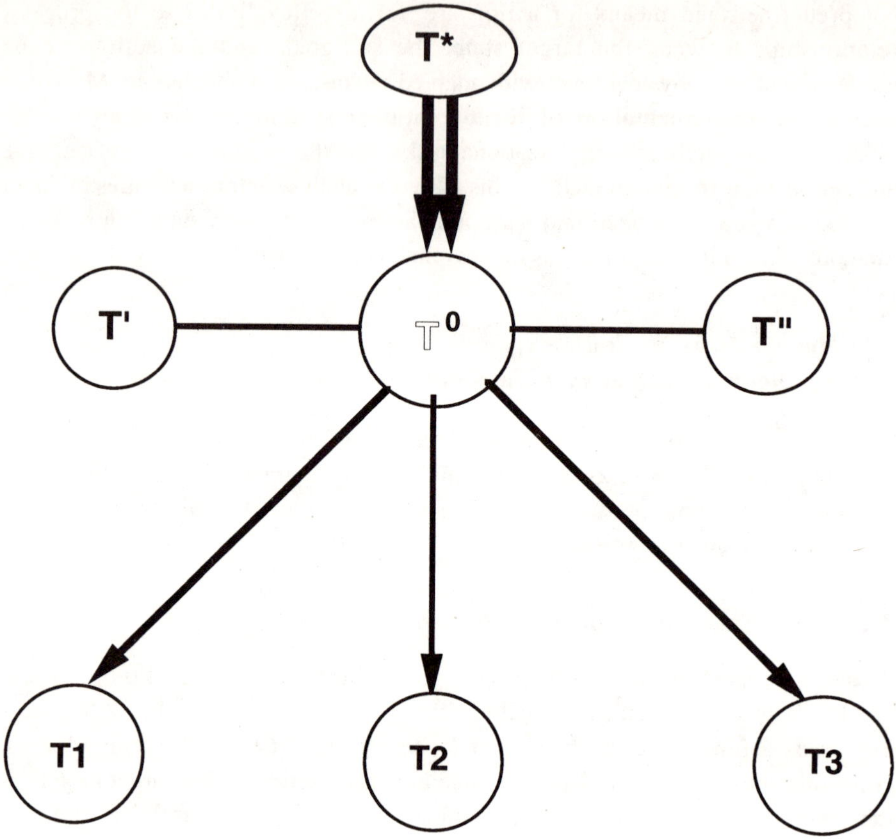

Fig. 2. Local relationships in a tasks network. The "target" task T^0 is considered as a node in a network of tasks. This network is partly hierarchized: tasks embedding is related to a hierarchy of sub-systems in the operational system. The target task T^0 is a sub-task of T^*. Relations to T' and T" represent dependence in achievement. T^0 has as sub-tasks T1, T2 and T3, which are said to be embedded in T^0. These last tasks are not in any hierarchical relationship.

"granularity" or level of task analysis (the network may be refined until elementary actions, that increases the number of embedded task levels).

3.2.2. The Task: Three Components and a Contract

Task may be structured in three components or fields. The task components are:
1) A goal to be reached; this goal is embedded in a system of goals defined at short or long-term, which may be specific goals or general purposes, objectives and

intentions: this is the teleological field. There is a duality in goal definition: it can be expressed as an action to be performed – for instance: prepare decision; it can also be expressed as a target state: "set of possible choices with their pros and cons". This duality is present in the well known Rasmussen's "ladder" modelling process control: each cognitive operation (such as "diagnosis") produces a state of knowledge, used as a starting point for the following operation (Rasmussen, 1986).

2) Criteria for evaluating if and how well the goal is reached; these evaluation criteria are themselves organized in a system of criteria and values: this is the axiological field. Criteria are partly hierarchized; they may be combined for decision making. Values concern high level purposes. They intervene in and a priori, implicit selection of possible goals or sub-goals, and for ordering evaluation criteria (for instance safety by respect to productivity).

3) Resources and constraints: schematically a set of resources are allocated for task performing, under constraints (time is a major one): this may be seen as the operative field (field of "ways & means"). In fact, resources and constraints may be two faces of a same element: for instance, prescribed procedures are resources as they propose solutions; at the same time they are constraints because they do not allow other ways for managing the present situation.

For similar reasons, instruments are also at the same time resources and constraints: they support operator's activities and extend the scope of his/her possible actions; but internal properties of the artefact itself constrain and restrain operator's activities (Rabardel, 1993), because the "logic of functioning" of artefacts generally differs from the operator's "logic of use" (Richard, 199).

This embedding of components is related to a systemic organization for action (operational system). Inside this system, task is object of a "contract" between a prescriber and an operator, or between a sub-system of prescribers (for instance Methods Office) and a sub-system of operators (a work team) inside the whole system. This contract is partly explicit and included in the global work contract. For a large part it remains implicit: a system of mutual expectations, adjustments for avoiding or solving conflicts, functions through "rules-in-action".

Short-term interactions intervene between task components during task performing: for instance a constraint as time pressure may decrease the weight of the safety criterium. Conflicts between productivity and quality criteria may conduce to ask for other resources or to modify goals ordering. Mean-term trade-offs occur between prescribers and operators concerning task components: a frequent trade-off deals with time constraints and material resources. Evaluation criteria may be modified or reordered for instance for insuring at the same time product quality and work safety. The last component to be modified concerns goals themselves: changes in goals may be observed in open situations like Rescue Operations or design situations (Walz et al., 1987). Long-term regulation mainly

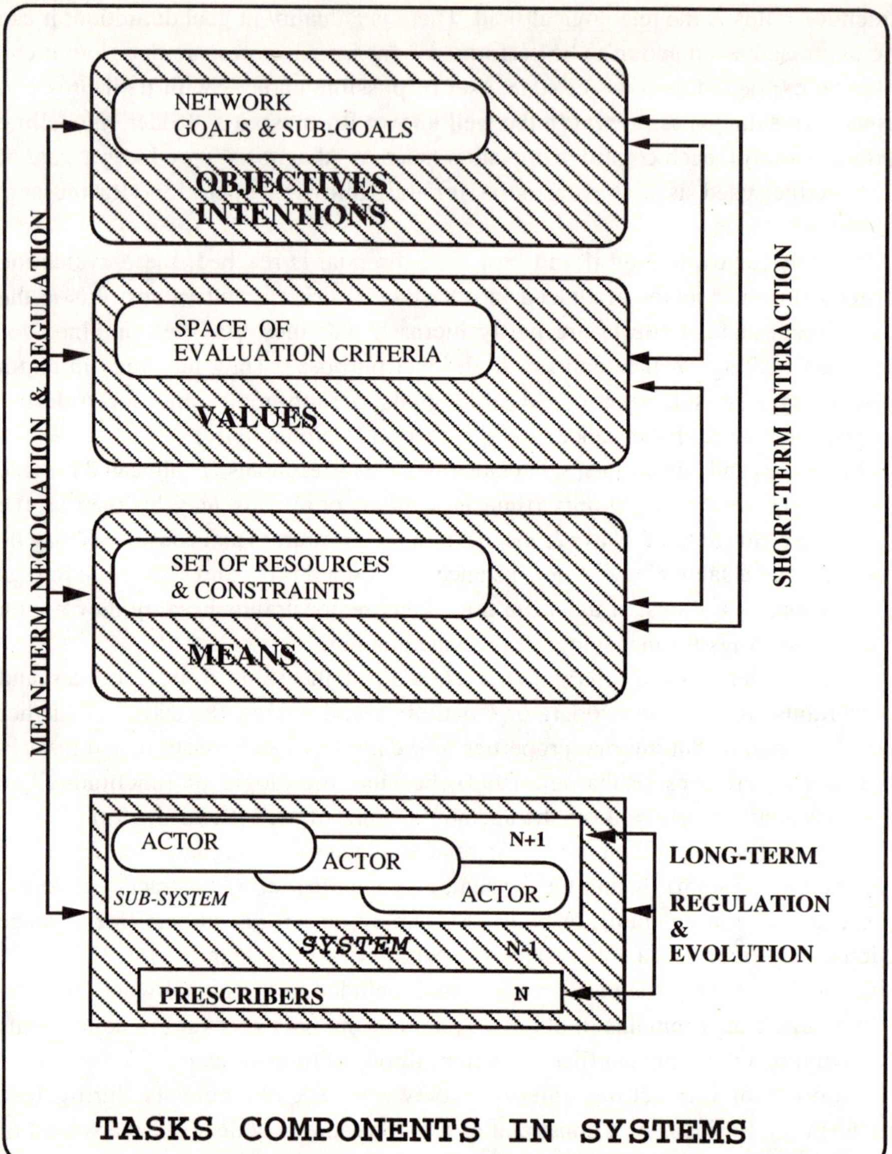

Fig. 3. Schematic representation of tasks components in a systemic approach

concerns the contract itself, and at another level the operational system -hence the organizational setting itself.

Transformations affecting the working process, through technological changes, such as automation and developpement of computer systems, and through organizational changes, modify all task components. They also modify the tasks network and the relationships between prescribers and operators (moreover, automation introduces new "actors": automatized systems). Identifying such a task architecture as we proposed above for defining a given task for an actor (or a system of actors) may be useful to articulate issues oriented toward cognitive ergonomics and sociology of organizations (for instance).

Moreover, such an architecture allows to specify questions about collective activities: not only the tasks of the various actors involved in collective work may be mutually situated, but issues about the specific task components for interacting actors may be specified: Are they sharing the same overall purposes? Are they using the same evaluation criteria, with the same high level values? Is their system of resources and constraints the same? And so on. All these questions are concerned with actors "task representation" (or effective task) and their effects on collective activities, whatever their organization.

3.3 Task Analysis: Virtual Operator and Virtual Operational System

Two notions of **virtual operator** for a task (where any task is defined by a situation requiring decision making and by an associated target-state to be reached to solve the problem) and of **operational system** where introduced for ODEM using a "Method for Tactical Reasoning" (MTR) as an interface between task and activity (Samurçay & Rogalski, 1991; Rogalski, 1991) The first notion represents an omnipotent (fictive, hence virtual) operator in charge of the task. Cognitive requirements on this virtual operator are infered from MTR in order to "implement" the virtual operator into a virtual operational system. Sub-systems embedding in the virtual operational system are related to the levels and functions identified by task analysis. The number of subordinates for mediatization tractability, and the number of parameters to be processed for "variety" tractability (Mélèse, 1992) are taken into account as two types of constraints.

3.3.1 Task Analysis and Virtual Operator

A virtual operator is a theoretical construct: It is a (fictive!) operator without any cognitive nor physiological limitations, without any specific values nor general intentions outside the goal of the target task. No time pressure nor cognitive load require consideration. It is also defined without any organizational consideration: there is no prescriber and no contract for a virtual operator. A virtual operator is defined by the task to be performed, which is characterized by the fact that two components remains empty (the axiological and the operative ones) and the teleological component is reduced to the target state (task goal). Naturally, this target state has to be related to the actual state of affairs for the situation to be managed.

This notion of a virtual operator allows a common approach for individual or collective activities. Moreover, it can be used for designing effective organization, when specifying possible resources and constraints and possible tasks networks.

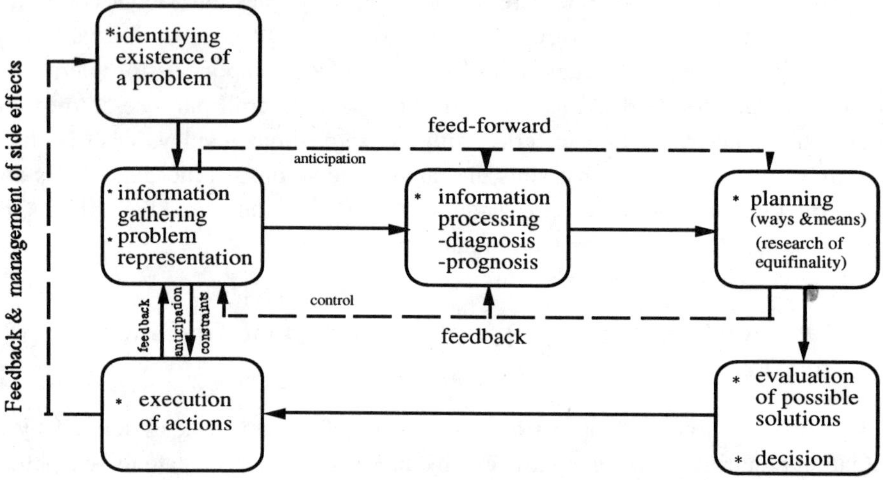

Fig. 4. General schema for goal-oriented situation management, with a twofold regulation: feed-forward (anticipation) and feedback (control). Side-effects of actions are taken into account at a global level.

Figure 4 expresses a logical task analysis for a virtual operator having in charge open dynamic environment management (ODEM). This model can be seen as a minimal a priori task decomposition, whose field of validity is wider than dynamic environment management: under the condition of suppressing the specification "diagnosis/prognosis" in the information processing unit. The planning unit involves designing possible “ways and means" or, in other words, possible “solution paths". In fact, this decomposition fits for every ill-defined problem-solving situation (no finite research space).

For analysing possible "effective implementation" of the virtual operator linked to this task, a more detailed task model is needed. For ODEM we took as a basis a method designed by experts actors in Rescue Services (high ranking firemen Officers) for rationalizing decision making in complex situations. This Method for Tactical Reasoning (MTR) was transposed from a similar military method. (It is taught to Officers in vocational training sessions).

In fact task decomposition is refined along two dimensions. One concerns the processing units (as an example, state diagnosis can be decomposed into information processing about the contextual environment and information processing about the dynamic event itself). The other is related to "granularity" in analysis: processing units can be decomposed until reaching "elementary" objects or operations (with respect to an hypothetical state of knowledge of the virtual operator). These decompositions are partly intertwined. Information processing about the contextual environment can be decomposed into analysis of spatial, human, meteorological contextual variables; this can be seen as a decomposition of the processing unit or as a more fine grained analysis of the contextual environment.

We observed in-service training to MTR: presenting MTR from its logical point of view, without any organizational constraints or time pressure, put trainees in a position similar to those of a virtual operator. As a result it was almost impossible to define efficient stopping rules for decomposition into more fine-grained levels. The adopted solution was to define a virtual operational system expressing the "granularity dimension", with embedded decision levels, and to specify a given level as trainees position.

3.3.2 Virtual Operational System

The approach of the virtual operational system related to ODEM is developped in (Rogalski, 1991). With respect to Stoelwinder's approach, concerned with the whole tasks network in an organization (Stoelwinder & Charns, 1981), our approach is "orthogonal", dealing with a main task refinement. The "task field model of organization [proposed by Stoelwinder & Charns] provides a framework for handling the multiple goals and their associated organizational requirements. It can be used in organizational analysis, design and changes programs". The virtual operational system (related to a task for a virtual operator) provides a framework for handling the dimension of hierarchical task analysis (granularity). Its domain of use is operational organization, defined "locally" in terms of task, time and actors.

There exist relationships between vertical and horizontal organizational constraints. It was observed that if a group works with high "horizontal" interdependence (and low "vertical" interdependence), it needs a decentralised system to be efficient. If the interdependence between the group and the vertical organization is strong, then the group needs centralization (Cannon-Bowers et al., 1992). In the first case, the adapted organizational functioning is on-line self-coordination, in the second case the group needs either a prescribed coordination or an internal centralized control (or even both). Interdependence may concern the three task components: for instance, in flight management, time pressure (as a constraint shared by all interacting sub-systems) and use of space as a shared resource (strictly limited resource for taking-off and landing) introduce high

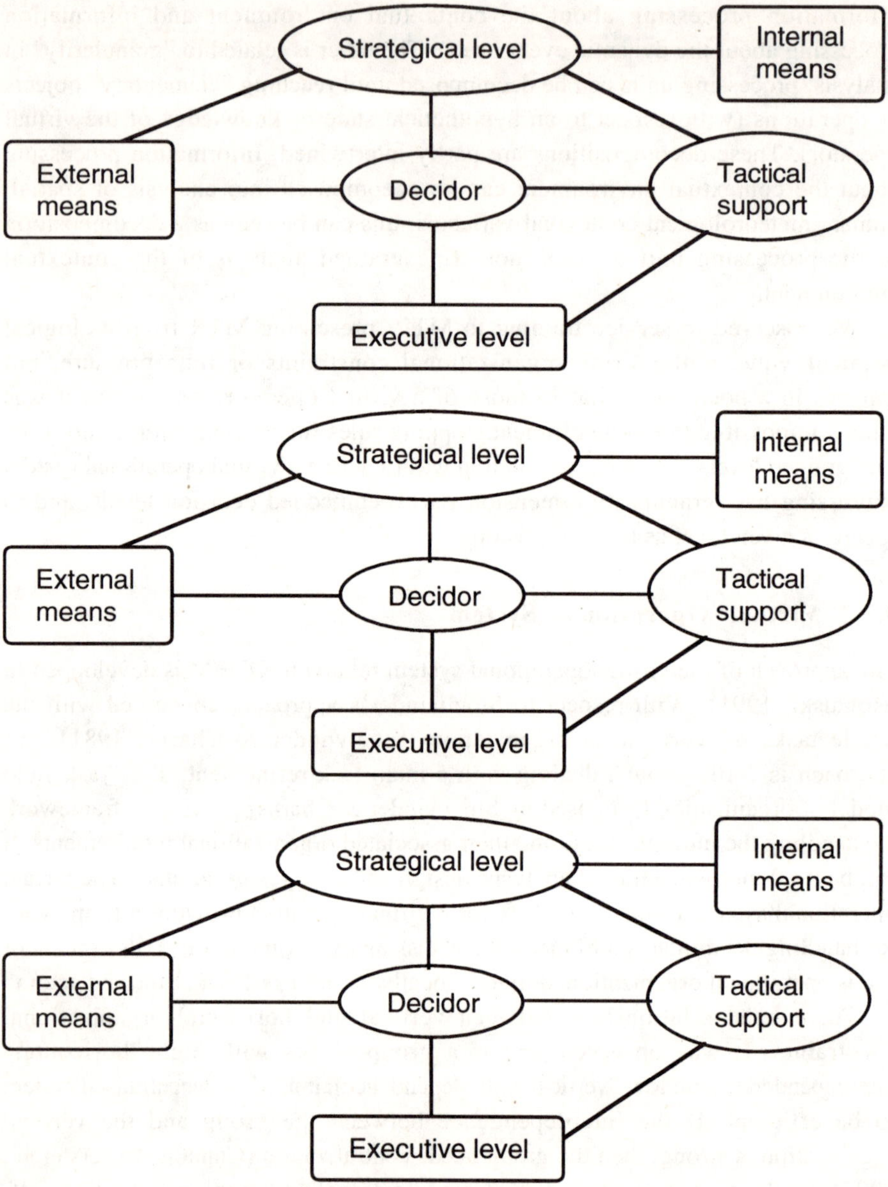

Fig. 5. A model for distributed decision making in ODEM, with a vertical decomposition (for information and action) and a horizontal decomposition with decision support systems (staffs at various levels). This virtual operational system can be used to analyse effective operations or to design appropriate operational structures.

interdependence between sub-systems (hence a high level of prescribed coordination and centralized control).

Modeling a task for a virtual operator or designing a model for a virtual operational system can be considered from two different points of view: Firstly, it is a framework for analysing existing operational systems. Secondly, it can be taken as a process aiming at organization rationalization, where organization is considered at the fine grained level of individuals and collective activities of individuals, that is from the point of view of psychological subjects. The first orientation was used 1) for studying distributed decision making along the horizontal dimension: group efficiency was compared in a simulated situation, using MTR as a framework for analysing the dynamics of collective activity on elaboration phases -from diagnosis to choice of a decision, (Samurçay & Rogalski, 1993); 2) for analysing distributed decision making along the vertical dimension in effective operations -large forest fires in South-East of France, using the model of virtual operational system as framework for collective activity analysis (Rogalski & Samurçay, 1993). The second orientation was taken as a basis for defining a "reference knowledge" about the organizational task in ODEM, for training Officers on Operations Management (Rescue Operations Chiefs and Staff Chiefs).

3.3.4 Tasks Temporalities and Organizational Levels

Embedding tasks in a network may also involve the mean and long-term temporal dimension, which is not taken into account in the preceding models of task and operational system. Figure 6 presents an example of functions embedding - including the level of Operations - for risk management in Rescue Services.

The previous models concern the operational level, where actors have to manage a specific ODEM. The operational system is a model for local organization to face this specific task. However, analysing distributed decision making leads to identify activities which cannot be related only to this direct operational level. Constraints and resources are depending on upper levels (prevision and prevention). Organization of means affects temporal means availaibility on a given operational field. Action structuration has also drastic effects. Action structuration may deal with 1) the "object" on which to "act" when implementing distributed decision making: previous analysis of types of dymanic events is here a crucial point; 2) the effective operational system which can be considered as an instrument for distributed decision making: current organization may be more or less well adapted to specific operational systems. Human training for specific types of risks is also a key point in operational efficiency: debriefing may be considered as linking operational activities to mean-term development of collective competences.

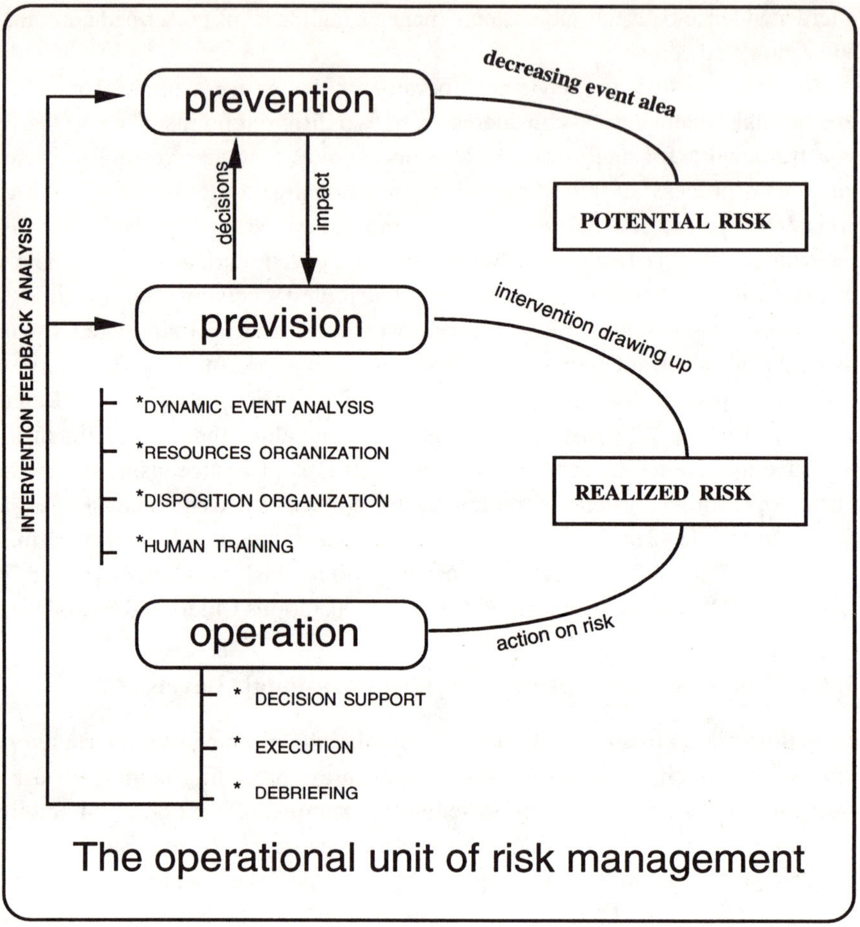

Fig. 6. The operational unit of risk management represents logical and temporal links between various levels of activity, at different temporal horizons. Prevention and prevision concern the "potential risk", and aim mainly at decreasing event alea and possible consequences. Prevision and operation are oriented towards a "realized risk" and strongly related by organizational decisions about means and operational device (tasks distribution and communication), and about actors training. Post-operational analysis concerns the activity of the whole operational device and introduces feedback at all levels, allowing the system to progress.

This example briefly illustrates relationships between questions dealing with task performing and operational system at the level of short-term actors' activities, and issues concerning organizational learning itself, at the level of mean and long-term interactions (here risk management in Rescue Services). Case studies on process control provide a number of important suggestions about how different forms of production and working conditions in industrial settings lead to different forms of cooperation and communication in distributed decision making (van Daele & de Keyser, 1991).

The previous models concern the operational level, where actors have to Analysing actors' activities indicate that action is not limited to task performing. From the one hand, a task goal interacts with other actors goals or purposes; there exists a "metafunctional activity" oriented towards knowledge and control on ones own activity (Valot & Amalberti, 1992; Valot et al., 1993). From the other hand, any human action always produces other effects than the aimed task target state. These "by-products" in task performing or "side effects" of activities may contribute to individual and collective learning as well to organizational learning; their temporalities differ, and it remains a research issue to better identify when and how these effects are taking place.

Our approach was developed on Open Dynamic Environment Management. It could be generalized for other dynamic environments such as process control. Figure 7 presents a schema of possible interactions between systems and human actors for integrating various situations of process management in a common frame.

4. Discussion and Conclusion

We have proposed elements for a comprehensive framework, with as a main purpose to articulate analysis of actors activities, where individual or collective actors are considered as psychological subjects, and analysis of conceptual entities defined as a more global level such as collective actor or organization itself. This framework consideres a task from three point of view: 1) Firstly, it is a node in a tasks network, which is partly hierarchized by a relation of mediatization. 2) Secondly, a task is object of a contract between prescriber(s) and operator(s). 3) Thirdly, a task is defined by three components: teleological (goals), axiological (values) and operative (resources and constraints). This general task model can be used for Tayloristic organizations (strong tasks decomposition related to hierarchical organization) and for flexible organization (the relationship "prescriber/operator" being here those of "client/fournisseur"); handcraft could be taken into account by a notion of self-prescription (prescriber = operator).

Limits of this framework mainly concern: 1) The temporal dimensions, and the embedding of tasks in functions defined at more global level, until the general purpose of a firm or service. 2) The choice of a priori "rational" analyses for task and operational systems is not aimed at representing "organistic" evolutions due to feedbacks from task performance and cooperation itself. 3) Modelling operational systems (for a given task) should be considered as designing under constraints,

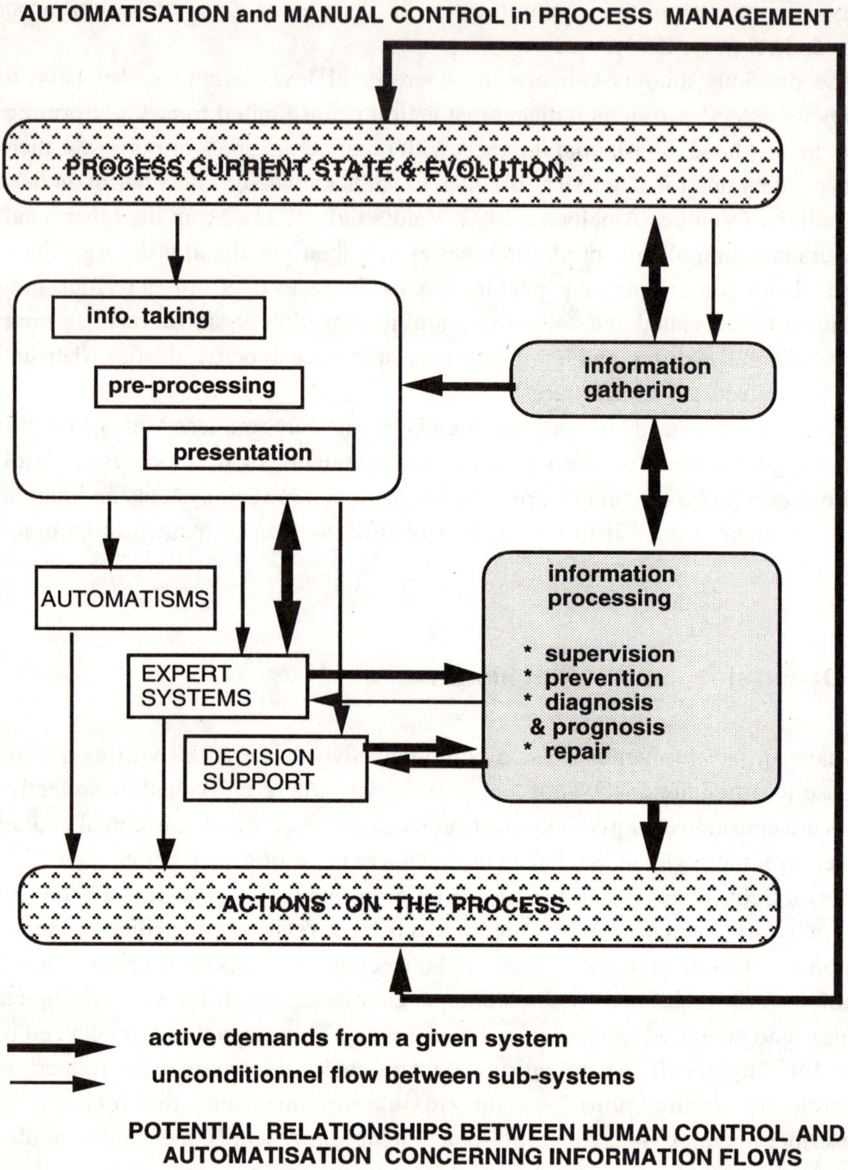

Fig. 7. Interactions between human activities and systems in process management. The shadowed areas on the right represent human activities on process supervision, while the areas on the left represent the automatized processing with possible support systems: automatisms, expert systems and decision support systems. Interaction involves two types of communications: inconditional flows or active demands (from a sub-system to another for information or for action). With respect to automatisation,

without any idea of solution unicity (Charue & Midler, 1992), as it was the case in our example. However our hypothesis is that open dynamic environment, characterized by high event variability and uncertainty, could not be managed through flexible organization, because temporal possibilities for organizational learning are strongly limited.

Intrinsic task analysis – from the point of view of a virtual operator and for the point of view of levels of granularity – offers possibilities for designing operational systems in taking into account specific possibilities and constraints related to human actors, human communication and/or interactions between human actors and automatized systems. This framework was used for analysing distributed decision making in various settings, for Open Dynamic Management. It could be developped in other field, by using task analysis based on the same rationalization principle; Methods for Problem Solving as they exist in many vocational fields (Rogalski, 1993) are good candidates for supporting this orientation.

The choosen approach also aimed at rationalizing design of operational systems embedded in a global organization, rationalizing task allocation, and actors training. It was used in a vocational context for defining a "doctrine" for Operations Management in Rescue Services and for designing adapted training situations. In fact, organizational change and actors' expertise were also shown as dependant variables for increasing efficiency and safety in nuclear power plants (Masson et al., 1993).

In fact, articulation between actors activities and organization was considered at the primary level of interactions between individual activity and inter-individual cooperation and the operational system related to a complex task, requiring distributed decision making. The proposed elements for a comprehensive framework should be embedded in a broader frame, to tackle higher articulation levels.

Acknowledgements

This work was financed by a grant of the Ministery of Research (91D03401) about "Work and learning". We wish to thank the Direction, the trainers and trainees of the Ecole Nationale Supérieure des Officiers Sapeurs-Pompiers (ENSOSP, National Institut of Civil Safety, Nainville-les Roches) and of the Centre Interrégional de Formation de la Sécurité Civile (CIFSC, Civil Safety, Valabre) for their welcome and participation.

---

there are two poles: on the one hand, fully automatized process control, with unconditional flow between sub-systems (dominant left areas) and, on the other hand, "open dynamic environment management", with mainly or only active demands betweens humans operators (dominant right areas).

References

Amalberti, R. (1992). Safety in flight operations. In B.Wilpert & T.Qvale (Eds.), New technology, safety and systems reliability. New York: Lawrence Erlbaum

Amalberti, R., Falzon, P., Rogalski ,J., & Samurçay, R. (1993). Communication et coordination dans les cockpits automatisés. Rapport intermédiaire SFACT-DGAC/ CERMA, CNAM, CNRS-Paris 8

Bærentsen, K.B. (1991). Knowledge and shared experience, Third European Conference Cognitive Approaches to Process Control. Cardiff: University of Wales, 217-232

Bagnara, S., Rizzo, A., & Stablum, F. (1991). Partners' credibility development and assessment in distributed decision making, International Journal of Human Factors in Manufacturing, 2, 109-119

Bagnara, S., Failla, A., & Polo, P.A. (1991). Banche, informatica e risorse umane, Sonzono: ETASLIBRI

Bagnara, S., & Failla, A. (1994/1995). Organizational models, profession roles, decision making and errors. Le Travail Humain (to appear)

Baron, M., Rogalski, J., & Samurçay, R. (1989). Un système de traitement de situations dynamiques complexes: analyse de la décision avec une méthode de raisonnement tactique (MRT); conséquences pour l'élaboration d'aides informatiques et la formation, European Congress on System Science. Paris: AFCET, 697-709

Brehmer, B. (1991). Organization for decision making in complex systems. In: J. Rasmussen, B. Brehmer, & J. Leplat (Eds), Distributed Decision Making: Cognitive Models for Cooperative Work, pp. 335-347. London: Wiley

Cannon-Bowers, J., Oser, R., & Flanagan, D.L. (1992). Work teams in industry: a selected review and proposed framework. In: R.W. Swezey & E. Salas (Eds.), Teams. Their training and performance, pp. 355-378. Norwood, NJ: Ablex

Cellier, J.-M., & Mariné, C. (1984). Expérience professionnelle et gestion des communications dans une tâche de régulation de trafic, Psychologie et Education, VIII, 2, 21-39

Charue, F., & Midler, C. (1992). Mutation industrielle et apprentisage collectif. In: G. de Terssac & P. Dubois (Eds.), Les nouvelles rationalisations de la production, pp. 187-225. Toulouse: Cépaduès Editions

Cicourel, A.V. (1990). The integration of distributed knowledge in collaborative medical diagnosis. In: J.Galegher & C.Kraut (Eds.), Intellectual Team Work, pp. 221-244. Hillsdale, NJ: Lawrence Erlbaum

Curtis, B., Krasner, H., Shen, V., & Iscoe, N. (1987). On building software process models under the lamppost. In: Proceedings of the 9th International Conference on Software Engineering, pp. 96-103. Washington: IEEE Computer Society

Dailey, R.C. (1978). The role of team and task. Management Sciences, 24, 15, 1, 1579-1588.

de Keyser, V., de Cortis, F., & Feree, F. (1985). La conduite collective dans un système automatisé appréhendée à travers les communications verbales. Psychologie et Education, IX, 1-2, 21-61 de Terssac, G., & Chabaud, C. (1990). Référentiel opératif commun et fiabilité. In: J. Leplat & G.de Terssac (Eds.), Les facteurs humains de la fiabilité, Marseille: Editions Octarès/Entreprises

Favaro, M., Guillermain, H., & Guyon, C. (1991), Identification, estimation et représentation des risques: l'exemple d'un atelier de centrifugation. Actes du 5ème Congrès Européen de Psychologie du Travail et des Organisations, Rouen, 24-26 March

Faverge, J.-M. (1966). L'analyse du travail en terme de régulation. In: J. Leplat (Ed.), L'analyse du travail en psychologie ergonomique Recueil de textes, pp. 61-86. Toulouse: Octarès (1992)

Green, R. (1190) Human errors on the flight deck. In: D.E. Broadbent, J. Reason, & A. Baddeley (Eds.), Human Factors in Hazardous Situations. Phil. Trans. Royal Soc. Lond. B 327, 503-512

Hollnagel, E., & Woods, D.D. (1983). Cognitive systems engineering: New wine in new bottles. International Journal of Man-Machine Studies, 18, 583-600

Hutchins, E. (1990), The technology of team navigation. In: J. Galegher & C. Kraut (Eds.), Intellectual Team Work, pp. 191-220. Hillsdale, NJ: Lawrence Erlbaum

Jansens, L., Grotenhuis, H., Michiels, H., & Verhaegen, P. (1989). Social organizational determinants in nuclear power plants: operators training in the management of unforeseen events. Journal of Occupational Accidents, 1,1, 121-129

Kuklan, H. (1988). Perception and organizational crisis management. Theory and Decision, 23,3,259-274

Lee, J., & Moray, N. (1992). Trust, control strategies and allocation of function in human-machine systems. Ergonomics, 35, 5, 1243-1270

Leplat, J. (1990). Relations between task and activity: Elements for elaborating a framework for error analysis. Ergonomics, 33, 10/11; 1389-1402

Leplat, J. (1991). Organization of activity in collective tasks. In: J. Rasmussen, B. Brehmer, & J. Leplat (Eds.), Distributed Decision Making: Cognitive Models for Cooperative Work, pp. 51-73.London: Wiley

Marek, J., Iversen, M.J., & Hellesoy, O.H. (1987). Risk, organization and safety on an oil-platform. In: W.T. & J. Hovden (Eds.), Risks and decisions, pp. 123-137. Chichester, UK: Wiley

Masson, M., Malaise, N., Housiaux, A., & de Keyser, V. (1993). Organizational change and human expertise in nuclear power plants: Some implications for training and error prevention. Nuclear Engineering and Design, 144, 361-373

Mélèse, J. (1991). L'analyse modulaire des systèmes (AMS). Paris: Les Editions d'Organisation

Navarro, C. (1987). Communications fonctionnelles et complexité des tâches dans le pilotage d'un avion de ligne. Le Travail Humain, 50, 4, 289-304

Navarro, C. (1991). Une analyse cognitive de l'interaction dans les situations de travail, Le Travail Humain, 54, 113-128

Norros, L. (1989). Responsibiliy and risk control as elements of process operators professional expertise. Second European Meeting on Cognitive Science Approaches to Process Control, Siena, 11-28

Petermann, F., Neubauer, W., & Grünheidt, B. (1992). Trust in the relationship between superiors and subordinates: managers' subjective theories of trust. European Review of Applied Psychology, 42, 3, 209-216

Rabardel, P. (1993). Les activités avec instruments. Habilitation Thesis. Paris: University Saint-Denis

Rasmussen, J. (1986). Information processing and human-machine interaction: An approach to cognitive engineering. Amsterdam: North-Holland

Rasmussen, J., Brehmer, B., & Leplat, J. (Eds.) (1991). Distributed Decision Making: Cognitive Models for Cooperative Work. London: Wiley

Rogalski, J. (1991). Distributed decision making in emergency management: using a method as a framework for analyzing cooperative work and as a decision aid. In: J. Rasmussen, B. Brehmer, & J. Leplat (Eds.), Distributed decision making: Cognitive models for cooperative work, pp. 291-301. London: Wiley

Rogalski, J. (1992). Rationalisation de l'activité et gestion de la compatibilité cognitive: les méthodes de résolution de problèmes. In: P. Dubois & G. de Terssac (Eds.), Les nouvelles rationalisations de la production. Toulouse: CEPADUES

Rogalski, J., & Samurçay, R. (1993). Analysing communication in complex distributed decision making. Ergonomics, 36, 11, 1329-1343

Samurçay, R., & Rogalski, J. (1991). A method for tactical reasoning (MTR) in emergency management: analysis of individuel acquisition and collective implementation. In: J. Rasmussen, B. Brehmer, & J. Leplat (Eds.), Distributed decision making: Cognitive models for cooperative work, pp. 291-301, Chichester, UK: Wiley

Samurçay, R., & Rogalski, J. (1993). Cooperative work and decision making in emergency management. Le Travail Humain, 56, 1, 53-77

Savoyant, A. (1984). Définition et voies d'analyse de l'activité collective des équipes de travail. Cahiers de Psychologie Cognitive, 4,3, 273-284

Savoyant, A., & Leplat, J. (1983). Statut et fonction des communications dans l'activité de travail. Psychologie Française, 28, 3, 247-253

Schmidt, K. (1991). Cooperative work: a conceptual framework. In: J. Rasmussen, B. Brehmer, & J. Leplat (Eds.), Distributed decision making: Cognitive models for cooperative work, pp. 75-110. London: Wiley

Sorensen, J.H., Mileti, D.S., & Copenhaver, E. (1985). Inter- and intra-organizational cohesion in emergency. International Journal of Mass Emergency and Disasters, 29-52

Stoelwinder, J.U., & Charns, M.P. (1981). The task field model of organization and design. Human Relations, 34 (9), 743-762

Susman, G.I. (1972). The impact of automation on work groupe autonomy and task specialization. In: L.E. Davis & J.C. Taylor (Eds.), Design of jobs, pp. 357-373. London: Cox & Wyne/Penguin Books

Valot, C., Grau, J.-Y., & Amalberti, R. (1993). Les métaconnaissances: des représentations de ses propres compétences. In: A. Weill-Fassina, P. Rabardel, & D. Dubois (Eds.), Les représentations pour l'action. Toulouse: Octarès

Valot, C., & Amalberti, R. (1992). Metaknowledge for time and reliability: Luxury or necessity? Reliability Engineering and System Safety, 36, 199-206

van Daele, A., & de Keyser, V. (1991). Distributed decision making and time in the control of continuous processes. In: J. Rasmussen, B. Brehmer & J. Leplat (Eds.), Distributed decision making. Cognitive models for cooperative work, pp. 261-274. Chichester, UK: Wiley

von Cranach, M., Ochsenbein, G., & Valach, L. (1986). The group as a self-active system: outline of group action. European Journal of Social Psychology, 16, 193-229

Walz, D.B., Elam, J.J., Krasner, H., & Curtis, B. (1987). A methodology for studying software design teams: an investigation of conflict behaviors in the requirements definition phase. In: G.M. Olson, S. Sheppard, & E. Soloway (Eds.), Empirical studies of programmers: second workshop, pp. 83-99. Northwood, NJ: Ablex

Learning Strategies and Organizations

Sebastiano Bagnara[1], Antonio Rizzo[1], and Oronzo Parlangeli[2]

[1] Dept. of Communication, University of Siena, Via del Giglio 14, 53100 Siena, Italy
[2] Dept. of General Psychology, University of Padova, Piazza Capitaniato 3, Padova, Italy

Abstract. Actual organizations refer to two predominant models, the Tayloristic or "mechanical" model, and the "organic" or "network" model. Each model implies distinctive qualitative differences in the jobs and in the cognitive work required. These differences will be illustrated in the paper with the aim to show that learning strategies have to vary depending on the organizational setting they are to be put into action.

Keywords. Organizational analysis, job profile, cognitive work analysis, error analysis, learning strategies.

1. Introduction

Theories and studies on organizational learning have stressed either the importance of social and organizational factors or the relevance of cognitive mechanisms (Olson et al., 1993). In the first case, the focus is on the organization, its goals, how its internal agents come to share a common view of these goals (see, e.g., March, 1988). And learning is characterised as a social process that takes place in a community of practice, where the socio-technical relationships play a fundamental role in shaping both the activities and the acquisition of the skills and knowledge needed to perform them. The interest is directed to identify the socio-cognitive processes by which individuals learn to act so that the organizations they are part of develop and survive in a competitive environment. The basic underlying principle is that the socio-organizational context forces the individuals to learn to behave as the organizations they enter require.

In the other case, the prevalent view is that the organizations are collections of individuals each of whom has a set of built-in strengths in learning, problem solving, decision-making, accuracy in performance, and motivation. The focus is on the mental structures, processes, and mechanisms the individuals in general activate and develop in order to acquire and use skills and knowledge (see, e.g.,

Anderson, 1986, 1987). The basic idea is that each organization requires individuals with a specific set of cognitive, social, motivational and personality characteristics. The key for the development and the efficiency of a competitive organization is in selecting the proper individuals. However, until recently, this approach has paid scant attention to the organizations. It has implicitly assumed that the organizations differ as for the skills and knowledge required, but has not conducted an in-depth analysis in order to differentiate and classify them in terms of what (skills and knowledge) the human resources should provide. This approach has not yet fully recognised that organizations may differ also as for the learning processes they privilege.

This contribution moves from the idea that there are different types of organization, each of them implying qualitative differences in the actual use of cognitive processes and mechanisms, but also in the learning strategies. It aims at identifying the learning strategies that better fit with the different types of organizations and the cognitive work they require. The reasoning will follow a path where, first, the organizations will be differentiated as a function of the model they refer to, then, the profiles of the main jobs and the cognitive work they have to perform will be illustrated. It will be shown that jobs and cognitive work qualitatively vary depending on the organization where they take place. It will be concluded that also learning strategies have to be tuned with the organisational settings they are to be put into action.

2. Organizational Models and Job Profiles

The actual organizations of work can be positioned along a continuum whose poles, at one extreme, have the Tayloristic or "mechanical" model, and at the other, the "organic" or "network" model (Butera, 1990). The distinctive features of the mechanical model are a hierarchical and pronounced division of labour, the concept (and the practice) of personnel as expendable spare-parts within the organization and the culture of dependence and of quantity. Responsibility and authority are clearly and unambiguously defined, delegated or controlled along the hierarchical line. Upkeep and innovation are assigned to the board, while co-ordination and execution are assigned downwards. Everything works, or should work, like a "clock". This model fits with stable internal and external socio-economic environments, long-lasting technological innovations, economy of scale, mass production, and markets where competition is on the cost, rather than on the added quality value of a product.

It is not by chance that the metaphor employed to illustrate the alternative model is biological: the organization is conceived as a complex organism where the individual parts are sub-systems which, while carrying out specialised functions and working autonomously, are embedded in a network of relationships. The system is continually redefined on the basis of reciprocal influences among components and is modified following processes of adaptation to the external environment as well as the characteristics of the internal processes.

The network model is mostly effective when the external environment is turbulent and unpredictable and it is difficult for management to control events. It requires an aptitude for planning, diagnosis, problem setting, and the assumption of decision-taking responsibility, both at an individual and at a wider level. Tasks and activities are not rigidly defined, but have blurred boundaries and are interdependent. Many specializations are required. The relationships and integration among activities, roles, processes and structures are dynamic.

Within the "network" organization, individuals are agents who know how to take on the role of signal interpreters and generators of hypotheses and alternatives. They are not mere resources exclusively employed to work upon established and defined processes. The relationship between workers and the system is defined by a continuous interaction of co-operation and conflict, in which involvement in work on the basis of appreciation of one's importance filters through from professional experience (Failla and Bagnara, 1989).

The working of a network system provides for the centralisation of strategic control and the decentralization of operative control, a competitive strategy based on innovation, the predomination of teamwork. As a rule, groups are self-regulating, typified by a multiplicity of specializations and by a high degree of co-operation for the attainment of agreed goals.

The mechanic and network organizations provide for qualitatively different jobs, who are also paid in a substantially different way: in the mechanical organization, on the basis of time and of working conditions; in the network organization, on the basis of results.

At the operative level, for the mechanical and network models respectively, the "manual" or "data worker" and the "process operator" are provided for. Manual and data workers are the employees who carry out simple operations, often manning machines, within a hierarchic-functional organization. The tasks they carry out are usually of an individual nature and concern the fulfillment of a segment of a procedure. The tasks are, therefore, predefined, with spheres of variation reduced to the minimum. They take very short time to be learned, and can be automatically carried out after extensive practice. The required work is geared to production in terms of quantity. Autonomy and responsibility are severely limited, if not non-existent.

In the network model, the manual and data workers tend to disappear and are replaced by the process operators, appointed to control and supervise, independently of the sector in which they are located. They usually work interactively with information and automation technologies and understand the system with which they interact and guarantee its efficient management by anticipating events and rapid malfunction diagnosis, thereby preventing possible breakdowns (Hacker, 1993). They are occupied in the upkeep of the process and in its flexible adaptation to ensure the quality of the product and they are capable of planning complex actions even in unfamiliar and highly variable situations. Then, learning is continuous and automatisation of knowledge plays a minor role. It is centered on the refinement of the mental model of the system processes rather than on the improvement of speed and accuracy in performing a few overpracticed actions.

At the intermediate level, the mechanical organizations provide for "middle-management" and the "specialist technicians", while the network organizations provide for the "professionals". Within the mechanical organization, the specialist technicians deal with the solution to technical problems which arise within their zone of competence. They have a thorough knowledge of a clearly-defined category of machinery and of a portion of the work process. Middle management, instead, interacts very little with the machinery. Its task is primarily that of medium-term planning of the activities and co-ordination of human resources related to given objectives. It therefore possesses a general understanding of the work processes and a deep competence on how to control and manage the human resources.

On the same level, in the network organization, the professionals carry out a more articulate and less pre-determined role. There is a core of competence common to all professionals, independent of their function, their position and their specific role (Failla & Bagnara, 1989). This "core competence" comprises the ability to use and co-ordinate all the opportunities provided by new information processing and management systems. They have a multi-disciplinary preparation: they understand the work procedures, but are also able to make effective use of the human resources. However, the professionals always have specific knowledge and abilities, allied to an ability to analyze and summarise. They claim and have right to autonomous decision-taking in their area of competence, they are efficient in horizontal communication and can manage their own schedule. They work for set aims.

The professionals, however, are undergoing transformations and two distinct figures are emerging: "the decision-taking support professional", characterised by an ability to anticipate and resolve problems within defined contexts and by a disposition for an adaptability to changing circumstances; and the "professional innovator", more dynamic, capable of initiatives and assumption of responsibility.

Finally, at the highest level, the mechanical organization provides for the "director", while in the network organization there is the "manager". The main aims of the directors are the running and upkeep of the system, seeking to rationalise the relationships between the parts with the scope of maximising global efficiency. Their activity requires a deep understanding of the internal structure and is characterised by hierarchical and formal communication. Choices and decision-taking are based upon the scheme of co-ordination among the various substructures of the system.

Managers' activities are mainly directed at problem setting: they seek new opportunities, based on their knowledge of the system, of the internal and external environments and of the interdependent relationships between the various systems involved. The managers tend to differentiate themselves and three types of manager can be identified: a) The "social" managers, who prevail in phases of technological and organizational changes, are typified not so much by their technical knowledge, but rather by their competence in human relations which apply inside the traditional framework of corporate culture. Their focus is on motivating human resources to keep up with the changes. b) The "super-professional" managers, who have a remarkable technological competence and are oriented towards the use of information technology, "know their way around" it with the same level of competence as (or, ideally, superior to) that of their collaborators. They direct themselves to the discovery of new processes, products or services. c) The "entrepreneurial" managers, who pay particular attention to the coherence among technology, organization and human resources. As their goal, they have not only the discovery of opportunities, but also the use of those opportunities within a dynamically evolving system, which however, must keep its identity. They propose and carry through new goals, but they also create and handle the conditions to achieve them.

3. Jobs and Cognitive Work

The jobs in the two organizational models present qualitative differences in cognitive work (Bagnara et al., 1989; Schuler, 1989). This can be shown by using the well-known approach to cognitive work analysis originally proposed by Rasmussen (1983), and further developed by Reason (1986) and Rizzo et al. (1987). This approach moves from the idea that human work can be differentiated not so much according to the kind and number of visible actions which make it up, but according to the pattern of control of cognitive processes that mediate upon it, and of which the observable operations are often a very partial product.

It is well-known (Norman and Shallice, 1985) that mental operations vary with regard to the level of conscious control necessary to carry them out. The large majority of mental processes, those for which one needs a lot of practice and which are at the basis of skills, are "automatized": they are carried out rapidly in a standardised way without attention and conscious control. They are triggered off by other mental processes or by the simple detection of the appropriate environmental conditions. Other processes are conducted under conscious control and are relatively slow. They are not carried out in a standardised way, they have to be kept under constant attention and require the active use of knowledge possessed by the agent. The use of feedback is essential.

With regard to the quality of control on the cognitive processes which govern and mediate human activities at work, it is possible to arrive at their classification on four levels of quantitatively and qualitatively increasing cognitive control (Bagnara, 1990) (see Fig. 1).

a) **Skill-based activities**. They include all the operations which are carried out smoothly and with timing and form almost constant. The actions tend to be ballistic: once the proper conditions are detected, they begin and cannot be stopped before they are finished. Obviously, the width of the ballistic unit of action tends to increase with practice. While they are being carried out, one may easily produce a verbal report, referring to the purpose of the actions, but never to how they are carried out.
The cognitive processes which mediate these actions are automatic in nature or automatized through practice. Only in the case of a pronounced discordance between the mental model of the environment and the real state of the same environment, the cognitive processes which produce these operations are placed under scrutiny. The typical errors are slips. A slip appears as a disruption in the execution. The intention is satisfactory, but the actions are not carried out as planned. It normally occurs in routine situations characterised by automatic and over-practised behaviour. These activities and errors prevail in the work of manual and data workers. Indeed, these jobs often exclusively require activities which do not call for any cognitive control. However, one should remember that behavioural components based on skills are present in all the jobs, even at top management level, but the routines of automatized actions and processes are usually quantitatively less important and are also qualitatively "richer".

Organizations		
Cognitive work	**Mechanical**	**Network**
a) Skill-based	Manual and Data Worker	
b) Rule-based	Middle-Management Specialist	Process operator
c) Problem-solving	Director	Professional
d) Problem-setting		Manager

Fig. 1. Organizations, jobs, and prevalent cognitive work

b) **Rule-based activities**. They are sequences of actions which follow a known procedure. In verbal reports on current activity, both the starting conditions and the state which is to be achieved are reported precisely but, above all, the temporal order is privileged. The structure of reporting are essentially "scripts". Cognitive control is not so much dedicated to the carrying out of single units of action, as to the temporal dimension, which organises and regulates the course of the activity. The typical errors are lapses and rule-based mistakes. The lapses derive from the loss of control of the intention that is currently under execution, or from the failure to trigger the intended action at the proper moment. The rule based-mistakes usually consist of the wrong application of well-known rules or procedures which have been consciously acquired, and depend on biases in the selection or to quasi-automatic overriding by more familiar rules onto the appropriate ones, both in identifying the situation and in adopting the plan of action.
In the mechanical organization, this type of cognitive activity covers a substantial part of the work, both of middle-management and of the specialised technician, but also of the director. In the network organisation, it is the mental work typical of the process operators, even if it does not exhaust their activity, which tends towards higher levels of control. In fact, the process operators are also asked to assume a determining role in those situations which deviate from the norm. They must, in fact, as far as possible, be able to anticipate and deal with evolving states of the system and with potential future configurations. It is a question of planning and diagnostic activity. In this case control is regulated by feedback and decisions cannot be taken automatically, indeed they are often the product of mental simulations of hypothesis verification on the functioning of the system conducted within the mental model of the said system. The performance, therefore, should not be evaluated on the basis of the ability to apply skills, but on the basis of speed and effectiveness in diagnosis and prediction (Hacker, 1993).

c) **Activities based on the active use of knowledge and directed towards problem solving**. These activities take place in unfamiliar or complex situations, when the goals have a notable degree of certainty but it is necessary to discover how to proceed to their fulfillment. They undergo continuous attentive control. Almost every plan of action is consciously and repeatedly evaluated, both with respect to the goals to be reached, and to the mental model of the environment, when the solution hypothesis is mentally simulated to test it. When the hypothesis is subject to empirical verification, in as much as it is considered reasonable on the basis of mental simulation, or because one goes to find out additional information to refine the mental model, feedback from the environment is actively searched for.

The typical errors are knowledge-based mistakes: the selected plans, or even the intended goals can turn out to be not well-fitted to the problem to be solved. People are not able to properly recognise the relations among different aspects of the problem or to achieve an adequate diagnosis of it, because of lack of completeness of the mental models used and/or a fault in causal thinking.

In the mechanical organizations, this type of cognitive activity above all belongs to the director, but also, even if in a limited way and domain, to the specialist technician and to the middle-manager. In the network model, instead, it is prevalent and typical in the professional who supports decisions, whose work is characterised by processes of "problem solving" and inference from data. However, within a network organisation, a shifting of activities towards the area of "problem setting" is needed: the decision is less and less a technical fact and increasingly the product of a process of evaluation of alternatives and innovations compatible with the values and strategies of the organization. The professional producer of innovations therefore tends to shift the level of cognitive control to the highest level.

d) **Activities based on the active use of knowledge and directed to problem setting**. These activities are centered on to "discover" problems. The active use of knowledge is qualitatively different. In the case of problem solving, it is directed towards an explicit purpose, and consists of searching out and following through the way to achieve it. In the case of problem setting, cognitive work is directed at establishing aims, and evaluating the actual state of affairs, with respect to scenarios and values relatively independent of the situations in which one is immersed.

These two mental work patterns are related to two different aspects of decision taking. Decisions are in fact based upon two main components (Beach and Mitchell, 1987; Kahneman and Miller, 1986).

One component concerns the inferential process of prediction on the basis of available data: its products are the estimation of probability which an expected event has of happening, which a hypothesis has of succeeding, or the predicting of the evolution of a dynamic situation. This activity is based on the knowledge the decision maker possesses regarding a certain environment or domain, on the mental model of this domain, on its states and on its evolution. Here, cognitive work is centered in cost-benefit analysis or in using short-cuts which, though often affected by well-known cognitive biases (Kahneman et al., 1982), can heavily reduce mental workload.

The other component is related instead to the role played in decisions by both individual and collective values such as those, for example, mediated by the "culture" of a corporate (Hampden-Turner, 1990), by its history and its long-term strategies. Here, a tacit selection takes place, by which only alternatives compatible with the principles the individual and social identities are based upon are made available for the subsequent comparative evaluation.

In problem solving, values have, as it were, already carried out their role: aims and objectives have already been established (often, within a mechanical organization, by others), and cognitive activity is directed to "discovering" the best way to achieve one or more given objectives. In problem setting, the crucial aspect regards the compatibility and the coherence between individual preferences and values, and the company values and strategies. The typical error is the failure. Failure is the result of decisions correct in theory, but wrong with regard to the context in which they are made because they are not consistent with the values, and the whole set of cultural and social objects that characterise the company system. This kind of activity and errors is typical of the entrepreneurial manager, but also, even if to a lesser degree, of the other professional managerial roles.

4. Organizations, Jobs, and Learning Strategies

The analysis conducted so far shows that, as cognitive work, errors are constrained and shaped by the organizational structures and the model they refer to. However, it has to be noted that many errors are not immediately detected and may become resident in the organization. This means that a distinction is needed between active and latent errors (Rasmussen & Pedersen, 1984).

Active errors and violations have immediate and visible negative effects, and occur right before a physical, social, or organizational accident or breakdown. Usually, the people involved in them operate near to the outcomes of a system process. Latent errors are usually organizational and design decisions, or undetected errors that yield conditions that may remain silent for a long time, until they combine with triggering conditions and then produce an accident. They are present in the system well before the onset of an accident sequence, and may facilitate the adoption of unsafe behaviour.

It is well-known that errors can have a positive function in learning (Bainbridge, 1989; Rasmussen, 1987; Reason, 1990) both at individual and at organizational level. At individual level (see, Frese, et al., 1991), they may induce an explorative cognitive attitude and, then, an improvement in work practices and in the related mental models, knowledge and skills, provided their management keeps low the motivational/emotional impact on self-esteem and anxiety. Thus, strategies aimed at mitigating the negative (for example, by increasing the awareness of their useful function), and to maximise the positive components of errors (for example, by providing suitable feedback in order to make errors become an opportunity for information seeking) have to be implemented.

Since errors signal mismatches between performance and organizational demands, they are of crucial importance in organizational learning. In fact, their detection displays the organizational dysfunctions and their handling allows to identify and test new and possibly more effective ways the organization itself can use to deal with the environment.

Of course, the positive function of errors in organizational learning implies a company policy that favours communication and discussion of errors and their handling rather than their encapsulation. To be effective, however, this general policy has to be tuned with the characteristics of the actual organization where it has to be implemented. The analysis of the relationships among types of organization, jobs and cognitive work they privilege allows to identify the learning strategy better suited for a specific organization, job, and cognitive work (see Fig. 2).

The skills, which mediate stable activities, are more efficiently acquired through repetition and practice, since learning should aim at the consolidation of knowledge schemata that allow automated reliable performance. Accordingly, here the traditional behavioristic framework can play a fundamental role, and sequential training strategies are the more adequate. Errors occurrence should be avoided (Carroll and Carrithers, 1984) in learning to facilitate the consolidation of the correct behavioural simple schemata to be used by the manual/data workers in the stable mechanical organizations.

By and large, sequential strategies are effective also for the development of rule-based activities, but are counterproductive for the development of competence in problem solving or problem setting. The learning of active use of knowledge is favoured by exploratory-style strategies. Then, for the mid-level roles of the mechanical organizations (specialist technician and middle management) the better learning strategy should be a combination of sequential and exploratory strategies. Errors should not be prevented but managed to facilitate the structuring of an adequate mental model of the system with which one interacts. The latter strategy should be the core of learning for the process operator, who is essentially involved in the prevention and diagnosis of unpredictable events.

For roles characterised by problem-solving (directors and professionals) exploratory learning strategies are the most suitable, since they must have a thorough comprehension of the problem space, so to allow, when needed, its restructuring. Here, the typical errors (knowledge-based mistakes) are a natural outcome both of work activity and of the learning process. Since it is well-known that their handling, detection and recover are highly facilitated by social interaction (Seifert & Hutchins, 1992), a social learning environment should be provided, both for training novices and updating experts' knowledge and competence.

Cognitive work	Error	Strategy
Skill-based	Slip, Lapses	Sequential
Rule-based	RB mistakes	Sequential, Exploratory
Problem-solving	KB mistakes	Exploratory, Social
Problem-setting	Failures, Latent errors	Social, Counterfactual

Fig. 2. Cognitive work, type of errors and learning strategies

Finally, failures are particularly likely to occur when one is engaged with problem-setting activities, but they can depend also from undetected slips and lapses and from tacit norms developed when violations are widely accepted in practice. For both overt and latent failures, the focus should be on the corporate culture and has to be based on an appropriate social learning environment where the methods of counterfactual thinking can play a major role.

The "corporate culture" is intended as a set of principles and values that are shared by the member of a company and displayed both through the symbols adopted by the company to mark its identity and by the patterns of behaviour accepted and rewarded (Schein, 1985). The corporate culture, as above noted, tacitly frames the decision making process both at collective and individual level.

When the organization and the environment are stable the coherence among symbols, behavioural patterns, and tacit framing of decision making is warranted through imitative cultural learning (for the distinction among imitative, instructed, and cooperative cultural learning, see Tomasello et al., 1993), while instructed learning is used only for the newcomers. Then, in the classical mechanical organization, latent failures are very likely not to be detected: in most of the cases they are coherent with the principles and values of the company. For example, if the principle to follow is quantity of production, any latent failure related to quality is very likely to remain covert (Bagnara et al., 1991).

When change and innovation characterise a company and its network structure and orientation to the market make it permeable to external values and principles, then the corporate culture of that organisation becomes less and less powerful in framing the decision making and overt failures become more and more likely. Network organizations are more prone to overt failures and coherence in decision making is less and less warranted by imitative cultural learning and has to be based on instructed and co-operative cultural learning.

Independently of whether a failure be overt or latent, socio-cultural norms, stereotypes, and compensation laws tend to put the responsibility of on the people that had to face the final deleterious outcome, rather than on those who established the conditions for its happening. Such norms are the social and institutional consequences of a very general cognitive bias. As shown by Girotto, et al. (1991), the human action is in general thought as if constraint-compatible alternatives to it were always available, whereas the organizational and environmental states are thought as binding, though hidden and implicit, attributes and properties of scenario. The actual consequence of this cognitive bias is that, in assessing the causes yielding to a deleterious result, people and organizations are biased toward the persons active at the outcome of the failure and under-evaluate the role played by latent conditions, unless a deep analysis of the causal scenario that yielded to the negative result is carried out.

To counter this bias, the organization should learn to make positive use of the rules underlying the counterfactual thinking (i.e., the retrieval or construction of alternatives to experience) (Kahneman and Miller, 1986). These rules determine which attributes of experience are taken as immutable (usually the organization itself) and which are likely to vary (the human actions) in the construction of counterfactual alternatives to reality. The organization and the people in it should learn to deal with all (either explicit or implicit) alternative states of the world that might have occurred, and to make apparent the hidden, tacit features of a scenario.

Such an objective can not be reached by imitative cultural learning, it needs instruction and cooperation learning environments and strategies, methods and techniques both at individual and organizational level tuned with the priority to be tackled, that is latent failures in the mechanical organizations and overt failures in the network organizations.

Such cultural learning environments are also required to counter a phenomenon common to both types of organization. The level of turn-over has become very high: only a few people will spend the whole working life within the same company. The time allowed for adopting a specific corporate culture is becoming shorter and shorter. People have to enter in various corporate cultures: instruction and co-operation based cultural learning are a forced choice.

Finally, the quantitative development of professionals in the company has brought about an issue previously of scarce importance. Professionals tend to belong to at least two organizations: the company they are part of and the professional group (the so-called "diffuse college") (Butera and Failla, 1993) they belong to, which is essential for the updating and evaluation of their technical and scientific competence and expertise. The value systems may refer to differential principles. Then, the professionals may often enter in situations of conflict, that can not be solved in terms of cost-benefit analysis, since the contradiction is on values, and there are not well-established norms and guidelines available. Again, both organizations and individual have to learn through a cooperative learning environment that facilitates negotiation and co-decision on principles, rather than mere information exchange. Coherent and effective behaviours have to be found, built, and convened on: there are not proved solutions to be tacitly imitated.

5. Conclusions

Organizations differ as for their aims and structures. The network organizations focus on quality and innovation and are decentralised and "flat". In contrast, the mechanical organizations are hierarchically structured and aim at maximising quantity. They differ also for the roles, jobs, competences, cognitive work, and errors they inhibit or favour. As a consequence, though there may be a general policy toward favouring communication and to avoid encapsulation of dysfunctions, the strategies to implement this general policy have to be tuned with the characteristics of the actual organization. Learning processes, strategies and objectives should vary depending from the organization they take place. To improve coherence, both individuals in the organizations and the organizations themselves have to apply learning methods and strategies coherent with the actual cognitive work activities in order to avoid dysfunctional expectations and competences of the human resources.

Of course, the degree of coherence can never be complete: even the more mechanical and Tayloristic organization needs to change and innovate, and change and innovation are founded on, and require individuals' and organizational learning. So, a certain degree of incoherence is physiological to improve and innovate the organization. What has been maintained is that each organization allows a quite limited range of variations in learning strategies, and that there is no better learning strategy and method *per se*. The same conclusion holds also for the educational technologies in as much as they refer to and embody learning strategies and theories (Bagnara et al., 1993).

These conclusions do not imply that the learning processes and strategies have to follow whichever organizational demand: there would be no need for a bridge between cognition and organization if the control is in the hand of the latter. Human cognitive processes are very much flexible but have their own constraints. For instance, when the organization requires the human resources to acquire low level skills only, they can adapt, but with two types of cost: stress and fatigue will appear very soon, and the potentiality to learn to perform complex tasks and to cope with turbulent and dynamic environments deteriorates very soon. As a consequence also the degree and the quality of organizational learning decrease. One major reason for the change in the model the organizations refer to (i.e., from mechanical to network model) can be found in the increasing costs the Tayloristic organizations incur to counter the effects of fatigue and stress and in their incapacity to innovate because they are no longer capable of learning (Lei & Goldhar, 1993).

References

Anderson, J.R. (1986). Knowledge compilation: The general learning mechanism. In: R.S. Michalski, J.C. Carbonell, & T.M. Mitchell (Eds.), Machine learning: An artificial intelligence approach, Vol. 2. Los Altos, CA: Morgan Kaufmann

Anderson, J.R. (1987). Methodologies for studying human knowledge. Behavioral and Brain Sciences, 10, 467-505

Bagnara, S. (1990). Ergonomia cognitiva e interazione uomo-calcolatore. In: D. Gilli & F. Grimaldi (Eds.), Interazione col computer e sistemi formativi. Milano: Angeli

Bagnara, S., Cambiganu, C., & Parlangeli, O. (1993). Organizzazione, tecnologie, educazione. In: A. Piromallo Gambardella (Ed.), Luoghi dell'apparenza: mass media e formazione del sapere. Milano: Unicopli

Bagnara, S., Di Martino, C., Lisanti, B., Mancini, G., & Rizzo, A. (1989). A human error taxonomy based on cognitive engineering and on social and occupational psychology. EUR 12624 EN. Ispra: JRC

Bagnara, S., Rizzo, A., & Stablum, F. (1991). Partners' credibility development and assessment in distributed decision-making. International Journal of Human Factors in Manufacturing, 1, 109-119

Bainbridge, L. (1989). Cognitive processes and training methods: A summary. In: L. Bainbridge & S.A. Ruiz Quintanilla (Eds.), Developing skills with information technology. Chichester, UK: Wiley

Beach, L.R., Mitchell, T.R. (1987). Image theory: Principles, goals, and plans in decision making. Acta Psychologica, 66, 201-220

Butera, F. (1990). Il castello e la rete. Milano: Angeli

Butera, F., & Failla, A. (1993). I professionisti d'azienda. Milano: Etas

Carroll ,J. M., & Carrithers, C. (1984). Blocking learner error states in a training wheels system. Human Factors, 1984, 377-389

Failla, A., & Bagnara, S. (1989). Manager e professional nell'automazione d'ufficio. Studi Organizzativi, 3/4, 149-174

Frese, M., Brodbeck, F., Heinbokel, T, Mooser, C, Schleiffenbaum, E., & Thiemann, P. (1991). Errors in training computer skills: On the positive function of errors. Human-Computer Interaction. 77-92

Girotto, V., Legrenzi, P., & Rizzo, A. (1991). Event controllability in counterfactual thinking. Acta Psychologica. 78, 111-133

Hacker, W. (1993). Occupational psychology between basic and applied orientation: Some methodological issues. Le Travail Humain, 56, 157-169

Hampden-Turner, C. (1990). Corporate culture. From vicious to virtuous circles. London: Hutchinson

Kahneman, D., Slovic, P., Tversky, A. (1982). Judgment under uncertainty: Heuristics and biases. Cambridge: Cambridge University Press

Kahneman, D., & Miller, D.T. (1986). Norm theory: Comparing reality to its alternatives. Psychological Review, 93, 136-152

Lei, D., & Goldhar, J.D. (1993). Implementation of CIM technology: The key role of organisational learning. International Journal of Human Factors in Manufacturing, 3, 217-230

March, J.G. (1988). Decisions and organizations. New York: Blackwell

Norman, D.A., & Shallice T. (1985). Attention to action: Willed and automatic control of behaviour. In: R.J. Davidson, G.E. Schwartz, & D.Shapiro (Eds.), Consciousness and self-regulation. Advances in research. Vol. IV. New York: Plenum

Olson, J.S., Card, S.K., Landauer, T.K., Olson, G.M., Malone, T., & Leggett, J. (1993). Computer-supported co-operative work: research issues for the 90s. Behaviour & Information Technology, 12, 115-129

Rasmussen, J. (1983). Skills, rules, and knowledge; Signals, signs, and symbols, and other distinctions in human performance models. IEEE Transactions on Systems, Man, and Cybernetics, 3, 257-266

Rasmussen, J. (1986). Information processing and human-machine interaction: An approach to cognitive engineering. New York: Elsevier

Rasmussen, J. (1987). Reasons, causes and human error. In J. Rasmussen, K. Duncan, & J. Leplat (Eds.), New technology and human error. London: Wiley

Rasmussen, J., & Pedersen O.M. (1984). Human factors in probabilistic risk analysis and risk management. In: Operational Safety of Nuclear Power Plants, vol. 1. Vienna: International Atomic Energy Agency

Reason, J.T. (1987). Generic error-modeling system (GEMS): A cognitive framework for locating common human error forms. In: J. Rasmussen, K. Duncan, & J. Leplat (Eds.), New technology and human error. London: Wiley

Reason, J.T. (1990). Human error. Cambridge: Cambridge University Press

Rizzo, A., Bagnara, S., & Visciola, M. (1987). Human error detection processes. International Journal of Man-Machine Studies, 27, 555-570

Schein, E.H. (1985). Organisational culture and leadership. San Francisco, Jossey-Bass

Schuler, R.S. (1989). Strategic human resource management and industrial relations. Human Relations, 42, 2, 157-184

Seifert, C.M., & Hutchins E.L. (1992). Error as opportunity: Learning in a cooperative Task. Human-Computer Interaction, 7, 409-435

Tomasello, M., Kruger, A.C., & Ratner, H.H. (1993). Cultural learning. Behavioral and Brain Sciences, 16, 495-552

Part III
Technology for Cooperative Learning and Working

Introduction

Cristina Zucchermaglio

Department of Social and Developmental Psychology
University of Rome "La Sapienza" Rome, Italy

All the analysis proposed by the socio-historic movement starts from the idea that the source of individual cognitive functioning should be search in the social context of life. To explain and describe the complex forms of human awareness we should go "beyond" the individual, analizying instead the social and historical conditions of human life. [Luria 1976]. The human world is characterized by the presence of material and simbolic artifacts, that are culturally, historically and socially determined. Since the cognitive activities are always carried out by those artifacts they are, by definition, social processes. And they are social, also if they are carried out individually. In this sense all the cognitive activities, also the construction of new knowledge, are social processes . [Scribner, 1990].

So in this view learning must be considered as a constructive, social, interactive and situated process: a process of enculturation, of entry into a culture of practitioners. Learning is a complex social action embedded in historically and culturally defined frames of activity (Vygotsky, 1966; Cole, 1989). Moreover it is necessary to evaluate how any symbolic system and artifacts allow the development of different types of competence, that could be defined as "skills in a medium". This means that the content of thinking is an important variable and that we think and learn in ways that are directly connected to instruments and symbolic systems of the cultural context in which we live (that is, the cognitive artifacts of Norman (1989) and the cultural amplifiers of Bruner (1989)).

Learning could be mainly described as a process of apprenticeship, that is as a graded, contextually embedded practice, inserted into a relevant frame of activity (Lave and Wenger, 1991). Learning about an activity depends on authentically participating in that activity - not in an "ersatz" approximation to it. In this sense learning does not always involve abstract, decontextualized subject matter, but regards also work practices, social rules and communicative behaviour of the organization. In this view learning is in part the process of socially constructing a communal understanding: learning is not individual and receptive at all, but a constructive and social process.

On the contrary in traditional educational contexts (also the organizazional ones) the learning process is mainly considered as a decontextualized and solitary process, that should be carried out without those supporting aspects (other persons, information, instruments, notes, books, etc.) typical of the informal contexts of learning (Lave, 1988). In fact the view of learning , schooling and training as an enterprise in "knowledge delivery" continue to dominate our society (Eckert, 1993). Most of workplaces embody a view of learning that focuses on the individual in isolation. It assumes that knowledge is a substance that can be delivered in the learner's mind. According to this view, learning is an individual activity that is separable from, even independent of, social engagement. In this sense the distance between learning "in" the educational contexts (intended as a specific ability) and learning "out" should be diminished and completely rethought (Resnick, 1987).

The Institute for Research on Learning (IRL) (1993) proposed seven principles of learning that summarize the new perspective described above:

1) Learning is fundamentally social.
2) Knowledge is integrated in the life of communities.
3) Learning is an act of membership.
4) Knowing is engagement in practice
5) Engagement is inseparable from empowerment.
6) “Failure" to learn is the normal result of exclusion from participation. Learning requires access and opportunity.
7) We already have a society of lifelong learners.

These principles present a serious challenge to the usual, "comfortable" view of learning that dominates our institutions and call for change in schooling and workplace practices. They also call for change in the way we develop and design learning and (also technolgical) learning environments.

Technology, Learning and Cognition

The problem of the relationship between technology and learning is embedded in the wider relationship between technology and cognition (Winograd & Flores, 1986). Technologies are cultural artifacts and amplifiers (both material and symbolic) that mediate our cognitive interactions with the environment: their presence and use in life contexts change our way of working, doing, thinking and learning substantially. For example, Simon (1988) highlights how the growth of management technologies (decision support systems, etc.) are changing the cognitive demands of working in the business workplace.

Indeed a cognitive artifact is an artificial device designed to maintain, display or operate upon information in order to serve a representational function (Norman, 1989). The notion of artifact concerns the inseparability of individual human growth from the instruments of the culture in which it occurs (Cole, 1989). Developments in the cultural psychology, also drawing on sociohistorical theory (Leontiev, 1981; Luria, 1976), provide researchers with an appropriate theoretical approach to analyse technology, intended as cultural tools. As Norman (1990) says "By understanding the ways in which cognitive artifacts serve human cognition, we may be better able to design new ones and improve old ones" (p.28). Technology design is compared with the need to take these cognitive research outcomes into account.

This means that we need to construct a perspective allowing a cognitive based design of new technology, not only to suggest new uses of old technology. We should construct a "design science" of technology for learning and working (Norman, 1989) or a "science of the artificial" as Simon (1969) defines it. Research on learning and instruction is always a science of the artificial: the study of how learning interactions work can never be far removed from the task of engineering them to work better. What is expected from the insertion of information and cognitive technologies in work contexts would be radical changes in traditional ways of conceiving and realizing learning. Designers are forced to integrate the kowledge of learning processes consciously into their design effort, contributing to create a new learning culture in educational contexts. The design of technology should be explicitly focussed on social interactions and communication and they are to be intended as cultural tools that mediate and support the social, distributed and situated construction of new knowledge. A technolgical system should not be merely revealing itself to a user, but has to be conceived as an instrument that help to reflect on the learning process and check it consciously, and supports communication and negotiation processes between members of a social community inserted in rich learning contexts.

An objective is to develop specifications that are applicable to a wide range of systems and that are concerned with increasing the ability of the system to support higher order learning and thinking abilities. To respect the constructive and social nature of the learning process it seems to be necessary to build and use "empty and open" technology to fill and construct in the head of the learner, rather than always and only proposing technology already "full" of knowledge that the learner must transfer into his own head (Zucchermaglio, 1993).

The main point treated in this part of the book can be summarized as follows:

a) **Integrating working and learning:** Following what we said before, there is no need to drastically separate working from learning, as this latter was a specialized activity. Organizations should recognize that *learning processes are continually present and active also in the workplaces.* Learning should not be an activity to be carried out a few times a year, but should be recognized and used constantly: organizations must not lose the contribution of these distributed innovation processes. There is a growing body of research (Seely Brown & Duguid, 1991; Sachs & Scribner, 1990) that supports the conclusion that work environments differently organized promote, as results, different kinds of learning. In this sense work contexts able to sustain the continually inte-gration between working and learning are needed: The design of the organizational learning context should take into consideration also the design of the social organization of work itself.

What emerges is the need to define a set of design principles for the construction of (technological) organizational learning contexts, in which it should be possibile to use and take advantage of the distributed and situated educational activities embeddeded in work practices, that is, such forms of non formal education as on-the-job training, guided practice with a mentor and learning through doing on the job as a participant in a cultural community.

b) **Communication as learning in the community of practice:** If learning is embedded in work itself (rather than abstracted from it) the working and communicative practices of the community could create a potential "curriculum ", in the broadest sense (Lave e Wenger, 1991). We prefer to use the word "could", because not all the community of practice are automatically "good" leaning environments. A good learning community makes itself transparent to its members: so members could find easy access to the resources they need in order to become full partecipants: "And with direct acess to people, places and activities, members' partecipation brings a sense of what the community is about, what possibilities it holds, what their own futures can be" (Eckert, 1992). The relevant aspects of practice (that are the community's learning resources) include activities, artifacts, places, and people and more specifically: the sociocultural organization of space into places of activity; the circulation of knowledgeable skill; the structure of access of learners to ongoing activity and the transparency of technology, social relations and forms of activity; about the segmentation, distribution, and coordination of partecipation within a community, etc. (Lave & Wenger, 1991). All these aspects should be intentionally designed as cultural tools for learning.

c) **Coherent technology design**: The cognitive and learning theories, often hidden, supporting most present educational and working technology, do not correspond to the cognitive processes through which we learn and work. Formal office procedures have almost nothing to do with how people do their jobs (Suchman, 1987).

It is necessary, just in the technology design phase, to use the results of ethnographic and basic research: studies of the development of competence in people who are becoming experts in their field (Resnick, 1989) and about how people actually work, learn and function cognitively in their normal (in - and out-of-school) lives. Just from the design phase it would be forecast to be flexible, explorative and open to social uses and they are therefore technologies able to incorporate at best those principles defined as a "correct" or "realistic" theory of learning, able to substitute common sense learning theory utilized mostly by technical designers.

A learning organization reaches a high level of innovation and efficiency when structured in such a way that eases (and takes advantage of) the social, distributed, participative, situated processes of learning developed by its members, also in interaction with the technological environment. The organizational structure, also with the technology, should support and allow members to learn while the organization itself learns from its members. In this perspective, the design of educational technology should not be conceived only at the human-system interaction level, but it should be conceived at a organizational level also.

It is necessary to pursue an organizational coherence between technologies and working context in which they should be used: otherwise their insertion could create confusing effects in the working context iteself. The designers ought to be sensitive to the socially organized settings in which technology might function: they should use the organizational change that technology produces ad a *mediator* that could amplify the possibility of social interaction and of organizational changes. It follows from this assumption that also the design of the technology must be related to the design of the organizational learning context (in terms of technology, reward systems, belief systems, communicative flows, availability of information, etc.)

The transparency cannot be viewed as a feature of an artifact in itself but as a process that involves specific forms of partecipation, in which the technology fulfills a mediating function. (Lave & Wenger, 1991). It should considered that the openess and transparency of a technological tool deeply affects its use as an instrument in instruction.

Reasoned Summary of Papers

All papers presented in this part speak around and about the themes described above. In particular we can divide the eight papers in two group: the first one (Newman; Cavalli-Sforza and Lesgold; Mac Mahon; Ohrun) are more directly concerned with the design, the implementation and use of technology in school or other educational settings. The second one (Schael; De Michelis; Yadzani and Mealing; Giannetti and Gerbino) are instead more directly concerned with the design, the implementation and moreover the use of technology in work contexts.

We will now analyze specifically how each paper contribute to the development of the above quoted themes. **Denis Newman** shows how technology could contribute to change school practices: the computer instructional system described in his paper is aimed to support project-based school work: an approach consistent with a pedagogy that puts learners in an active role and sustains social and distributed learning. **Violetta Cavalli-Sforza and Alan Lesgold** describe two computer tools that provide opportunity to learn while engaging in authentic activities: the first one, Sherlock, is an intelligent apprenticeship environment for learning by doing electronic troubleshooting; the second one is a computer environment to assist learners in practicing collaborative epistemic argumentation in science instruction. Although their obiectives and uses are rather differenr, the two systems share many similarities in their instructional approach, in particular in using argument as a coaching techniques. **Harry MacMahon** presents Bubble Dialogue, a computer-mediated projective role play technique that can be used in fostering the development of high complexity interpersonal skill The "philosophy" of this software is a constructivistic one and the role of technology in the training process is to stimulate dialogue between trainees, not replace it making trainees conform to a prepackaged informations of one side of an well established human-computer interaction. **Emrah Orhun** illustrates how the technical features of hypermedia could sustain organizational learning intended as a process of collaborative interaction: in this view hypermedia are considered as mediational means creating and shaping new and more effective work and learning practices. **Thomas Schael** emphasizes how "appropriate" technology could help organizations adopting a process organization of work: the paper shows how Workflow Management Technology, through an integration of data, information and communication flows, could support cooperative practices in work context. In the same line of work, **Giorgio De Michelis** analyze the problem of the threshold of sustainable complexity each member of an organizational cooperation network could manage as one of the factors affecting transaction costs of work processes. The paper shows how this problem can be overcome through "well designed" CSCW systems, instead of only organizational means. An analysis of the requirements that such a CSCW Environment should satisfy to face with the complexity of cooperation work processes is traced. **Masoud Yazdani** and **Stuart Mealing** present design principles for a visual language that could support multilingual communication. A system using iconic language for hotel booking is described. **Anna Giannetti** and **Francesco Gerbino** show how an "enterprise knowledge communication network" where organizational knowledge is captured, stored and retrieved could be one of the more effective means to sustains learning and innovation. An this is particularly true in a company producing software: in the software development/maintenance process tools support and methodological principles are needed for the acquisition and tha subsequent constitution and maintenance of knowledge bases. Two examples of this type of tool are illustrated.

References

Bruner, J. (1989). Culture and human development: a new look. Invited address to annual meeting, Society for Research in Child Development, Kansas City, MO, April

Cole, M. (1989). Cultural psychology: a once and future discipline? In: J. Berman (Ed.), Nebraska symposium on motivation: Cross cultural perspectives. Lincoln: University of Nebraska Press

Eckert, P. (1993). The school as a community of engaged learners. Unpublished manuscript, IRL

Lave, J. (1988). Cognition in practice: mind, mathematics and culture in everyday life. New York: Cambridge University Press

Lave, J., & Wenger, E. (1991). Situated learning. Legitimate peripheral participation. Cambridge, MA: Cambridge University Press

Luria, A.R. (1976). Cognitive development: its cultural and social foundations. Cambridge, MA: Harvard University Press

Norman, D. (1989). Cognitive artifacts. Workshop on Cognitive Theory and Design in Human-Computer Interaction, Kittle House Inn, Chappaqua, NY, June

Resnick, L.B. (1987). Learning in school and out. Educational Researcher, 6(9), 13–20

Resnick, L.B. (Ed.) (1989). Knowing and learning. Issues for a cognitive science of instruction. Hillsdale, NJ: LEA

Scribner, S. (1990). Reflection on a model. Quarterly Newsletter of the Laboratory of Comparative Human Cognition, XIII (2), 90–95

Scribner, S., & Sachs, P. (1990). A study of on the job training. Technical report, February

Seely Brown, J., & Duguid, P. (1991). Organizational learning and communities of practice: toward a unified view of working, learning and innovation. Organization Science, 2/1 (February), 40–57

Simon, H.A. (1969). The sciences of the artificial. Cambridge, MA: MIT Press

Simon, H.A. (1988). Informatica, Direzione Aziendale e organizzazione del lavoro. Milano: Franco Angeli

Suchman, L. (1987). Plans and situated actions: the problem of human-machine communication. Cambridge, UK: Cambridge University Press

Vygotsky, L.S. (1966). Pensiero e linguaggio. Firenze: Giunti-Barbera

Winograd, T., & Flores, F. (1986). Understanding computers and cognition: a new foundation for design. Norwood, NJ: Ablex

Zucchermaglio, C. (1993). Toward a cognitive ergonomics of educational technology. In: T.M. Duffy, J. Lowyck, & D.H. Jonassen (Eds.), Designing environments for constructive learning. NATO ASI Series F, Vol. 105, pp. 249–260. Berlin: Springer

The School as a Workplace

Denis Newman

Bolt Beranek and Newman Inc., Cambridge, Massachusetts, USA

Abstract. Considering the school as a workplace for both teachers and students leads to new approaches to the design of computer instructional systems. To support a project-based approach to teaching and learning, in which the students' role is similar to an apprentice, computer network systems must provide for students gradually assuming greater responsibility. Individual and group work must be supported both within the school and over wide area networks from which resources for projects can be brought into the school.

Schools are the workplace for both teachers and students. Traditional instructional technologies usually ignore the work of the teacher while transforming the work of the student into the absorption of information. The notion that individual students are receptacles for information is a fundamental premise of many American schools and may have outlived its utility as we move from the industrial to the information age. Some of the technology now entering American schools is based on a different pedagogy, one that sees the teacher and student as simultaneously active. In broad strokes, schooling can be made more like an apprenticeship in which students are productive under the supervision and guidance of teachers. Students are engaged in projects which do not necessarily have real world economic value but do typically have some tangible product that is worked on over an extended period of time often by a collaborative group.

A project-based approach implies a shift in the nature of school work as well as in the relationship of the computer to the organization of instruction. The fundamental design problem for the technology is to create a system that has a chance of helping an organization to change. American schools are notoriously conservative. Work skills of both teachers and students will have to change. How can technology help not only with the educational change process for the students but also with the organizational change that teachers and other school personnel must engage in as a continuous process?

We have begun developing computer systems to support project-based school work. This paper will describe the work of the Earth Lab and the Copernicus projects which are developing local area and wide area network systems for schools (Newman, 1990; Newman et al., 1989; Newman, Bernstein & Reese, 1992). This work is a large part of the basis for the technology design for the Co-NECT

School, a BBN project designing and creating technology-rich schools in which school work for both teachers and students will be transformed.

Keywords: Schools, computer networks, organizational change

1. Designing for a Changing School

The design of a school technology that promotes change is not as obvious as many advocates of school technology sometimes make it appear. They are heard talking in glowing terms about the revolution that is about to take place in learning. Technology, it is said, will transform classrooms, provide patient tutors for each individual student, enliven learning with graphics and interactive video. Seldom is it suggested that most school technology has either a fundamentally conservative influence or little impact at all. While technology can be designed to change the curriculum or the use of time and space for project-based or inquiry-based school work, most technology functions within the existing structures and supports the traditional compartmentalized curriculum. For example, for many years "individualization" has been considered the primary benefit of instructional technology. But individualization makes the development of a work group very difficult. The path for change becomes static. The school as a workplace resembles an assembly line with each worker devoted to a separate task.

The notion of student products has wide ranging implications for information technology. Instead of instructional or training systems, what becomes relevant are productivity software and even so called "computer supported cooperative work" tools. The use of communication systems, in particular, becomes of increasing interest as students extend their research out to resources on wide area networks such as the Internet and as teachers find support and resources for innovations in their own work. Communication on local area networks, while less common, is an area of tremendous potential for supporting apprenticeship by providing a channel for coordination among students and teachers within the school.

Except for the marginal use of telecomputing, network technology in schools is a mechanism for instructional delivery. Painted in the broadest strokes, the distinction that needs to be made is between systems that *deliver* individualized instruction from a central repository and systems that enable teachers and students to *access* and gather information from distributed resources and communities of peers. For example, will a high school teacher who is learning to make the most of a new simulation of the cardiovascular system in his biology class be able to communicate with peers who are trying to do the same thing or will the network be designed simply to deliver cardiovascular courseware to the workstations of the students?

The *delivery* approach can be traced to the instructional paradigm that has dominated school computing from the first computer assisted instruction programs to advanced intelligent tutoring systems. The associated pedagogy puts the student in a passive role of a receptacle for information. Often the instructional goals tend toward the basic skills, the usual items found in standardized tests. Individualization is the major contribution of the technology and the students are isolated from peers (advertisements often show students wearing headsets) and from other resources such as remote data-bases that are necessary if students are to become involved in inquiry skills or open-ended real world tasks not addressed well by the standardized tests.

The *access* approach is consistent with a pedagogy that puts the learners in an active role in exploring complex problems and favors constructing collaborative environments. Often the computer becomes a tool or a stimulus for projects that let students delve deeply into subjects and example problems. A simulation of the cardiovascular system makes a very complex system accessible to the classroom. Networks in this approach are channels for obtaining more information about that system, more peer collaborators, and additional materials that will help make most of the simulation available at the desktop.

2. LANs for Instructional Delivery

School LANs are dominated by the "integrated learning systems" (known as ILSs) that have evolved from the time-sharing CAI delivery systems of the 70s. ILSs are sold usually as a lab consisting of enough computers for each student in the class to work individually. Courseware covers all major elements of the curriculum (thus the term "integrated"). Scheduling, tracking student progress, and managing instruction are the major functions of the LAN. ILSs make it quite easy for schools to acquire and use computer technology without major changes in their goals or organization. A report by the EPIE Institute (Sherry, 1990), an independent consumer information organization, provides a detailed catalog of this class of school technology which are growing in popularity according to the EPIE report which estimates that sales doubled in 1989. The ILSs can be considered a successful implementation of computers: one that serves a perceived need and fits in well into the practices of the traditional school. We can examine three prominent features of ILS use: location, curriculum, and the time frame of the ILS tasks.

Space. The EPIE report notes that it is an all but universal practice to put the ILS in a lab. Although there is nothing inherent in the technology that requires this configuration, ILSs are sold as "labs" in units of about 30 which is sufficient to simultaneously accommodate a whole class assuming that students will be working at the computers individually. In many ILSs, students also wear headsets as part of their computer interactions so are further removed from peer interaction.

Having the ILS centralized in a lab means that the system can be more easily managed by a school computer lab teacher, or non-teaching paraprofessional. For a school starting with a low level of expertise, the centralization reduces the technical and subject matter training costs that would otherwise be necessary. It is also argued that having students occupied with computer tutorials offers teachers an opportunity for individualized dialog with students who are often ignored in whole class teaching (Schofield et al, 1990). While this is a fundamental change from the traditional classroom lecture style, it does not change the nature of the school work considered in terms of the curriculum goals and class scheduling already in place.

Curriculum and Assessment. Another feature that provides a good fit to traditional school work is the content organization of the courseware. It is straightforward to categorize the courseware into the standard school subjects: math, reading, language arts or science similar to the typical textbooks. Also, like the typical textbook, the content is, for the most part, presented as facts or procedures to be mastered in sequence.

The content and its sequencing is an integral part of the management and evaluation functions. Discrete tasks which result in a single correct answer can be evaluated by the system itself. More open ended tasks requiring, for example, formulation of the problem or research into sources outside the computer, or any kind of free form response cannot be handled by the system. Intelligent tutoring systems (ITSs) are a natural extension of the integrated learning system approach since they also operate best within very confined subject matter domains with similar level of control over student responses.

The division of topics into clearly defined subject areas also eliminates the need for teachers handling different subjects to collaborate or for a teacher in a self-contained classroom to consider the integration of learning across the subjects. In this respect, too, ILSs strongly support work practices in most schools where teachers are not expected to know much about what other teachers are doing.

Time. ILSs are designed on the assumption that each period will be self contained. The tasks do not require preparation prior to the computer lab period and call for a minimum of technical capability on the part of the student. Teachers can thus schedule the use of the computer lab into the preexisting slots in the day. The time frame assumed by ILSs is thus well suited to the structure of the school day which is broken down into discrete periods. Single tasks, such as an arithmetic problem, can usually be done in a matter of a few minutes. The short tasks of the ILS are also very similar to the kinds of tasks found on the standardized tests. The short, carefully constrained answer slots are ideal for automatic scoring in both instances.

3. LANs for Access and Communication: Earth Lab

The ILSs support traditional school work as highly compartmentalized into discrete subjects, grades, and classrooms. But LAN technology can also help to breakdown these barriers of time and space while providing a natural mechanism for extending access to resources outside the school. In the school environment, the effects may result from flexible access to files (including data, text, reports, diagrams, etc.) across physical contexts (classroom, computer lab, library, home) and time frames (periods, units, grades). In essence, because files can be stored and retrieved for a file server on the network, it becomes possible to make that data available wherever the student is working. This makes project work more flexible and provides a sense of unity to the work which is no longer tied to a particular time and place.

This effect is illustrated by the work of the Earth Lab project which has developed a local area network system for schools and conducted formative research on its functioning in several schools over the last four years (Newman, 1990; Newman, et al, 1989). The main school site is a public elementary school (grades 3 to 6) located in Central Harlem, New York City. The school population of approximately 700 students is predominantly African-American with a minority of Hispanic and other groups.

An AppleTalk network of approximately 50 Apple II and Macintosh computers includes a file server that allows for central storage of data, text, and programs. The environment contains a variety of word processing and data-base, desktop publishing and other tool software, including electronic mail which connects students and teachers both over the local and wide area networks. Along with the technology, we introduced a year-long earth science curriculum designed in collaboration with the teachers (Brienne & Goldman, 1989).

The primary means for supporting project work in Earth Lab is what we call "workplaces" which are folders on the file server and in which the work of the project, in the form of text, data-base, graphics and code files is stored. These workplaces, available to any computer on the school LAN, give groups a location for their work together. Figures 1 and 2 show the workplaces available to one sixth grader both at the school-wide level and within her class. At the school-wide

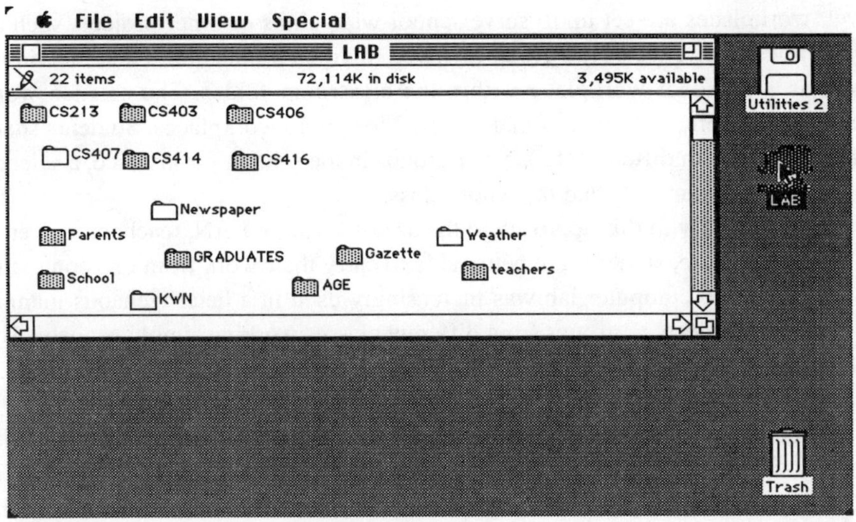

Fig. 1. Workplace available to student at the school-wide level.

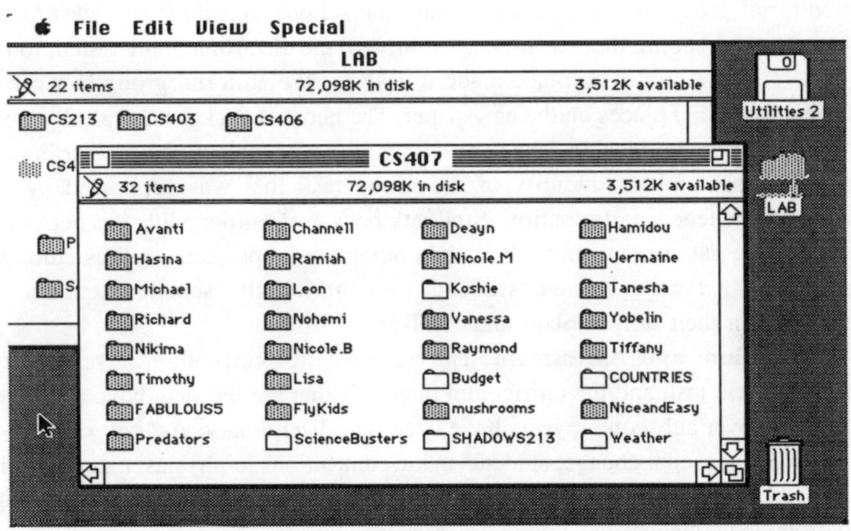

Fig. 2. Workplace available to student at the school-with-lee class

level, workplaces are set up to serve school-wide clubs or other projects such as Kid Witness News (KWN) a group involved in video production. Each individual also has a personal workplace within the classroom folder. The science work groups gave themselves names that are used for group workplaces. Students share different data with different students or groups in the school, for instance, a science group, a noon hour club, and the whole class.

We find that with this approach to the use of a school LAN, teachers are better able to collaborate, students are better able to carry their work from one context to another, and the computer lab was increasingly used in a heterogeneous manner with several projects or groups from different classes working simultaneously. The following examples taken from our observations at the school illustrate these changes.

Space. The school's computer network includes two separate labs, a satellite lab in a small room off one of the classrooms and network connections in several other classrooms into which computers can be moved as needed. The way the project workplaces were set up for groups and individuals helped to develop a sense of continuity.

For example, several students from different classes and different grades are editors for the school newspaper. The newspaper has a workplace on the network that students use for storing articles and other material for the newspaper. Beyond the editorial group, many students around the school contributed articles to the newspaper by sending them as messages through the electronic mail system to the editors. The common workplace made it easy for the editorial group to work at different times and places on the newspaper. The network makes the walls between classrooms more permeable. The ease with which any student could contribute to the newspaper and the identity of the group task that was supported by the workplace widened participation. Students became familiar with the network's function as a data organizer so that when other school projects such as editing a video newscast were started, students thought it quite sensible to create a workplace for their scripts, plans and edit lists.

Curriculum and Assessment. The earth science curriculum developed for the initial field test, and the curriculum materials that the teachers have continued to develop over subsequent years have been interdisciplinary. As they worked on weather and seasonal change, students made connections to physics, math, writing, and social studies. The network system made classroom projects easier to manage and promoted collaboration among the teachers.

For example, at the beginning of the first year of operation some teachers in this essentially traditional school had doubts about the students' capabilities for handling the autonomy involved in small group project work. Having the small group workplaces on the network helped communicate to the teacher community that students were expected to do collaborative work. Where interdisciplinary projects become a more common feature of the curriculum, the workplaces give the students a clearer group identity or sense of project continuity and thus help in the classroom management. Instead of greater centralized control of individualized instruction, as is common in integrated learning systems, control can be distributed to the students.

The Earth Lab network made no attempt to provide a technological solution to the problem of assessing student progress or grading student projects which is the central function of integrated learning systems afforded by the hierarchical nature of the courseware. We have, however, begun exploring the use of group and individual workplaces as portfolios of student work. The notion of a portfolio is receiving growing attention among educators as an alternative means of assessment in which the stages of work on a project collected in a portfolio can provide insight to both the teacher and the student about the state of their work, and in retrospect about the process of learning (Gardner, in press). The workplaces currently serve as archives of the group or individual project work and so can serve the function of a portfolio in this sense. Assessment systems such as TextBrowser (Kurland, 1991) which serve a data-base and analysis function for the teacher's own evaluations of the students work (their writing in the case of TextBrowser) provide a division of labor between teacher and technology that is appropriate for project based school work.

Time. The workplaces provide continuity over time as well as location. Projects involving collecting weather data and data on seasonal change extended over many months. In some cases, projects may extend over years as new cohorts of students move through the school curriculum.

Figure 3 shows a portion of a personal workplace of one sixth grade student as it appeared during the first week of school. Notice that her writing and other data from fifth grade is still available to her, providing the student with a powerful sense of the continuity of her own work that transcends the boundaries between grades.

The use of tool software requires a greater initial investment in order to bring students up to speed with the technology than is required for more traditional CAI programs which present small tasks and simple interactions with the technology. However, the availability of the Earth Lab system to students over a period of years and the consistency of the available tools has made it increasingly easy for teachers to introduce long-term projects as part of their curriculum. In the first year of operation, the sixth grade class spent several months on fairly simple introductory projects designed to familiarize them with the word processing, database and communication tools. Several years later, teachers are able to start immediately with substantial projects.

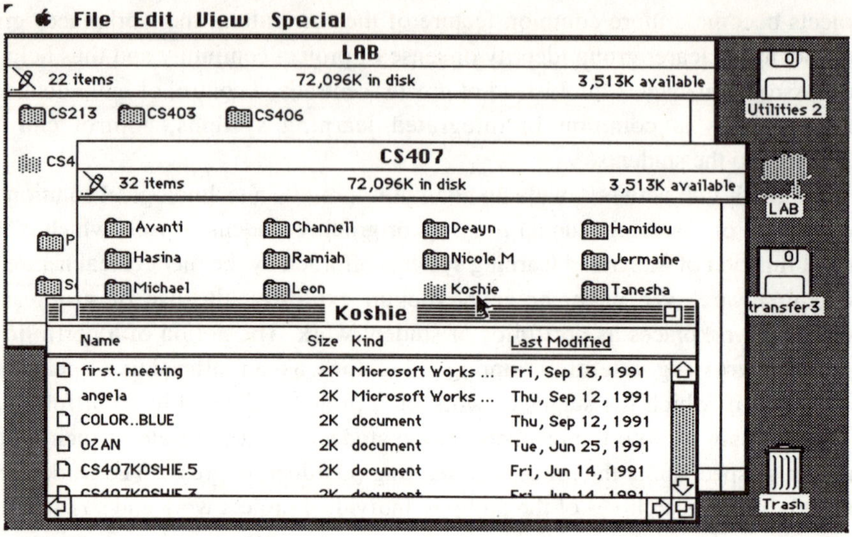

Fig. 3. Example of personal workplace of a sixth grade student

4. Principles for Design of a Project-Based School

The system that is in use at the Ralph Bunche School makes use of off-the-shelf technology and is not at all optimized for the task of organizational change that is called for by the school's new structure and goals. As we set out to provide very many schools with network systems and to create a whole new set of schools on the model of the Co-NECT schools, we have to start from scratch with some basic notions about school technology.

Students and teachers take increasing responsibility. The fundamental notion is that students as apprentices are expected to continuously take ever greater responsibility for their work, that is their projects, and the learning that is necessary for the projects and expected from the projects. Traditional ILS systems take exactly the opposite approach in centralizing all management and scheduling functions that are controlled by a system administrator. Teachers are not even expected to have any role in system management. In other systems designed for research or business environments, central control of the server systems is almost always assumed. We will want to distribute all management functions. There will be some system functions managed centrally, but teachers will have responsibility for such things as setting up work groups, creating new email users, providing software tools to individuals and work groups, and so on. We expect that many teachers will take on these responsibilities only after several years experience with the system. Until that time, a system manager may have to provide administrative support.

Students will also be able to perform many of these functions. Young students, or students who are just starting with the system may be provided with a ready made environment. Over time the expectation will be that they will take over many of the management and scheduling functions. Learning to use the technology environment as a tool for carrying out their individual and group projects is a central part of learning to do school work.

Our current design for what we are calling the Copernicus server calls for a database of people and groups containing a large amount of ancillary information that supports the work. The interface to this database will both act on the database and on the file systems that support the projects. The central administrator, the teachers, and the students will each have different levels of access to make modifications to the database. The goal will be to gradually increase each person's level of responsibility for modifying the computing environment.

We do not believe that a school environment in which students work on projects will be manageable unless students take a significant level of responsibility. The idea is not to create a more elaborate and detailed central management but to distribute the management to the students. That gradual change in responsibility is a major goal of the educational environment.

Workplaces are the unit of work and communication. We are elaborating the concept of a workplace that was illustrated in the discussion of the file server used in the Ralph Bunche School. In the Copernicus environment, we assume that students may be working on many different tasks at many different computers in the school as well as at home. Once they identify themselves to a networked computer, their work environment will be recreated such that workplaces that were open last time they were at the computer will be re-opened. Other personalizing effects will be recreated so that the computer becomes a familiar environment. When they choose a workplace their files and other work products are available. Importantly, application programs that are run from within the workplace context will be take their settings from that context so that each project or work situation is independently configurable.

The sensitivity of applications to the workplace context can be extended to communication applications such as email and bulletin board access. In most email system, a "user" is the unit of communication. It is a user that sends and receives email. In Copernicus, workplaces can also have that function. Each person will have a workplace. But so will groups of people who form work groups to tackle projects. With an email application taking its settings from the workplace, the workplace itself can send and receive email. The rationale for extending context-sensitivity to communication applications is that very often work groups will work together around a single computer and will be responsible for creating a single product. Take for example a newspaper as a class project. The newspaper will be assembled in the workplace. Students will be able to submit articles to the workplace and the editorial group, as a collaborative will be responsible for responding to the incoming information. Since students will also have personal workplaces, they will also be able to communicate as individuals.

Groups of workplaces could form email mailing lists just as groups of users can be addressed in conventional systems. The teacher could address all the project groups working on a certain topic, for example just as she would be able to address all the individual members of an after school club.

The workplace will most often exist on a file server but in some cases may have its home on an individual machine (e.g., a notebook computer) carried around by the student. We do not expect that students will have individual machines for quite some years but as that occurs, individual workplaces can reside there. Group workplaces will still have to be located on some common ground available to any member of the group or to the group when it begins to work on any computer. The Copernicus server can provide access to these workplaces from any machine on the school LAN.

Connecting the LAN and WAN. Many projects will involve access to resources and other peers outside the school as well as community access to the school. We note that a few schools are now beginning to take advantage of the packet-switched nature of the Internet to make network connections between their PCs and the network. As LANs become increasingly used for communication and as the WANs of interest to schools are distributed packet switched networks, the two technologies begin to converge.

There are two advantages arising from this convergence. First, with the LAN connected to the Internet, a student or teacher can sit down at any computer to call up a mail server or other remote resources. A second advantage of the convergence is the integration of the school community with widely distributed communities. With a mail server on the local school network, local and distant communications are supported by the same system from the students' and teachers' points of view. In these cases, we find local communications supporting local project groups that are accessing remote resources. Teachers use the local communications to coordinate their own work. The LAN is now a medium for supporting communication and sharing information among the local community, not just a means of access to the wide area resources.

5. Conclusion

In the U.S. there is growing dissatisfaction, even embarrassment, with the current schools as a grossly inadequate preparation for the information age. It is believed that American competitiveness cannot be reestablished if graduates enter the workforce with no sense of what work is and with attitudes that devalue solving problems and contributing to organizational change. Of course, the workplace will have to become more accepting of change to take advantage of a new crop of students, that is, both have to change simultaneously. At BBN we are concerned with the school side. We believe there is a contribution that technology can make if we take care to design it specifically for the task.

The key task is to put in place a system that supports and invites change. Just as we do not expect students to come to school with the skills and knowledge that they are there to learn, we cannot expect teachers and other school personnel to have in place the whole organization that is necessary to provide the conditions that will allow students to move through a productive apprenticeship. The technology system has to expect that teachers will be slow to appropriate it to the new instructional tasks such as projects. Teachers, themselves, must learn new approaches to their work and new ways of organizing the school. The technology has to be usable from the beginning but continually invite change toward giving students greater responsibility. In other words we have to design for the simultaneous change in two systems: the students and the teachers.

References

Gardner, H. (in press). Assessment in context: The alternative to standardized testing. In B.R. Gifford & M.C. O'Connor (eds.) Future assessments: Changing views of aptitude, achievement and instruction. Boston: Kluwer Academic Publishers

Kurland, D.M. (1991). TextBrowser: A computer-based tool for managing, analyzing, and assessing student writing portfolios. Paper presented at the annual meeting of the American Educational Research Association, Chicago

Newman, D. (1990). Opportunities for research on the organizational impact of school computers. Educational Researcher 19(3), 8-13

Newman, D., Bernstein, S., & Reese, P.A. (1992). Local infrastructures for school networking: Current models and prospects. (BBN Report No. 7726). Cambridge, MA: Bolt Beranek & Newman

Newman, D., Goldman, S.V., Brienne, D., Jackson, I., & Magzamen, S. (1989). Peer collaboration in computer-mediated science investigations. J. Educational Computing Research 5(2), 151-166

Schofield, J., Evans-Rhodes, D. & Huber, B.R. (1990). Artificial intelligence in the classroom: The impact of a computer-based tutor on teachers and students. Social Science Computer Review 8(1)

Sherry, M. (Ed.) (1990). The integrated instructional systems report. Water Mill, NY: EPIE Institute

Intelligent Learning by Doing Tools for Technical and Dialectical Knowledge

Violetta Cavalli-Sforza and Alan Lesgold

Learning Research and Development Center, University of Pittsburgh, PA 15260, USA

Abstract. New members entering productive organizations require considerable training. Computer tools can support such training by providing an opportunity to learn while engaging in authentic activities and receiving appropriate coaching. We describe two tools that incorporate this approach. Sherlock, an existing computer coach, is an effective environment for learning how to troubleshoot complex electronic devices. A newer research effort focuses on tools for supporting knowledge-building argumentation and scientific theory evaluation in post-elementary school science education. Both tools offer users opportunities for reflecting on their own performance and support individual as well as collaborative learning.

Keywords. Learning by doing, collaborative learning, intelligent tutoring systems, computer coaching, electronics troubleshooting tutors, argumentation environments

1. Introduction

Productive organizations are complex cultures into which new members continually enter. These new members require considerable training, since the roles within productive organizations are usually quite complex. At the Learning Research and Development Center, we have been developing one kind of tool for such training for several years, now. This is an intelligent coached apprenticeship environment for learning by doing electronic troubleshooting, called Sherlock. We have also begun to consider how tools might be built to train some of the interactive skills involved in group cognitive activity, especially dialectical knowledge-building (epistemic) activity. We feel that the complexity of modern productive organizations regularly requires that teams address complex arguments involving different people's expertise, multiple standards of validity, and difficulties of consistency and coherence.

In the first part of this chapter, we will briefly describe the current Sherlock environment and consider a number of ways in which it can be extended to support collaborative learning. Several features of Sherlock's design and existing components of the system make it feasible to contemplate an environment in which system-student and student-student teams can work collaboratively toward solving a problem and understanding how their solutions match or differ from an expert's solution to the same problem.

The second part of the chapter is devoted to our newer project, a computer environment to assist students in practicing collaborative epistemic argumentation by arguing with each other and, in restricted domains, with the computer. We began with the idea of constructing tools to support students as they discuss policy issues for the purpose of acquiring a broader and more informed perspective on these issues. While we intend to return to policy arguments in the near future, our research currently focuses on argumentation and argumentation-supporting tools in science instruction. We aim to expose students to a view of science in which knowledge is not simply proposed and accepted, but rather debated and scrutinized, requiring the proponents of new theories to provide certain kinds of support for their proposals and to be aware of the context of beliefs and practices in the domain.

Sherlock and the argumentation project have more in common than meets the eye. Both subscribe to the learning-by-doing instructional approach: task-specific as well as more general strategic knowledge are learned while working on a substantial and realistic problem. Both systems incorporate aspects of the cognitive apprenticeship approach (Collins et al., 1989), with the system providing progressively more support if the student requests it. Both projects aim at engaging the student in authentic activity (Brown et al., 1989) and at providing concrete reifications of cognitive activity, permitting later reflection on one's performance. Finally, in both systems, there is a movement toward addressing the importance for students of learning to construct explanations (Chi et al., 1989, Chi & Vanlehn, in press).

2. Sherlock: Coached Practice for Electronics Troubleshooting

The original goal of the Sherlock project was to develop a computer environment for training avionics technicians to troubleshoot a complex electronic testing device, the F-15 manual avionics test station, which is used for diagnosing aircraft navigation components. Ordinarily, when a navigation module from an aircraft is brought into the repair shop because of suspected malfunction, the technician attaches the module ("unit under test" or UUT) to a test station, and, by carrying out a series of test procedures, is able to locate the fault. However, sometimes the airman discovers that he continues to get an unexpected reading even when he replaces the UUT with a shop standard. When this occurs, the problem is not with the UUT; rather, the test station itself is malfunctioning and needs to be repaired, a much greater challenge than troubleshooting a suspect UUT. Now, instead of following a set of documented procedures with the substantial support provided by the test station, the technician is on his own and must engage in complex heuristic problem-solving (the test station contains about 1.5m^3 of circuitry) with the aid of minimal tools. The essential task confronting the airman is to construct a mental representation of the active circuitry - the active circuit path - and to search this path until he locates the fault. Sherlock's job is to scaffold the process of learning to construct these abstract representations and developing effective troubleshooting strategies.

We began with the concept of coached apprenticeship - learning by doing. We wanted trainees to learn by tackling, in a simulation environment, the hardest parts of the job for which they were being trained, receiving help from a computer coach when necessary. The resulting intelligent computer-based training system,

Sherlock I (Lajoie & Lesgold, 1991; Lesgold, in press; Lesgold, Lajoie, et al., 1992), was quite successful. Independent field tests by the Air Force showed that 20-25 hours of training on the system was roughly equivalent to four years of on-the-job experience (Nichols et al., in press), and retention after six months was about 90%. However, we realized that conceptual learning and the abstractions needed to support transfer probably could not occur while the trainee was overloaded with the problem-solving task, but that experience with the task could undergird that learning. So, in Sherlock II, we added opportunities for trainees to review their performance, have it criticized by the computer, compare their actions to an expert's, and receive explanations of the expert actions.

2.1 Sherlock II: The Environment

Sherlock II is a realistic computer simulation of the actual job environment.

Trainees acquire and practice skills in a context similar to the real situation in which they will be used, interacting with video displays of the test station control panels and with schematics of its circuitry and components. Sherlock presents students with problems chosen from its curriculum on the basis of the student's assessed competence and, optionally, on input from the student himself. It uses its expertise to coach the student during the problem solution process and offers feedback on the student's solution in the review phase that follows problem-solving activity. Sherlock II has a strong object orientation[1]. An example is provided by the organization of domain expertise. This is represented by a multi-level model of the test station: there are circuit-level objects (the active circuit path) whose definitions are based partly on component-level objects (electronic components within the active path). Problem-solving knowledge is distributed throughout the objects in this model. Each object has an associated goal structure for deciding whether that object is faulty, and it has strategies for achieving the subgoals of that goal structure. For example, a circuit-level goal is to verify the signal path between the aircraft unit and the measurement device. One strategy for achieving this goal is to perform tests each of which splits the space of remaining fault locations in half, ruling out one half or the other. We call this strategy "space-splitting." At the component level, the testing goal structure is more specific to particular functions, but there is still considerable generality among the component objects. A component-level goal is, for example, to verify the outputs of the component. This can be achieved by carrying out a sequence of measurements.

Circuit-level subgoals translate into sequences of component-level goals that must be satisfied. The object-oriented design of Sherlock has several advantages from the perspective of conceptual design, reusability, as well as software management.

2.2 How a Problem-Solving Episode Works

The fundamental activity within the Sherlock tutor has two parts. Trainees first solve a problem, receiving various types of assistance from Sherlock. Because learning from task situations requires significant cognitive effort (Owen & Sweller, 1985; Sweller, 1988; Sweller & Cooper 1985), some of the instruction is parceled out to a period of reflective 'review' that immediately follows the problem-solving activity and in which the student has an opportunity to evaluate his solution and compare it to an expert's solution for the same problem.

2.2.1 Problem-Solving

When the student has determined that the test station, and not the UUT, is faulty, he starts troubleshooting the test station. At this point, he has a number of options: He can test components by attaching probes to pins on a video display of an extender board, replace a suspect component with a shop standard and rerun the

[1] An object is a self-contained collection of procedures and data. The procedures respond to messages received from other objects. Object-oriented programming consists of designing collections of objects and orchestrating the messages that can pass between them.

failed checkout test, etc. The student can ask for help at any point while troubleshooting. Sherlock provides help at both the circuit and component levels.

Within each level, Sherlock offers functional/conceptual ("how it works") hints, and strategic/procedural ("how to test") hints. Coaching becomes increasingly directive if the student asks for repeated help with a particular aspect of a problem, and it is guided by the system's model of the student (Lesgold, Eggan, et al., 1992]. More directive hints recommend a goal to achieve next, and - at the furthest extreme - information about how to achieve the next goal (e.g., by specifying which pins to test, on a particular component).

In addition to providing advice, both tactical/procedural and strategic, on where to look and what to test, Sherlock II also provides advice on how to trace through schematics to develop a representation of the circuit path on which the fault being diagnosed must lie. The biggest advance has been in graphical support for coaching and explanations. At certain stages of circuit-level hinting, the student is shown a diagram of the active circuit path (Fig. 1). We have developed an intelligent schematic display routine that configures drawings on the computer screen to match the coach's current explanatory goal, the current problem-solving context, and information about the trainee. For example, more space and more

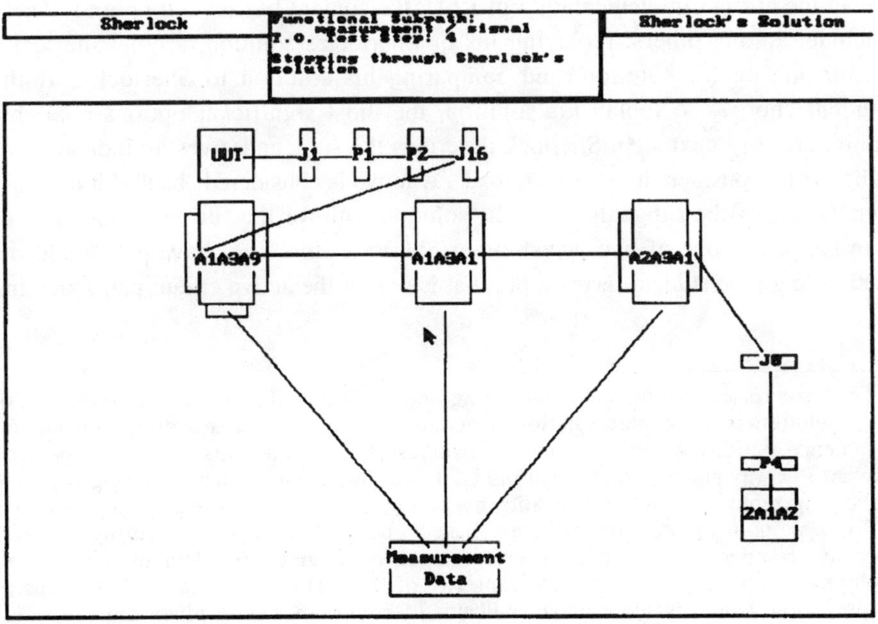

Fig. 1. Example of circuit path display

expanded circuit detail is provided in the part of the circuit on which an expert might now be focusing or the part the coach wishes to emphasize. Labeling and color coding are used to show what is known about circuit components and information paths from tests already completed and are filtered to assure that diagrams don't unintentionally give away too much information. Overall, we believe we now are doing a pretty good job of realizing our strategy of hint inertia, that is, trying to maximize the likelihood that a trainee solves an impasse by receiving almost enough coaching and then inferring the specific knowledge needed for the situation at hand.

2.2.2 Reflective Follow-Up

While the student is troubleshooting, Sherlock creates a record of the actions of thestudent, the goals he has achieved, and how he achieved them, at both levels of analysis (circuit and component). This goal/test record is the crucial input to the second part of a Sherlock II exercise: what we call reflective follow-up (RFU).

The goal/test record contains a "trace" of the student's solution. Sherlock also constructs its own "expert" solution to the problem, using the goal and strategy knowledge in its device model. Both expert and student solutions are scored before entering RFU and Sherlock updates the student model[2].

In the current implementation of RFU, the student has several options. These include, among others: replaying his or Sherlock's solution, seeing Sherlock's comments on his solution, and comparing his solution to Sherlock's. If the student chooses to replay his solution, the most significant options available tohim are: My next step. Sherlock describes the step, and gives an indication of whether he considers it "good" or "bad". A move is considered "bad" if it does not satisfy one of the subgoals along the solution path for the current problem. This can happen in one of two ways: either Sherlock would not have performed the action (e.g., the student tested a pin that is not on the active circuit path, yielding

[2] The basis for evaluating and commenting on a trainee's solution and for comparing of that solution to an expert solution generated by Sherlock is a scoring scheme for problem solutions within the manual avionics job domain. This was developed, by Robert Pokorny and Sherrie Gott of the U. S. Air Force, via a policy capturing approach involving ranking of solution quality by senior experts and subsequent justifications of those rankings. A point scheme, based on the Pokorny-Gott scoring scheme, permits comparing the student's performance on a given problem to the expert's solution, taking into account the complexity of the problem. The ratio of the expert's score to the student's score as an indicator of competence on a given problem. This ratio can be used (a) to decide on progress within the curriculum; (b) to decide on progress toward expertise; and (c) for special purposes during cross training (taking someone from a different job and training them in Sherlock's job). For more details see (Pokorny et al., in press).

a non-informative measurement), or the action was performed incorrectly (e.g., the student measured resistance instead of voltage.)

What would Sherlock have done? Sherlock tells the student which test it would have performed on the current card or indicates which card it would have tested, thereby enabling students to compare their performance with expert performance. What would Sherlock do next? Sherlock tells the student which test it would perform next on the current component, or indicates what component it would test next.

As steps are replayed, Sherlock color-codes the circuit diagram (see) to indicate what is known about the status of each component. Those components which have been verified as working are shown in green, for example, and pathways through which incorrect data is flowing (e.g., wrong voltages or resistances) are shown in red (black indicates no information yet obtained, and yellow indicates insufficient information to decide whether a component or data path is good or bad). RFU thus presents an Abstracted replay of the solution, using both verbal and visual displays, allowing students to see which components are considered part of the active circuit path for a problem and how their measurements constrain what can be concluded about the components of this path.

If a student notices that Sherlock was able to draw a conclusion that he disagrees with or did not draw, he can ask Sherlock for an explanation by "mousing" on the component in the diagram.

In addition to the trace of his actions, the student also can examine a scoreboard to see how he is doing on particular troubleshooting skills, according to the current state of the student model; he also can help Sherlock determine the next problem.

2.3 Opportunities for Collaboration in Sherlock II

We expect Sherlock II to be even more successful than Sherlock I, which had only a fraction of the new version's capabilities. However, as we have worked on Sherlock II, we have come to realize that, unlike many of the schemes used in training people for important jobs, it treats learning as a human activity performed alone, or at most with a tutor. Educational psychologists often mention motivational aspects of the collaborative side of education, but only recently has American psychology begun to realize the fundamental importance of social processes in bringing about learning. To some extent, Vygotskii anticipated this in his view that learning is the internalization of knowledge that is initially present in social interactions to which the student is a party (Vygotskii, 1978).

Collaboration in problem solving models skills of learning that may be quite useful when internalized by a solo learner. When two people bring different viewpoints to a task, they may achieve greater insights from trying to reconcile the two positions or at least to understand how they could both be valid (though perhaps not simultaneously). With practice, this dialectical activity may become internalized, and the participants may learn to search spontaneously for different viewpoints as a way of broadening their knowledge. Collaborative activities might be especially helpful if they teach participants to examine the problem situation from multiple viewpoints and then try to reconcile those viewpoints.

2.3.1 Directions for Collaborative Learning

We have barely begun work on collaborative learning in Sherlock, but we see three kinds of socially mediated learning activities in which our electronics diagnosis trainees might participate. First, they can simply work together to solve problems, talking to each other and manipulating a simulated task environment on the computer. In addition to affording opportunities to negotiate understanding and to compare alternative problem-solving procedures, this might also might reduce the mental overload that arises during the solution of difficult problems and thus remove an impediment to learning from practice.

Second, they can pose problems for each other. When one person thinks he has learned something important, he can test his knowledge either by using it to solve problems or by selecting problems that he thinks involve the new knowledge and posing them for a partner. By observing the partner's performance and, if necessary, using his new knowledge to give advice to the partner, a person can test and tune his new knowledge. Further, if the problem posed turns out not to exercise the new understanding, this itself helps in understanding the boundaries on the new knowledge. There is, of course, also a motivational side to this game of two people posing tasks to each other.

A third form of socially mediated learning is peer critique of problem solutions. We are now designing schemes whereby a trainee can critique the problem solution process exhibited by another trainee. The tools for doing this exist already, for the most part, since Sherlock can already critique a trainee's solution. What needs to be added is a facility that permits one trainee to generate a critique for another trainee in a sufficiently standard form. This is a problem on the same scale as the interface tools needed to permit a trainee to work in a simulated problem-solving environment at all, so we believe it is quite attainable.

2.3.2 Extensions for Collaborative Activity

Katz (Katz & Lesgold, in press) developed a preliminary scheme that demonstrates the possibilities for collaborative activity using Sherlock II with relatively small enhancements. The two episodes of a Sherlock exercise, problem-solving and reflective follow-up, correspond to the dual roles that students can play during collaborative sessions - producer and critic, using Collins et al.'s (1989) terminology. Computer-based collaboration can take place during both phases. The nature of the task (problem-solving or review) and who the collaborators are (a student and the tutor; or, a group of students) provide two dimensions along which to categorize the collaborative activities we plan to incorporate within Sherlock II.

Although there are collaborative opportunities in the problem-solving phase, we have focused our design efforts on collaborative post-problem reflection activities for two reasons. First, even though there have been several suggestions for on-line reciprocal critiquing (Brown, 1985; Collins & Brown, 1988; Collins et al., 1989; Scardamalia et al., 1989) little is known about how a system can support students in carrying out this task; research suggests that this kind of support is necessary (Chi et al., 1989), since some students cannot or do not generate explanations from which they or their peers can learn. Second, research also suggests (Collins et al., 1989; Palincsar & Brown, 1984; Schoenfeld 1983) that building students' explanation skills may result in more effective peer collaboration and improvements in self-monitoring skills, as well as in task performance gains. Whether explicit training in explanation will indeed induce these effects, however, remains an empirical question.

In teaching the learning strategy of performance review, we plan to follow the three phases of traditional apprenticeships identified by Collins et al. (1989): modeling, coaching, and fading. During the modeling phase, students will solve problems and go through reflective follow-up as described above, examining Sherlock's evaluations, comments, and explanations. This will allow students to get used to reflective follow-up, and see a number of examples of good explanations. Then, students will work through another few problems, this time reviewing their solution themselves, but with support from Sherlock. In one scheme we have in mind, Sherlock will flag steps as "good" or "bad", but the student (rather than Sherlock) will have to articulate why a move was marked "bad". Sherlock will give students hints about how they can find out what they did wrong, and will scaffold their formulation of explanations by providing a vocabulary for review, via commenting menus. We expect that by actively using this vocabulary, students will come to internalize both the vocabulary and a set of criteria for review. After reviewing a few of their own solutions, students will review peer solutions. Coaching will be there when students need it and fade when students can review problems on their own.

2.3.3 A Concrete Scenario

To make our ideas more concrete, here is a scenario in which a student works with Sherlock to review another student's solution. A student (the "producer") would first would solve a problem and then, instead of reviewing the solution himself or having Sherlock review it, he would choose to have another student do it, selecting the "critic" himself or asking Sherlock to find a suitable one for him[3]. The student selected to be critic would perform the review the next time he accesses the sherlock system. Once in review mode, he would have similar options as in the current post-problem reflection mode, with a few additions. Descriptions of the producer's steps would be flagged as "good" or "bad" by Sherlock, or the critic would need to diagnose moves himself. In either case, the critic would be instructed to explain why each "bad" step is inappropriate. He would also need to recommend an alternative move, using the color-coded display to examine the goal status at that point in the solution, or to determine the goal status on his own.

If the critic knows what the producer did wrong, he can select the option Comment on Last Move from the menu, which expands into a layered sequence of submenus, each layer prompting the student for more specific information than the previous one. Figure 2 provides a simplified idea of what we have in mind. The first submenu provides the main error category - i.e., a circuit-level strategic error, pertaining to the goal status of the functional area of the test station in which the current move was made (Functional Area); a component-level strategic error, pertaining to the student's decision to test a component (Choice of Component); a tactical error, pertaining to how the student carried out a measurement (Measurement-taking), etc. These options make the curriculum goals of Sherlock II explicit to the student. The second submenu prompts the student to choose between two general functions of an explanation: describing the error and suggesting an alternative (both may be chosen). The third level of options specifies errors and associated fixes for each of the categories in the first submenu.

The critic might supply even more specific information by using a final level or by demonstrating a correction on the appropriate video display.

If the critic could not determine why Sherlock flagged a move as "bad", he might have two forms of assistance available. He could select Help with Finding Error to receive progressive hints that mirror the increasingly directive structure of the commenting menus, essentially seeing progressively more detail of the comment Sherlock would have generated had it been given this responsibility.

Alternatively, the student might select Experiment with Alternative Moves and temporarily be shifted into problem-solving mode, at the point of the solution trace immediately preceding the move he is trying to criticize. There, he could assume the original problem solver's role and try to understand what was wrong

[3] As "manager" of the review process, Sherlock will try to select as the critic a student who is not already too much in demand, and who is at least at the first student's ability level, according to their student models. There is also the intriguing possibility of having a set of simulated "learning companions" with different strengths and weaknesses in their troubleshooting skills.

Comment on Last Move		
Functional Area	Describe Error	UUT not Verified First Stimulus not Verified Measurement Signal not Verified
	Suggest Fix	Verify UUT First Verify Stimulus Verify Measurement Signal Verify Load
Choice of Component	Describe Error	Tested Component Earlier Cleared by Previous T.O. Test Verified by Testing Other Component Not Central in Unverified Path
	Suggest Fix	Specify Test Component
Measurement-taking	Describe Error	Probes Reversed Wrong Pin for High Wrong Pin for Low Meter not Set to Ohms Meter not Set to DC Oscilloscope not Set to DC Oscilloscope not Set to AC
	Suggest Fix	Reverse Probes Correct Pin for High Correct Pin for Low Set Meter to Ohms Set Meter to DC Set Oscilloscope to DC Set Oscilloscope to AC

Fig. 2. Preliminary design of comment menus

will be preferable to simply being told (or given hints about) appropriate critical comments. If, by using the Experiment with Alternative Moves option, the critic with the original move by trying to discover a better move. This latter approach were to discover a different way of solving it than the original student solver generated, he could record this alternative solution, and the original student would be able to replay it.

Sherlock would record the critic's corrections and, when the critic selected the Register Comment option, would assemble the pieces of the student's menu selections into a coherent English critique, to be displayed in a separate window[4]. We might also ask the student critic to rate the original problem solver on specific skills, and on overall troubleshooting proficiency. This might prompt useful interactions among students about each other's critical views.

2.4 Status of Sherlock II

Sherlock II is in final development. A full release of the system is scheduled for March of 1993. Feedback on intermediate products has been encouraging and has given much useful input into the design of the system and of the reflective follow-up phase in particular. Work on extensions for collaborative activities is still in the design stages. We expect that future research and development of extensions for collaborative learning will benefit and be benefited by our other project on tools for supporting dialectical activity, to which we now turn.

3 Tools for Supporting Dialectical Epistemic Activity Argumentation is an important part of many intellectual activities. However, the term argumentation connotes several different types of interaction, ranging from the adversarial exchanges of law courts or college debates, to the persuasive discourse of advertisements or negotiations, to the reasoned exposition and dialogue of public science. We focus our attention on what we call knowledge-building (epistemic) collaborative argumentation, that is, argumentation as critical discussion and dialectical inquiry in which arguers collaborate in resolving or gaining greater understanding of an issue, in explaining a natural phenomenon, or in understanding how some artifact works or fails to work. Participants in collaborative arguments are neither trying to win nor attempting to persuade by any means available. Collaborative argumentation occurs in some tutoring interactions (Cavalli-Sforza & More, 1992), in group design or decision-making (Coklin & Begeman, 1988; Lee & Lai, 1991; Mac Lean et al., 1991), and in high-productivity team work. It helps to uncover both the common ground of different viewpoints and the areas of fundamental disagreement, suggesting starting points for building consensus and areas in need of further investigation.

3.1 Supporting Collaborative Argumentation

Arguing in ways that make progress toward achieving collaborative argumentation goals is a skill that needs to be learned. Studies conducted by members of our group (Keith et al., 1989; Keith et al., 1990; Lesgold, 1989) and other researchers

[4] The critique will be constructed using a model of explanation and an algorithm for generating explanations being developed by our colleague Johanna Moore, extending the research she did on her dissertation (Moore & Paris, 1989).

(Kuhn, 1991; Perkins, 1985; Voss et al., 1983) have shown that novice arguers have difficulty generating and responding to arguments and handling the complexity of extended arguments. Apart from weaknesses in argument analysis, generation, and evaluation skills, some arguers have perceptions of argumentation that may be well grounded in their personal and social experience but are inappropriate for epistemic argumentation. It is not surprising to find shortcomings in argumentation skills among the subjects in these studies, many of whom were high-school and college. Secondary education provides few opportunities for practicing argumentation in general, and collaborative argumentation in particular. Where opportunities do exist - commonly in the form of essay writing, informal logic courses, and debate teams - they promote too narrow a view of argument (too often as a contest of people or ideas rather than potentially an integrative activity) and may be responsible for some of the behavior we have found in our studies.

The weaknesses exhibited by subjects in our studies suggest that novice arguers could benefit from a computer environment that reifies the structure of an ongoing argument and suggests ways of contributing to it. The environment needs to provide the means for visualizing the overall argument structure, inspecting its components at different levels of detail, and displaying the extent of support received by the different positions in the argument as arguments and counterarguments are advanced. In order to assist the user in constructing and responding to arguments, the system must know how to construct different types of arguments and which types of dialectical moves are acceptable or desirable in response to arguments advanced. In sum, the system needs to be able to act as an expert arguer. While some arguing expertise varies with each domain of argumentation, it is possible to identify a core body of argument knowledge that serves collaborative argumentation and is shared across domains. In doing this, we were led to a conceptualize argument as having three levels of representation, each of which addresses different concerns: the MICRO, DIALECTICAL, and EVALUATIVE levels (Cavalli-Sforza, 1991)[5].

3.1.1 The MICRO Level

The most detailed level of argument representation focuses on individual inference steps (microarguments) in a reasoning chain. We use a model of argument loosely based on the argument analysis framework introduced by the philosopher Toulmin (Toulmin, 1985; Toulmin et al., 1984), in which claims are supported by grounds or data via the warrant, a linking principle whose authority is established by field-

[5] The names of the 3 levels of representation differ from those used in (Cavalli-Sforza, 1991). We feel these names better reflect the aspect of argumentation each level is focusing on.

dependent means (Fig. 3). We are presently developing a computational representation of microarguments that associates descriptions of sound argumentation and common fallacies with different types of microarguments.

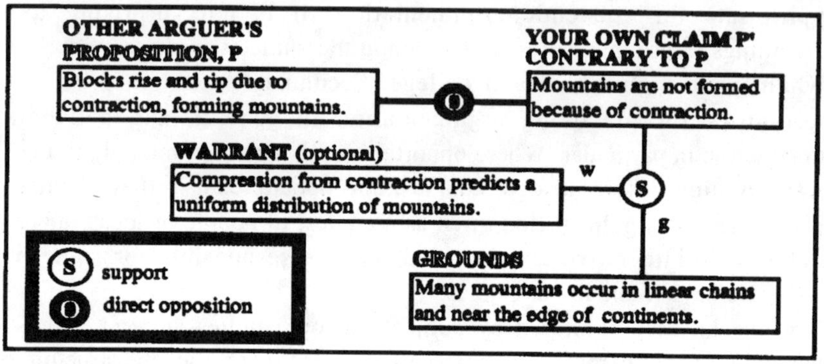

Fig. 3. A counterargument template, with microargument and counterclaim

Arguments and argument chains displayed in full detail at the MICRO Level usually give rise to complex graphical displays. Once the microarguments have been examined, much of this detail can collapsed and hidden from view, leaving behind display objects that serve both as reminders of their content and as handles into it, should the details need to be brought into full view again. Figure 4 shows one way in which the display can be simplified.

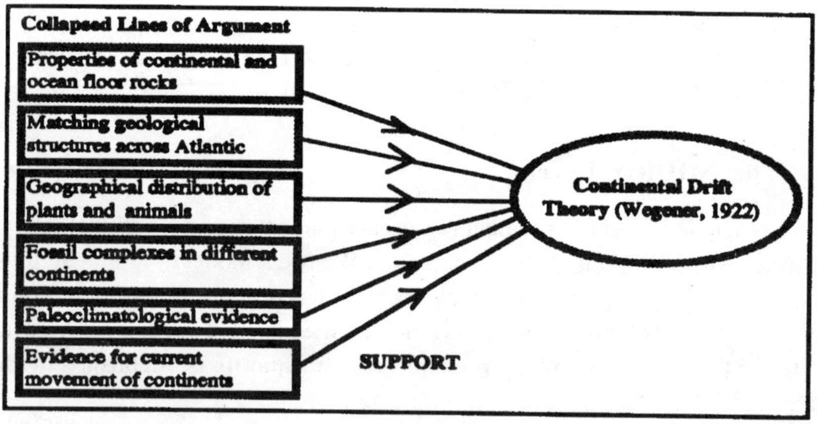

Fig. 4. Summarizing a complex argument by line of argument

3.1.2 The DIALECTICAL Level

The second level of argument representation addresses the structure of dialectical arguments. An argument, represented internally and displayed as a graph, includes support for a claim, opposition of competing claims, and attacks on the reasoning chains adduced in support of claims (Fig. 5). Help in contributing to an argument comes from a taxonomy of argument moves and a set of tactics and strategies for choosing possible moves. Argument moves (e.g., counterargue, support, etc.) are derived from the MICRO Level. Some actions, for example the counterargument shown in Fig. 3, are performed by filling out a graphicaltemplate (as in (Streitz et al., 1989)) that details the necessary components and the criteria they must individually and jointly satisfy. Students can also mouse on parts of the template to get more information on the terms used in the template descriptions.

Tactics suggest possible ways of responding to an action by the other arguer. A different set of tactics is provided for each argument action. For example, the tactics for responding to the action "support CLAIM i" include:

- counterargue CLAIM i
- attack-support SUPPORT i, by one of:
  - challenge GROUNDS i.1 through i.n, by one of:
    - counterargue one of GROUNDS i.1 through n
    - request-support-for one of GROUNDS i.1 through i.n
  - challenge WARRANT i, in one of several ways
  - attack-warrant-use for WARRANT i

Strategies are a set of preferences and rules that suggest, censor, or applaud particular actions or responses. A few of the ones we use are summarized in Table 1. Tactics and strategies are intended to maintain continuity (focus) in the discussion, to avoid inefficient maneuvers, to take actions that are most effective in achieving acceptance for each position, and to manage inconsistent argumentation. Some strategies (the Good Conduct Rules) are also designed to prevent the arguers from evading the responsibility of justifying contested positions and accepting positions they do not favor but are unable to contest. We have used these strategies to simulate dialectical arguments regarding the disposition of the West Bank (Cavalli-Sforza, Lesgold, & Weiner, 1992).

3.1.3 The EVALUATIVE Level

DIALECTICAL level tactics and strategies also draw from the third, or EVALUATIVE, level of argument representation. Here, each proposition in the argument is viewed as having an associated level of support, which is determined by whether it has been justified, accepted without justification, or contested by the

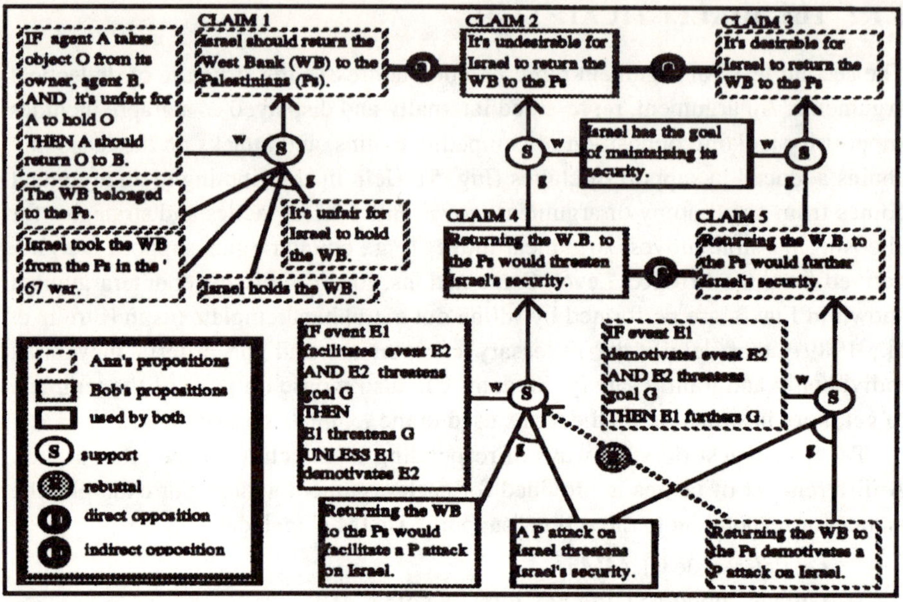

Fig. 5. Graphical representation of an extended dialectical argument

other arguer. Level of support for a claim is calculated as a function of the support that flows to it through arguments in its favor and the support that flows away from it due to competing claims and arguments in their favor. The ultimate source of support for a position can be either factual evidence or beliefs held by the arguers. The student can manipulate the strength of support associated with one or more propositions (often evidence or beliefs) and see the effect of such manipulations on the strength of support for other propositions. Thus, in addition to showing the extent of support received by different positions as the argument develops (see Fig. 6 below), this level provides a way of representing strongly held viewpoints (collections of propositions or beliefs) on an issue. It also helps to highlight areas of agreement, from which arguers can build consensus, and areas of disagreement, on which they can focus further inquiry.

3.2 Applying Argumentation to Science Teaching

In our current research, we are combining work on argumentation with tools for learning science. The goal of this research is create a computer environment in which students can learn about a scientific domain, and simultaneously learn about the kinds of reasoning and activities that are part of doing science and are

Table 1. Strategies for contributing to a collaborative argument

Strategies for Determining *Where to* Address a Response

- **Recency Preference**: preference for responding to the most recent proposal, question, or challenge.
- **Good Conduct Rules**: rules to prevent arguers from changing the topic or shifting attention to another part of the argument when at a loss for how to defend their position.
- **Level of Support (LOS) Heuristic**: a preference for supporting one's own weakest positions, if challenged, or weakening the other arguer's strongest positions that compete with one's own.

Strategies for Determining How to Respond

- **Eliminating Response Class+Target**: some actions are inappropriate no matter how they are carried out. So an arguer should avoid:
  - *Responses inconsistent with one's beliefs,* e.g., counterarguing a proposition she believes.
  - *Responses inconsistent with one's past actions,* e.g., challenging a proposition she used earlier in the argument.
  - *Unnecessary actions,* e.g., supporting or attacking propositions that have been conceded or abandoned.
- **Eliminating Response Instantiations**. Some actions are inappropriate if carried out in particular ways. An arguer should avoid:
  - *Responses inconsistent with one's beliefs,* e.g., relying on propositions she believes to be false.
  - *Responses inconsistent with one's past actions,* e.g., rely on propositions she attacked earlier in the argument, unless she conceded them.
  - *"Irrational" responses,* e.g., counterargue a claim unless she can attack all support provided for the claim.
  - *Dangerous responses,* e.g., responses on which, given her knowledge of the other arguer, she can expect an attack, unless she is can meet the challenge.
- **Selecting the Best Response**. One should prefer responses that:
  - Appeal to beliefs of the other arguer.
  - Help the arguer's position in more than one way.
  - Use stronger or more acceptable forms of support.

commonly not part of science courses. We focus on helping students develop the ability to articulate, evaluate, and criticize hypotheses offered as explanations of natural phenomena by examining and recreating the vicissitudes of a scientific theory from introduction to acceptance (Cavalli-Sforza et al., 1992), or by comparing different theories that have been proposed by real scientists or by their peers. Driving the design of this environment are three important ideas that relate science to argument and are largely missing in high school science education.

First, we want to convey the point that a great deal of scientific activity is justificatory, even in hypothesis generation. The linking of hypotheses to observations through sound explanation or argument is a key part of generating plausible hypotheses, as well as convincing others of their plausibility. In addition, there is accumulating evidence that elaboration of the material being learned, and especially explanation construction, increases retention of the material and performance on the task (Chi et al., 1989; Chi & Vanlehn, in press). We point out that, although argument and explanation are not identical, they have many similarities. An explanation of a statement can often serve as an argument for that statement (although not vice versa) (Cavalli-Sforza & More1992). In addition, to say that a scientific theory "explains" some observations is also to say that the observations support the theory.

A second point we want to get across to students is that an ongoing scientific controversy is a powerful determinant of scientific research. The arguments marshaled in support of or against a theory, and the arguments set forth in favor of competing theories, often shape the context within which scientists carry out their research activities. The goal of challenging a competitor's position or strengthening one's own suggests which new data to gather or how to re-interpret existing observations.

Third, students should understand that scientific knowledge is, in principle, at all times tentative and subject to revision. When new theories are introduced, they are likely to encounter critical examination and opposition and will often be dismissed; but there are times when the weight of the arguments in favor of a new theory will be sufficient to convince the majority of scientists to accept it.

Finally, we want to convey that actual scientific practice is collaborative. Not only do scientific projects often require team work from several specialists, but the scientists who argue against each other's positions in such an outwardly adversarial manner can be seen as working toward a common knowledge goal, with the criticisms of the one helping to refine and strengthen the reasoning of the other.

To summarize, a student of science should be aware of the importance of argument in science, should be well practiced in both generating cogent arguments and situating new work within the context of ongoing debates, and should be prepared to recognize that certain theories are better than others.

3.2.1 Resources for Scientific Argumentation

Learning argumentation in the context of science gives students on opportunity to practice their argumentation skills in a meaningful setting. However, textbooks and other instructional materials used in high school science courses are currently not designed to support this type of activity. Hence, in addition to general knowledge of argumentation, our environment will need to provide resources to support learning specifically about science and scientific reasoning.

One resource is a rich on-line information base of theories and hypotheses, data, and domain principles relevant to a scientific controversy[6], indexed by author and time at which they were advanced and further organized using argumentative and other rhetorical links. Examples include exemplification links relating principles or generalizations to specific exemplary observations, support/explanation links between evidence and hypotheses, and opposition links connecting competing hypotheses. Domain information will be accessible to students in the form of text (including excerpts from historical documents), tables, graphs, and bitmap images, with an underlying system-understandable representation[7]. This representation will allow the system to provide intelligent coaching in response to student requests for advice as the students construct arguments by "copying and pasting" the domain information into argument templates.

Another resource is a set of general strategies and criteria used by scientists in conducting research and evaluating hypotheses. For example, a general strategy for supporting one of two competing hypotheses is hypothesis splitting, i.e., searching for evidence that is predicted by the one but cannot be explained or even runs contrary to the predictions of the other. A common criterion in selecting among alternative explanations of a phenomenon is simplicity: In general, scientists prefer explanations that require fewer independent assumptions. Domain-specific and domain-independent scientific reasoning knowledge will be merged with knowledge of epistemic argumentation practices, such as the tactics and strategies mentioned above, to coach argumentation at different levels of domain specificity. We give an example of this type of interaction in the following section, where we describe how the progressive hinting strategies we already saw in Sherlock and knowledge of argumentation structures, tactics, and strategies can be employed to elicit scientific arguments from students.

3.2.2 Scenarios for System Use

Our science environment is intended to support a variety of uses. In one scenario, the student uses the system as an on-line textbook, freely browsing the data and hypotheses that were available to scientists at different times, examining the

[6] The scientific controversy we are currently working with is the emergence of the theory of plate tectonics, from the introduction and rejection of Alfred Wegener's theory of Continental Drift in the 1920s, through the evolution of ideas and availability of crucial new evidence during and after World War II, up to the acceptance of plate tectonics theory the 1970s.

[7] Eventually we want to add interactive graphics and simulations.

arguments linking them, exploring the structure of the debate by displaying it graphically and traveling along argument-related links (e.g., support, contradictory evidence, etc.), and viewing domain knowledge by traversing other rhetorical links (e.g., exemplification, elaboration, etc.). A variation on this scenario constrains the student to follow the historical evolution of the controversy. In both these scenarios, the system's main role is to assist the student in exploring its information base. Students are active browsers but passive viewers of argument structures; they take the initiative in exploring the system's scientific knowledge, while becoming familiar with argument concepts and seeing them applied to the material they are learning.

When a student feels more comfortable with both the scientific and argument concepts, she can begin practicing argumentation by playing the role of active disputant in a scientific controversy. In the role of advocate for a particular theory, she must produce arguments in support of her theory and respond to arguments and criticisms produced by another student (or the system) who is supporting a competing view. She must show how the theory she is supporting explains the observations she is using as evidence, and why her theory is preferable to the one espoused by her partner in the argument. The intention is not to have the students faithfully recreate the timing and details of the controversy, but rather to practice constructing arguments, applying scientific reasoning strategies, and searching on-line for data that achieves their dialectical goals. The role played by the system can range from active arguing partner (Cavalli-Sforza & More, 1992), to passive bystander or coach in an argument between two or more students.

Figure 6 shows an example of the kinds of interaction we have in mind. It depicts a small portion of the controversy existing in the mid 1920's between supporters of Continental Drift and its contemporary, well-entrenched Contraction Theory. Both theories attempted to explain the present distribution of continents, oceans, and mountains. Starting from the state of the argument depicted in Fig. 6 (a), in which neither theory is significantly better supported than the other, the student acting as a supporter of Continental Drift is asked to strengthen her position. She applies the strategy of hypothesis splitting, i.e., she finds an observation – the observed distribution of mountains – that is explained by one theory and not the other: Contraction Theory would predict a more even distribution of mountains than is actually observed. The results of this action are shown in Fig. 6(b). The counterevidence link used in the figure is a compression of the counterargument move in Fig. 3.

Figure 6(c) shows how we can use the graphical representation of an argument to illustrate different perspectives on an issue. It displays how the supporters of Contraction Theory might have viewed this controversy, considering not only empirical observations but existing domain presumptions. Many of them were Uniformitarians, i.e., they had a strong belief in the principle that the present state

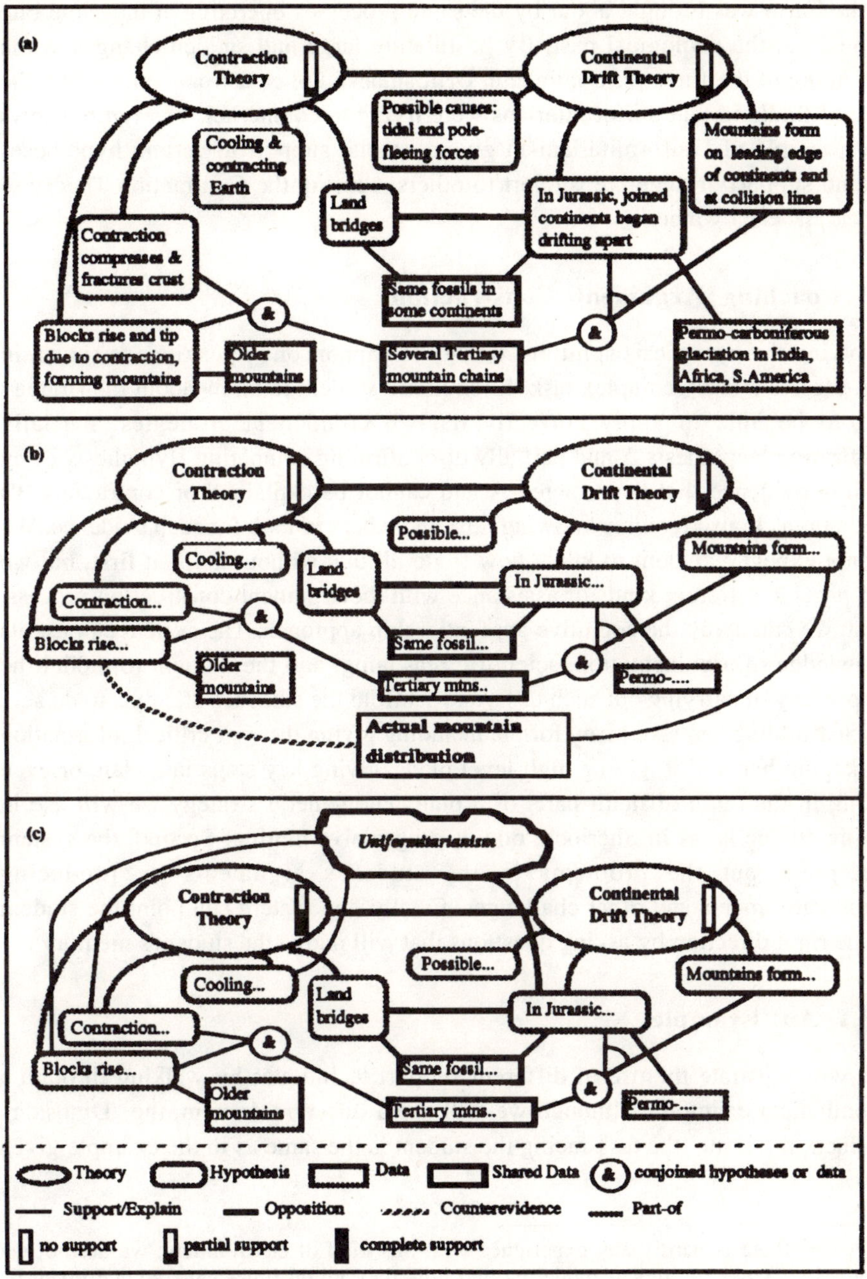

Figure 6 Displaying and interacting with a scientific controversy

of the Earth was brought about by the same processes operating at the same rate throughout the geological past. By postulating large and sudden changes at the beginning of the Jurassic, Continental Drift smacked of catastrophism, exactly the type of thinking that uniformitarians were trying to counteract. The figure shows the principle of Uniformitarianism giving special support to certain hypotheses and no support or negative support to others, causing the Contraction Theory to appear the clear winner.

3.3 Coaching Argument Construction

Applying the hypothesis splitting strategy to support one's position, as shown in Fig. 6(b), is a fairly complex task. It requires a student to know about the strategy and to be able to apply correctly its two component strategies: Partially confirming Hypothesis A and partially disconfirming competing Hypothesis B, by finding evidence that is predicted A and cannot be explained or contradicts B's predictions. It also requires knowing how and where to look for such evidence. We cannot expect a student to know how to do all this without help, at first, but we can provide different kinds of assistance with the argument construction process. First, we can apply the cognitive apprenticeship approach. The system can use its knowledge of argumentation, scientific reasoning, and the domain to model the proper way of carrying out such tasks and scaffold the student's attempts to do so.

Scaffolding can take many forms, including giving the user critical information or helping her find it, giving high-level plans, giving key steps in a plan, or even filling in the more difficult parts of a plan. The general strategy we will use in giving advice is, as in Sherlock, one of progressive hinting. Second, the system can point out the problems in the student's arguments by producing counterarguments and other challenges. Finally, the system can point the student in the right direction by asking questions that will nudge the student's memory[8].

3.3.1 An Example

We will illustrate the use of different support techniques by walking through a hypothetical example. Although we will use a different domain, the "Dinosaur" Extinction Debate[9] the task facing the student is the same as in the example given above.

---

[8] Because there is hardly any experience with this kind of environment, we are not yet addressing issues of student modeling and how they might impact the advice given by the system.

[9] This is the debate surrounding the cause of the large scale extinctions, including that of the dinosaurs, that occurred about 65 million years ago at the boundary of the Cretaceous (K) and Tertiary (T) geological periods. The main contenders in the explanation of this event are the Meteor Impact Hypothesis (or simply Impact) and the

The dialogue begins with the system asking the student to find evidence that supports the Impact Hypothesis. When the student responds by requesting help, the tutor replies by suggesting that a good way to support a hypothesis against a competitor is to apply the hypothesis splitting strategy. If the student requests further assistance, the system, using its argumentation, scientific reasoning, and domain knowledge, must develop a plan for carrying out the task (Fig. 7), which it will use to provide increasingly more detailed advice upon request from the student. Figure 7 also shows that the system has found a complete solution and a partial solution, i.e., a solution that works for "confirming" the Impact Hypothesis, but not for "disconfirming" the Volcanism Hypothesis (we mean, of course, incremental or partial, not absolute, confirmation/disconfirmation.). In response to the student's request for assistance it will suggest looking for some material in rocks that can be dated to the Cretaceous-Tertiary (KT) boundary and whose source might be a meteor impact but not volcanism. This advice corresponds to giving the student the high-level steps in the plan.

If the student needs more assistance in carrying out the first plan step, the system again tells her the necessary steps. It also gives her a hint of what to look for (a material present in both the KT deposits and meteorites). Armed with that information, she browses the information base and discovers that iridium, found in the KT layer in high concentrations, is a common element in meteorites. She selects iridium as the material (?m) with the system's approval, but she still needs help on the remainder of the subplan.

When the system cannot break down a subplan into finer steps, it switches to a question and answer mode and attempts to elicit at least part of the information from the student[10]. For example, it can get the student to select iridium concentration as the property (?p1) to use in the predictions (?ep1) that can be made for Hypothesis M. Once it has obtained the necessary information for constructing the argument in support of Hypothesis M (and therefore of the Impact Hypothesis), it will model the construction of the microargument using the graphical argument representation (Fig. 8, Claim 1).

Volcanism Hypothesis. The first holds that a large extraterrestrial body collided with the earth, setting in motion a number of reactions that affected the atmosphere and surface of the Earth that resulted in the extinction. The second holds that the cause was very large scale volcanism occurring prior to and during the time of the extinction, resulting in the extinction through environmental changes that were possibly quite similar to those caused by the meteor impact. Recently, some scientists have proposed that the two events may be linked.

[10] To simplify the natural language input problem we will rely heavily on the use of mouse-sensitive text to allow the student to select phrases in the system generated explanation and to ask further questions. This approach is based on the dissertation work of our colleague Johanna Moore (Moore & Swartout, 1990).

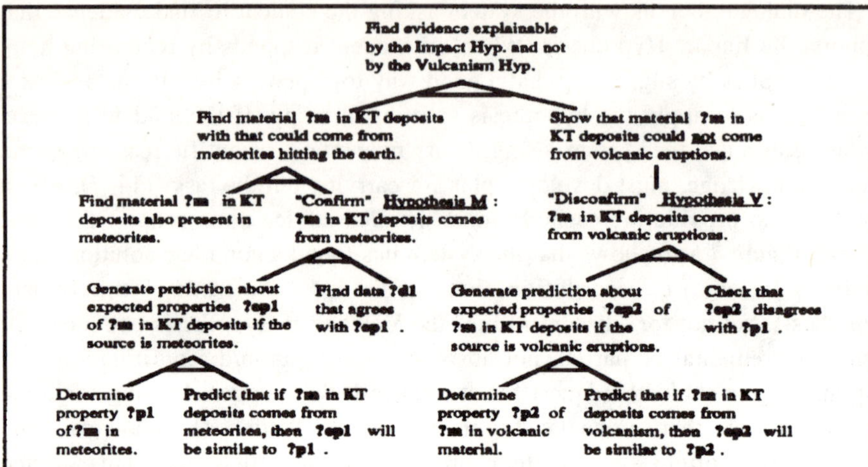

System Bindings

?m = iridium

?p1	?ep1	?d1	?p2	?ep2	check
Concentration in meteors is high	Concentration in KT deposits is high	KT clay (Italy)	Concentration in volcanic mat'ls is high	Concentration in KT deposits is high	*fails*
Iridium/gold ratio in meteors is low	Iridium/gold ratio in KT deposits is low	KT deposits (Denmark)	Iridium/gold ratio in volcanic mat'ls is high	Iridium/gold ratio in KT deposits is high	ok

Fig. 7. A summary of the system's plan for performing the task

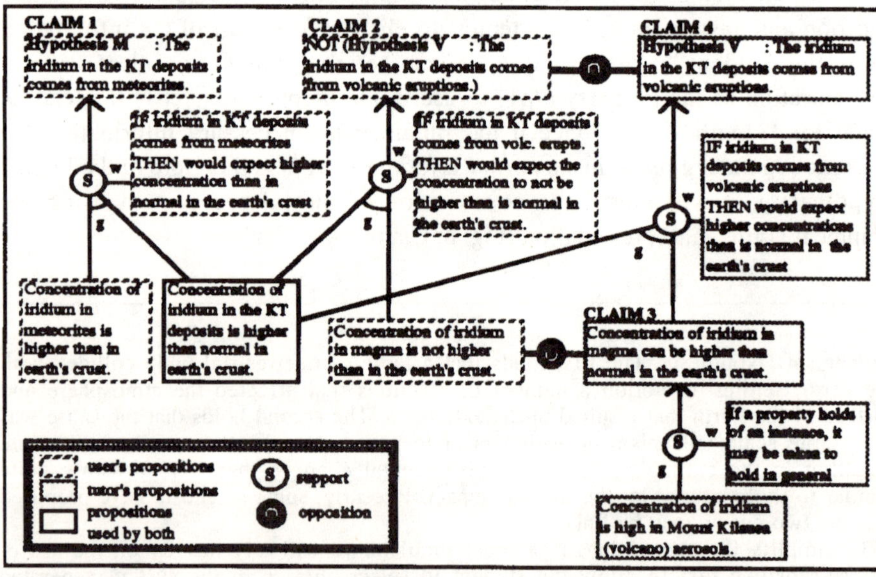

Fig. 8 Arguments constructed by the student and the tutor

Now, suppose that the student, having watched the system construct an argument to support the Impact Hypothesis, tries to construct an argument against the Volcanism Hypothesis (in favor of "NOT Hypothesis V"), using the previous argument for Claim 1 that she constructed with the system's help (Fig. 8, Claim 2). She uses the correct form of the argument but does not check whether the data she uses is actually correct. The system can respond by arguing against her data (Fig. 8, Claim 3) and showing that, in fact, iridium concentration cannot be used to disconfirm Hypothesis V but rather provides some support for it (Fig. 8, Claim 4). According to our rules for collaborative epistemic arguments, the student must either find another way of supporting "NOT Hypothesis V", or must be able to find fault with the system's counterargument.

If she chooses to resupport "NOT Hypothesis V", she must search for another suitable property of iridium, possibly relying on the system for advice.

3.4 Status of Argumentation Tools

Although, for simplicity, we have used the present tense in describing the argumentation tools and their intended use, we emphasize that this work is in its early stages. Some components of the system have been prototyped (Cavalli-Sforza, Lesgold, & Weiner, 1992) and others have been used in previous systems built by members of our research group, but the system overall is currently in the design and early implementation stages. As we develop these tools we are also talking with middle school science teachers and other education researchers in order to receive early feedback on how our tools might best be fitted into the existing curriculum and classroom structure.

4. Concluding Remarks

We have described two computer-based environments, at various stages of development, that emphasize learning by doing. Although their task domains, electronics troubleshooting and scientific argumentation, are rather different, the projects share many similarities in their instructional approach. This is perhaps not surprising, given their common genesis and the success of Sherlock I, which incorporates many of the ideas on which Sherlock II is also based.

The argumentation project borrows from Sherlock the approach of letting the student take responsibility for his or her own learning, providing advice only upon request, relying on progressively more detailed hinting, and using modeling and scaffolding as coaching techniques. The reification of argument structures and the effects of argument actions provides a vivid explanation of why certain moves are better than others, simplifying the reflective task addressed by Sherlock's review activity. The inherently dialectical nature of argumentation suggests that the system can coach students to argue against each other using knowledge very similar to the one it uses to coach a single student. Our focus on epistemic argumentation makes the system's knowledge of argumentation suitable for using argument as a coaching device. Sherlock's reflective follow-up activity and its scheme for evaluating a student's solutions against an expert's similarly provide opportunities for moving from tutor-critiques-student interactions to student-critiques-student interactions.

Both Sherlock and the argumentation project aim at providing students with an opportunity to learn concepts and skills specific to a task domain, as well as more general (and transferable) strategic/tactical knowledge, by practicing using the concepts and skills while performing an authentic activity. They provide the student with a vocabulary of actions that can be used to perform and clarify the task, and whose meaning can be acquired by the student through engaging in task-related activity. Finally, both systems will allow us to explore empirically whether practicing the construction of explanations - be they argument chains demonstrating why some observation supports a scientific hypothesis, or reasoning chains showing why a troubleshooting action was suboptimal - does indeed have the effects of improving performance, peer collaboration, and self-monitoring skills, imputed to it by other researchers.

Acknowledgements

Work on Sherlock has been funded by the Air Force, through subcontracts from Advanced Technology, Inc., Hay Systems, and Metrica, Inc. Sherlock II has been a collaborative effort that has included Marilyn Bunzo, Roberta Catizone, Dennis Collins, Richard Eastman, Gary Eggan, Maria Gordin, Linda Greenberg, Sherrie Gott, Ellen Hall, Edward Hughes, Sandra Katz, Susanne Lajoie, Alan Lesgold, Thomas McGinnis, Johanna Moore, and Dan Peters, Bob Pokorny, Rudianto Prabowo, Govinda Rao, Rose Rosenfeld, and Arlene Weiner.

The James S. McDonnell Foundation has funded our empirical studies of argumentation skills; the Andrew W. Mellon Foundation and, more recently, the National Science Foundation have funded research on tools for supporting argumentation. John Connelly, Gareth Gabrys, William Keith, Jarmo Levonen, Johanna Moore, Victor Schmidt, Mike Smith, Fred Sukitsch, Dan Suthers, Rich Thurlow, and Arlene Weiner have collaborated on this project.

References

Brown, J.S. (1985). Process versus product: A perspective on tools for communal and informal electronic learning. Journal of Educational Computing Research 1(2), 179-201

Brown, J.S., Collins, A., & Duguid, P. (1989). Situated cognition and the culture of learning. Educational Researcher 18(1), 32-41

Cavalli-Sforza, V. (1991). An environment for tutoring argumentation skills. Working Notes of the AAAI Spring Symposium Series: Argumentation and Belief, pp. 71-84

Cavalli-Sforza, V., Gabrys, G., Lesgold, A.M., & Weiner, A.W. (1992). Engaging students in scientific activity and scientific controversy. Workshop Notes of the AAAI Workshop Program: Communicating Scientific and Technical Knowledge, pp. 99-106

Cavalli-Sforza, V., Lesgold, A.M., & Weiner, A.W. (1992). Strategies for contributing to collaborative arguments. Proceedings of the Fourteenth Annual Conference of the Cognitive Science Society, pp. 755-760. Hillsdale, NJ: LEA

Cavalli-Sforza, V., & Moore, J.D. (1992). Collaborating on arguments and explanations. Working Notes of the AAAI Spring Symposium Series: Producing Cooperative Explanations, pp. 61-68

Chi, M.T.H., Bassok, M., Lewis, M.W., Reimann, P., & Glaser, R. (1989). Self-explanations: How students study and use examples in learning to solve problems. Cognitive Science 13, 145-182

Chi, M.T.H., & VanLehn, K. (in press). The content of physics self-explanations. Journal of the Learning Sciences

Collins, A., & Brown, J.S. (1988). The computer as a tool for learning through reflection. In: H. Mandl & A. Lesgold (Eds.), Learning issues for intelligent tutoring systems, pp. 1-18. New York: Springer

Collins, A., Brown, J.S., & Newman, S.E. (1989). Cognitive apprenticeship: Teaching the craft of reading, writing, and mathematics. In: L.B. Resnick (Ed.), Knowing, learning, and instruction: Essays in honor of Robert Glaser. Hillsdale, NJ: LEA

Conklin, J., & Begeman, M.L. (1988). gIBIS: A hypertext tool for argumentation. ACM Transactions on Office Information Systems 6, 303-331

Katz, S., & Lesgold, A. (in press). The role of the tutor in computer-based collaborative learning situations. In: S. Lajoie & S. Derry (Eds.), Computers as cognitive tools. Hillsdale, NJ: LEA

Keith, W.K., Weiner, A.W., & Lesgold, A.M. (1989). Argument comprehension: Some preliminary findings. In: B. Gronbeck (Ed.), Spheres of argument. Proceedings of the 6th SCA/AFA Conference on Argumentation. Annandale, VA: Speech Communication Association

Keith, W.K., Weiner, A.W., & Lesgold, A.M. (1990). Toward computer-supported instruction of argumentation. In: F.H. van Eemeren, R. Grootendorst, J. A. Blair, & C.A. Willard (Eds.), Proceedings of the Second International Conference on Argumentation, pp. 1144-1153. Amsterdam, The Netherlands: SICSAT

Kuhn, D. (1991). The skills of argument. Cambridge, UK: Cambridge University Press

Lajoie, S., & Lesgold, A. (1991). Apprenticeship training in the workplace: Computer coached practice environment as a new form of apprenticeship. Machine-Mediated Learning 3(1), 7-28

Lee, J., & Lai, K-Y. (1991). What's in Design Rationale?. Human-Computer Interaction 6, 251-280

Lesgold, A.M. (1989). Skills of Argumentation on School Subjects: Year 1. Report to the James S. McDonnell Foundation, LRDC, University of Pittsburgh

Lesgold, A. (in press). Assessment of intelligent training systems: Sherlock as an example. In: E.L. Baker & H.F. O'Neil, Jr. (Eds.), Technology assessment: Estimating the future (tentative title). Hillsdale, NJ: LEA

Lesgold, A., Eggan, G., Katz, S., & Rao, G. (1992). Possibilities for assessment using computer-based apprenticeship environments. In: W. Regian & V. Shute (Eds.), Cognitive approaches to automated instruction. Hillsdale, NJ: LEA

Lesgold, A. M., Lajoie, S. P., Bunzo, M., & Eggan, G. (1992). SHERLOCK: A coached practice environment for an electronics troubleshooting job. In: J. Larkin, R. Chabay, & C. Scheftic (Eds.), Computer assisted instruction and intelligent tutoring systems: Shared goals and complementary approaches. Hillsdale, NJ: LEA

MacLean, A., Young, R.M., Bellotti, V.M.E., & Moran, T.P. (1991). Questions, options, and criteria: Elements of design space analysis. Human-Computer Interaction 6, 201-250

Moore, J.D., & Paris, C.L. (1989). Planning text for advisory dialogues. Proceedings of the Twenty-Seventh Annual Meeting of the Association for Computational Linguistics, Vancouver, B.C., Canada, pp. 203-211

Moore, J.D., & Swartout, W.R. (1990). Pointing: A way toward explanation dialogue. Proceedings of Tenth National Conference on Artificial Intelligence, Boston, MA, pp. 457-464

Nichols, P., Pokorny, R., Jones, G., Gott, S. P., & Alley, W. E. (in press). Evaluation of an avionics troubleshooting tutoring system. Special Report. Brooks AFB, TX: Air Force Human Resources Laboratory

Owen, E., & Sweller, J. (1985). What do students learn while solving mathematics problems? Journal of Educational Psychology 77, 272-284

Palincsar, A.S., & Brown, A.L.(1984). Reciprocal teaching of comprehension-fostering and comprehension-monitoring activities. Cognition and Instruction 1, 117-175

Perkins, D. N. (1985). Postprimary education has little impact on informal reasoning. Journal of Educational Psychology 77(5), 562-571

Pokorny, R., Jones, G., Gott, S., & Alley, W. (in press). Evaluation of an avionics troubleshooting tutoring system. Special Report. Brooks AFB, TX: Armstrong Laboratories, Human Resources Division

Scardamalia, M., Bereiter, C., McLean, R.S., Swallow, J., & Woodruff, E. (1989). Computer- supported intentional learning environments. Journal of Educational Computing Research 5(1), 51-68

Schoenfeld, A.H. (1983). Problem solving in the mathematics curriculum: A report, recommendations and an annotated bibliography. The Mathematical Association of America, MAA Notes 1

Streitz, N.A., Hannemann, J., & Thuring M. (1989). From ideas and arguments to hyperdocuments: Travelling through activity spaces. Proceedings of Hypertext Conference, pp. 343-364

Sweller, J. (1988). Cognitive load during problem solving: Effects on learning. Cognitive Science 12, 257-285

Sweller, J., & Cooper, G. (1985). The use of worked examples as a substitute for problem solving in learning algebra. Cognition and Instruction 2, 59-89

Toulmin, S. (1958). The uses of argument. Cambridge, UK: Cambridge University Press 1958

Toulmin, S., Rieke, R., & Janik, A. (1984). An Introduction to Reasoning, 2nd ed. New York: Macmillan

Voss, J.F., Tyler, S.W., & Yengo, L.A. (1983). Individual differences in the solving of social science problems. In: R.F. Dillon & R.R. Schmeck (Eds.), Individual differences in cognition. New York: Academic Press

Vygotskii, L.S. (1978). Mind in society: The development of higher psychological processes. Cambridge, MA: Harvard University Press

The Use of Computer-Mediated Role Play in Management Training

Harry McMahon

Language Development and Hypermedia Research Group
University of Ulster at Coleraine, Coleraine, Northern Ireland

Abstract. This paper explores the potential of a computer-mediated role play technique called "bubble dialogue" for use in management training. The pedagogy involved, which has been tested in educational settings but not in the training sector, is outlined and related to the tenets of constructivism. Possible applications in management training are described and possible benefits suggested. An example of its use in developing high complexity helping skills is presented in an Appendix.

Keywords. Computer assisted learning, computer-mediated training, role play, character play, management training, constructivism, constructivist, dialogue, bubble dialogue , interpersonal skills, helping

1. Introduction

For a long time, training systems developers have pursued the dream of creating technology-based training of such character that the interaction between the computer and the user would be indistinguishable in terms of its quality from that between a trainee and an experienced, capable, informed, patient, and sympathetic trainer. Each new wave of computer technology, bringing with it more power, more memory, more peripherals, more whatever, has augured the arrival of the dream. And each wave has brought with it claims that computer-based training in tutorial mode is now, at last, authentic, effective and efficient, even fun.

Perhaps others don't feel the same way, but I feel that the dream of the machine-based intelligent tutor is a very long way from being realised. Indeed, I wonder if it ever can. For me, programming the current generation of computers to dialogue directly with humans, even to the limited degree of expecting the machines to ask sensible questions and judge specific answers in context, often produces human-computer dialogue which is stilted, either obtuse or simplistic, limited in scope and in vocabulary, straight-jacketed and more often than not, just plain boring. Of course, there are exceptions to this general "rule", which I have conveniently invented. But in these exceptional cases, the investment in developing interactive systems were the human-computer dialogue is natural, realistic and informative continues to be very high. I wonder: as the dream of the machine-based intelligent tutor continues to be pursued, could it be that it is the wrong thing to be trying to do in certain sectors of training - in particular those where the development, understanding and fluid use of interpersonal communication skills lie at the core of effective performance and where the context of use of these skills is changing very rapidly?

My concern is that, particularly in these times in this area of performance, it is dangerous to be captured by the notion that one can engineer a stable solution to a training problem and pass across to a computer the responsibility to effect the solution. The idea of the educational engineer being able to construct a machine (by programming a computer) which can deliver instruction on interpersonal skills as one might deliver a candy bar or a packet of cigarettes may be attractive to one seeking efficient solutions to definable problems. But training goals in the area of interpersonal skills are notoriously slippery concepts to hold on to during the training process itself. They can be defined in behavioural terms, of course, but so often our experience is that the formal language of behavioural specification fails to capture the reality of everyday experience, especially where that experience is constantly being modified by the pressures of the market place. Somehow the rapidly changing context of on-the-job performance and the contrariness (or humanity) of the trainees seem to combine to demonstrate the inadequacy of the notion that all the trainee has to do is to pay up by turning up, select to taste, press the right keys and wait for a package of skill to be delivered by the machine.

In the face of these problems some trainers have either refused to contemplate or abandoned altogether the thought that computers might be able to help in interpersonal skills training. Others have taken a different course. They have stuck with the technology, but kept it in its place - as a stimulus to dialogue between humans. For these people it has been the off-screen dialogue stimulated by the shared screen that has really counted as learning experience. For them, the quality of software has to be judged by the quality and relevance of the debate it helps generate among trainees and between the trainees and their trainer, not by its ability to make trainees conform to a prepackaged notion of one side of a preordained human-computer interaction.

The type of software I am talking about offers an alternative to the embodiment of behavioural and information processing theories in prepackaged instruction. It does not derive its power from the elegance with which a package of knowledge is defined, structured and delivered to the individual at the keyboard or the ease with which a basket of skills is encapsulated and transferred from the heart of the machine to the head of the user. I am pointing to the emergence of a new strain of software sympathetic to the so-called constructivist view of learning. Central to this view is the acceptance that, above all, learning is a social process which takes place in a context in interaction though dialogue with others. In the constructivist view, all instruction and learning (whether it is experienced in a working, training or educational environment) is less a process of communicating specified knowledge to learners and more a matter of nurturing the processes whereby learners come to understand, share and construct multiple perspectives and interpretations, develop and learn to defend their own positions while recognising those of others and become self aware in the sense that they can recapture their situated experience and exploit their understanding of the knowledge construction process itself (Knuth and Cunningham, 1993). For example, a management trainee does not become an effective manager by learning the theories and practices of management as Abstracted knowledge. He or she becomes effective by practising the art and science of management in an environment in which their understanding of managing is constructed and reconstructed through dialogue embedded in purposeful action, and critical reflection upon that dialogue, with superiors, subordinates, clients, consultants, trainers, or whoever, with people rather than with books or computers. If the machine is to play a role in this process, it must be to stimulate dialogue between humans, not replace it. And if it can stimulate dialogue about dialogue, so much the better.

2. Bubble Dialogue

The subject of this paper, bubble dialogue, belongs to this strain of software. Bubble dialogue is a computer-mediated projective role play technique which, I believe, could be used by management trainers to produce conditions where trainees are motivated to engage productively in the constructivist processes outlined above. Through a combination of elements of role play, dialogue creation and a process called reflexive dialogue analysis, trainees can be expected to engage actively with their own ideas and those of others, including those of their trainer. Through first the projective creation of bubble dialogue in context and then reflection about that dialogue, they have the opportunity to express personal (perhaps naive) views of key performance areas of their working environment, to contemplate multiple perspectives in both public and private domains and to accommodate their own thinking to contrary views. By sensitive selection of the roles and the topics explored with the aid of the bubble dialogue program, the trainer can ensure that the learning is focused on a particular set of skills required or issues likely to be encountered in realistic contexts which are relevant to contemporary working life.

The technique can in principle be applied in any situation where learners or trainees can usefully undertake role play exercises designed to facilitate the development of high complexity interpersonal skills, such skills as the ability to carry out interviews, to negotiate, to resolve or reduce conflict, to explore an issue or a concept, to help others through personal difficulties which are influencing work performance. These skills are broad ranging and complex, but they can be defined and exemplified, as, for example, the art of helping is defined by Carkhuff and Anthony (1979), Egan (1990) or Okun (1982).

In bubble dialogue, still graphics or photographs of two characters talking to each other are placed on the computer screen and users get access via four hot icons to empty "think" and "say" bubbles which are presented as selected, above the heads of the characters. The users, who can be trainees working individually or in groups, with or without the immediate involvement of the trainer, take on the roles of the characters and create conversations between the characters about topics of concern to the trainees. For example, in a retail management training course, one screen character could be an assistant manager in the retail sector and the other a dissatisfied customer. A dialogue creation protocol pushes the conversation forward, one "say-write" and optionally one "think-write" per character per turn. The users, anything from one to four seems to work, cluster around the screen and individually or collaboratively create the "play" between the characters on the screen. Evidence accumulated over three years suggests that users identify with the characters, get caught up in realistic dialogues and at the same time feel themselves able to take risks with what the characters are made to publicly "say" and privately "think". (McMahon and O'Neill, 1993) A review facility allows the users to retrace the dialogue either on the screen or in either of two text forms: a comic strip or a play script in rich text format. A comment facility allows commentary to be added to the dialogue when it is being reviewed. (McMahon, O'Neill and Cunningham, 1992)

In the bubble dialogue presented in the Appendix, a researcher acting as a management trainee is attempting to follow Carkuff's helping model (Carkhuff and Anthony, 1979) as he plays the part of the helper on the screen. In this particular example, the part of the helpee was played by the trainer, but it could just as well have been played by a fellow trainee or trainees, provided that the trainer's intention was to provide input during the review process.

3. The Need for Affordable Just-in-Time Training

It has been acknowledged that constructivist approaches, such as I advocate here, have much to offer managers seeking to preserve or recover a company's competitive edge. For example, AT&T have reported their broad conclusion (in a collaborative study with Indiana University of the design and development of enhanced learning and information environments) that there is a need for "a rather fundamental reconceptualisation of instructional design" in the training sector:

> "Our task becomes one of designing learning environments: which nurture... experiential processes; which nurture the contexts in which the knowledge will be used; which illustrate the constructive nature of knowledge and allow... workers to experience multiple perspectives from colleagues, experts, interactive media, and so on."
>
> ELIE Project (1990)

One of the more specific findings of AT&T's study was that sales account executives, working in high pressure circumstances, were of the view that the knowledge and skills most critical to job performance were sales process oriented rather than topic oriented, even in a rapidly changing technological environment. For them, it was the constant upgrading in a competitive market of the ability to build and maintain relationships, use business communication skills, understand people and negotiate that really counted. AT&T's conclusion was that to satisfy this training need an enhanced learning and information environment would need (among many other things) to "provide opportunities to gain knowledge and practice sales process skills through computer-based simulations". They saw the deployment of such training support as an element of "just-in-time" training, delivered when needed as close to the sales site as possible.

Using currently available interactive videodisc technology, such training would typically be delivered by a computer-based simulation of the type first explored by Rushby et al. (1987). In a project involving collaboration with Lucas CAV, Kodak and the Trustee Savings Bank, the Centre for Staff Development at London University looked at the problems involved in simulating a 30 minute interview dialogue. Their interactive video allowed the trainee to interview a single simulated applicant for one particular job, initially selecting questions from a menu but in later phases using a voice recognition system. The computer-driven applicant was simulated using short video sequences selected from a bank of 300-400 hundred on the basis of choices made by the trainee, who takes the part of the interviewer, from a bank of predetermined questions.

I wonder what the development to delivery time ratio for such a video might be - probably much greater that the 100:1 commonly quoted for computer-based training. I also wonder about the effect that such a high ratio might have on the ability of a training organisation to produce simulation packages at a rate high enough to satisfy the need for "just-in-time" context specific training of the type seen necessary by AT&T. For example, what would be the cost of rapidly producing a set of interactives which would allow trainees to simulate next month's interviews of a range of candidates lining up for newly defined positions which are subject to new employment legislation? If, in order to ensure context embeddedness of this type, companies want to design their own interactive video role simulation exercises then they will have to invest heavily in front end design.

But would doing that make sense? The rate of change in the market place is accelerating and changes are becoming more and more complex. The changes are affecting virtually all aspects of most businesses - products, processes, systems, employment legislation, organisational structures and the market itself are all on the move, to such an extent that job complexity is increasing and coping with change is beginning to be seen the highest order skill of all. Is it not likely that context specific simulations are going to be too expensive to deliver in these rapidly changing business conditions if the chosen mode of delivery is interactive video of the type described by Rushby et al.?

One of the characteristics of bubble dialogue is the speed with which it can be adapted for use in specific contexts, in role playing exercises where the characters being played and the dialogue encountered are situated in the contemporary scene as it is experienced in the workplace by the trainees. It takes only five to ten minutes work with a master program to select a pair of *characters* from a graphics library, name them, write a short *prologue* which describes the context of the dialogue and compose an *opener*, an opening statement by one of the characters which launches the dialogue. This just-in-time design ability means that trainers can respond rapidly to the newly defined needs which inevitably emerge as a corporation changes in response to the changing market place. Thus, rapid adaptability to changing circumstances and to the idiosyncrasies of local circumstances helps ensure that this form of computer-mediated training is effectively embedded in the world of work as actually experienced by the trainees.

I postulate that it might be possible to deliver just-in-time, context embedded training at affordable cost if the mode of delivery incorporates the bubble dialogue technique.

4. Story-Embedded Training

Our experience as teacher trainers is that inservice teachers prefer training that is somehow embedded in a story that is realistic and recognisable. The publications of the UK National Council for Educational Technology suggest that this is a widely held view; see, for example, the use made by North (1991) of training simulations embedded in case studies of school management. I have noted also that there is evidence that the same is the case in corporate training. For example, Hawkridge, Newton, and Hall (1988), report that trainees react favourably to training that reflects the current corporate culture. This is not in the least surprising given the degree to which companies make use of myth and metaphor (a form of story telling) in building the corporate image. (Sculley, 1987). In the light of these observations I see it as entirely appropriate that training in business and industry be embedded in the "story" of the company itself, in the corporate culture in which the trainees find themselves. This intention, to situate the training in a meaningful, recognisable and already experienced context, reflects one of the central tenets of constructivism and follows the recent thinking of Brown, Collins, and Duguid (1989) on situated cognition and Jonassen, Mayes, and McAleese (1993) on contextually mediated learning.

Suppose, for example, that instead of using the computer to drive one side of a simulated conversation, with the high front end costs in elapsed time and development that that entails, and the necessary focus on a single predetermined and hence "frozen" event in the corporate story, one were to produce at much lower cost a straight movie, or even more cheaply, a storyboard of a negotiation, interview, consultation, or whatever, with a number of key decision points where the differential effects of deploying up-to-date or out-of-date expertise would be critical to the outcome of the exchange. If the video/storyboard was then frozen at each of these critical moments (for example, when a salesperson is on the point of responding to a customers' claims about a competitor's recently announced product or service) and the trainees were branched into bubble dialogue, then one might be able to provide immediacy and embeddedness at greatly reduced cost. With the trainer in a position to choose to play one of the characters or to review and comment on dialogues produced by trainees working on their own, it could be assured that key issues and problems are encountered. The essential difference between the two approaches (apart from cost and timeliness) is that in the former the instructional systems designer has to predict all the permutations of question and answer within a predetermined conversational framework, all the possible branches in the interaction between the characters have to be in place from the beginning, while in the latter it is the trainees (or, by choice, the trainer) who invent the branches during the training process itself by creating bubble dialogue which springs from the frozen frame in the video anchor or storyboard. (I have borrowed the term 'video anchor' from the work of the Cognition and Technology Group at Vanderbilt (1993) on anchored instruction.) This way, bubble dialogue taps into the trainees' current view of the world as well as the trainer's, increasing the chances of accessing and reconstructing out-of-date or inappropriate views. By freezing the video or storyboard action and focussing on the trainees' projective simulation of possible/probable interaction, we cause trainees to concentrate on how the characters, standing in for themselves and their customers, create meaningful worlds through dialogue.

By following through with bubble dialogue in which the characters are even more closely identified with real people, in terms of sex, ethnicity, level of experience and responsibility, the trainees could "rehearse" the future deployment of newly acquired skills in collaboration with their peers and their trainers. In the extreme, which might prove a valuable extreme to use from time to time, a rehearsal dialogue could be created between characters based on digitised photographs of the actual protagonists.

The recent appearance of cameras which produce digitised stills means that the time needed to place real people on the bubble dialogue screen can be a matter of a very few moments. With this technology available, one can imagine many different kinds of training situations where trainees could be helped by bubble dialogue to reflect on and reconstruct their performance skills. For example, several still camera shots are taken of trainees as they role play a key interpersonal performance skill; the digitised stills are transferred into the bubble dialogue graphic library and the trainees choose one of the stills as representing the moment when things started to go wrong; they write their own prologue and produce a bubble dialogue in which an alternative approach to achieving the desired goal is explored.

My colleagues and I have have used the term "reflexive dialogue analysis" to describe this process whereby users of bubble dialogue are able to amend and extend, reflect on and narrate the story of the bubble dialogue they have themselves created or helped to create. We think that reflexive dialogue analysis, particularly if it is carried out in the presence of their peers and trainers, involves the trainee in the constructivist processes I described in the introduction and progressively nurtures their ability to construct and reconstruct contextualised knowledge and skill. Further, and of great importance, it nurtures the ability to monitor the processes in action. By this last point we mean that in training, awareness about the importance of meta-level skills is developed. Trainees grow to understand the processes by which knowledge is constructed in context and are thereby better able to act reflexively, to monitor, assess and manage their own performance, in real life situations. Thus, the training is of a type which is likely to increase the trainee's capacity to monitor and reconstruct performance in the field and so reduce the need for further training.

5. Rewriting the Story and Redistributing the Costs

Bubble dialogue lends itself to updating story-embedded training. Characters on the screen can be the same characters as those encountered in a case study based on the corporation itself or on a fictitious but closely analogous company. They can have typical backgrounds, qualifications and experience and face typical problems and opportunities in the workplace. What is more, the speed with which it is possible to create new bubble dialogue scenes, by creating prologues and openers which fit the opportunity or requirement of the moment, allows the storyline, the embedded life experience of these characters, to be developed and explored within the training session itself. One is not bound by the prepackaged case study and its prepared exercises, so the training can be developed as the company itself develops. The analogous company and its employees can be presented with the latest opportunities and problems facing the company itself. This last feature, the speed of adaptation of bubble dialogue to newly defined needs, allows existing training materials to have a longer working life and could also allow the adaptation of existing but not fully suitable training materials to fit the needs of a company not able to invest in the development of its own case study materials.

The use of bubble dialogue transfers cost from one part of the training cycle to another. Usually in technology based training, after training needs have been identified, the bulk of the cost goes into front end development of the training package, which may cost up to a couple of orders of magnitude more than the front end cost of conventional training. In theory, with this achieved, the trainer steps back from direct interaction with the trainees and consequent savings accrue. In contrast, with bubble dialogue, the front end cost is small, but the technique will only work if the trainer is involved in the creation and review of bubble dialogue. The technique is a stimulus to trainer-trainee interaction, not a substitute, so the cost of the trainer has to be added to the delivery cost. I have pointed earlier to the constructivist's view that delivering pre-structured packages of training via the machine may be much less effective than stimulating the reconstruction of knowledge and skill through human dialogue in contexts closely relevant to job performance. Keeping the trainer in the loop ensures that expertise is offered when it is needed, during the reconstruction process itself.

6. Time for an Admission

I have said enough for the time being. Readers with experience in the training sector will be better able than I am to imagine how the bubble dialogue tool might be used in particular forms and areas of training. I have to admit it - as if it hasn't been patently obvious already - I am largely speculating about the ease with which the undoubted success of bubble dialogue in the education sector might be replicated in the training sector. In isolation I can do nothing else. My colleagues and I need others to help us find out if there is any truth to be found in or constructed from my speculations.

References

Brown, J. S., Collins, A., & Duguid, P. (1989). Situated cognition and the culture of learning. Educational Researcher, 18 (1), 32-42

Carkhuff, R. R. and Anthony, W. A. (1979). The skills of helping Amherst, MA: Human Resource Development Press

Cognition and Technology Group at Vanderbilt (1993). Designing learning environments that support thinking. In: T.M. Duffy, J. Lowyck, & D.H. Jonassen (Eds.), Designing environments for constructive learning, NATO ASI Series F, Vol. 105. Berlin: Springer-Verlag

Egan, G. (1990). The Skilled Helper. Belmont, CA: Brooks/Cole

ELIE Project (1990). The design and development of enhanced learning and information environments. Indiana University, AT&T University of Sales Excellence, AT&T Bell Laboratories

Hawkridge, D., Newton, W., & Hall, C. (1988). Computers in company training. London: Croom Helm

Jonassen, D, Mayes, T., & McAleese, R. (1993). A manifesto for a constructivist approach to uses of technology in higher education. In: T.M. Duffy, J. Lowyck, & D.H. Jonassen (Eds.), Designing environments for constructive learning, NATO ASI Series F, Vol. 105. Springer-Verlag, Berlin

Knuth, R. A., & Cunningham, D. J. (1993). Tools for constructivism. In: T.M. Duffy, J. Lowyck, & D.H. Jonassen (Eds.), Designing environments for constructive learning, NATO ASI Series F, Vol. 105. Berlin: Springer-Verlag

McMahon, H., O'Neill, W., & Cunningham, D. (1992). "Open" software design: A case study. Educational Technology (February)

McMahon, H. and O'Neill, W. (1993). Zones of engagement in learning. In: T.M. Duffy, J. Lowyck, & D.H. Jonassen (Eds.), Designing environments for constructive learning, NATO ASI Series F, Vol. 105. Berlin: Springer-Verlag

North, R. (1991). The Role of the IT Coordinator. London: National Council for Educational Technology

Okun, B. F. (1982). Effective Helping. Belmont, CA: Brooks/Cole

Rushby, N.J., Weil, S., Schofield, A. and Delf, G. (1987) The ubiquitous trigger: a flexible resource. In F. Percival, D. Craig, & D. Burglass (Eds.), Aspects of educational technology XX: flexible learning systems. London: Kogan Page

Sculley, J. (1987). Odyssey: Pepsi to Apple. Glasgow: Fontana/Collins

Appendix

The scenario explored in the following bubble dialogue is based on one of a number of role play situations which are explored by inservice trainee managers attending a management course in the Corporate Learning Centre at Vanderbilt University, Nashville, Tennessee. The author, having read about half of one of Carkhuff's books on the art of helping (and scanned the other half) takes the part of the trainee and role plays the helper, George, while Professor Olin Campbell, tutor on the course, plays the part of Jack, the helpee, who has a problem about working in groups. The screen graphic, not reproduced here, showed two young men, one white (Jack), the other black, facing each other in attentive posture. They are dressed in casual clothes. A third character, not visible on the screen, is introduced in the prologue. Jack's opener is presented on the screen to launch the role play. The rest of the dialogue was created over a period of about 45 minutes. The notes in italics were entered into bubble dialogue by the author two days after the role play was completed.

One of Jack's best friends, Jody, is giving Jack a hard time in his project group. He turns to George for advice.

Jack says: I wish Jody could understand. My grandmother died very unexpectedly and I had to go home. I just couldn't do my share on the project and keep up with everything else. I'm getting a bum rap. It's not fair for Jody to be so mad at me.

George says: I'm sorry to hear that things are tough for you at the moment. You must feel strongly about your work to be so upset.

At this stage I was very conscious of trying to follow a helping model, so here I am doing my best to operate at the "Responding to Feeling" level.

Jack says: You're very perceptive. I've been working my tail off on this project. When I take a high priority trip to be with my family, everybody jumps on me. I wonder if something else is going on.

I was put off by the first sentence a bit. At first I didn't know whether to acknowledge the "compliment" or not; I decided to let it pass unremarked.

George thinks: Sounds as if he's got something else to say about that "something else going on". I won't get into that just yet but I must remember to come back to it.

This "something else going on" struck me as a very important statement but I decided to stick to "the facts". I didn't want Jack to start theorising too soon about WHY things were going wrong but I internally declared, in George's mind as well as my own, my intention to return to this remark - provided I could remember to do so!

George says: Are the others in the group under the same kind of pressure? Keen to do well and working their tails off, I mean?

Here and there in George's utterances are sprinkled my best attempts to use American idiom. People do not "work their tails off" in Northern Ireland, they just work wild hard.

Jack says: Pressure, yes. Working their tails off, no. At least not until recently. I like to get my work done early, so I've been really hitting it hard. Now it's the end of the semester and they finally got religion!

George says: Do you feel bad about Jody in particular. She's the leader and she has a lot of responsibility; is she putting all the pressure on you?

Jack says: Well I like Jody. And we've worked together well in several classes. At first I was mad at her, but now it seems others are giving me a hard time too, and I wonder if she's just reflecting their feelings - or did I really do something wrong?

George thinks: Now's the time to start exploring some possible action - but I'll ask about past action rather than suggest future action.

George says: Are you sure Jody knows about your grandmother and how much you were worrying about her?

Jack says: No I'm not sure. When we talked I was so bummed out that I didn't want to say ANYTHING to her. I just walked out.

George thinks: He's certainly going to have to talk to Jody soon. But those other things he was saying seem to suggest that the real problem may be with the others.

George says: What about the others? Do they know why you had to take off at the crucial time in the whole project?

Jack says: Well it was the crucial time, but I'd already done MORE than my assignment. Yes they knew why I took off, but maybe not how important my family is to me.

George says: Maybe it's a good idea to share your feelings when things like this happen.

Jack says: Yeah, but I feel a little like an outsider -- like they're not particularly interested in my feelings.

George says: I think I see what you mean - as far as they are concerned you are the workhorse and the workhorse is not allowed to go out to graze; sorry for the metaphor, but you know what I mean?

I worried about this metaphor as soon as I made George use it. Equating the visit to his grandmother to "grazing" wasn't a good thing to do; hence the apology.

Jack says: That's close. I do my work alright. But "workhorse" implies they really need me, and I'm not so sure they think that. Jody is a friend of sorts to me, but she also likes to butter up the others who are in her sorority.

George says: How did you divide up the work? Did Jody assign different tasks to each of you or did you choose what your contribution was going to be?

Jack says: We had a planning session and discussed what work had to be done. I said I like to gather information, so everybody said that would be o.k. I did it, but then when they started working on their parts they wanted different information than I found.

George says: Are you saying that what started out as an agreed set of equal tasks turned out to be unbalanced to your disadvantage?

Jack says: Well not exactly. You always need to get more information as you get into a project. And I was feeling fine about getting it. It's just that I had to leave for the funeral when they wanted more.

George thinks: Here we go; he's not suggested any action of any sort - probably because my questions haven't invited it. I think I'll have to start getting into possible futures.

This was the point were I decided that I had enough "information", and Jack had had enough opportunity to "have his say", for me to start moving towards defining action to be taken.

George says: If you had another planning meeting right now what do you think might be achieved?

But I crept up on the idea of "action" pretty slowly and carefully. I wanted the idea if possible, and certainly the commitment to action, to come from Jack.

Jack says: It's a little late to be planning. The project is due in three days. On the other hand I've felt really estranged from the rest of the group and maybe I could offer to do some of the finish-up work.

George says: That's a good idea, I'd say - but only if you feel up to it. And you never know, maybe the others have been making progress without you realising. Maybe "planning" was the wrong word to use. How about a "review" meeting?

I suppose this could be classified as the "Facilitating Helpee Acting" stage in the helping model.

Jack says: Your idea makes sense but I'm not sure I'm up to it.

George says: You might surprise yourself - especially if you don't go into the meeting with expectations that are too high. Do you think Jody might be prepared to see if the group is ready to wrap things up as a team rather than a set of individuals.

George thinks: Maybe that's an over statement! The others may be a team and Jack the isolate - but I'd rather not get into that too deeply unless it comes out even more strongly.

Jack thinks: I'm feeling a little pushed by George on this. He wants me to go right into the group or at least to Jody, but I'm afraid to do it -- what if they ostracise me?

Jack says: Well, that's possible, George. I don't know. Maybe I need to think about it a little first.

George says: Do you remember saying that you wondered if something else might be going on - I remember you saying something like that. Did you have anything in particular in mind?

For a moment I didn't know what to say next. (Note how I've expressed this!! There was no question but that at this stage of the dialogue I was fully identified with George, wanting him to say the right things to help Jack out of the hole he was in.) Then I remembered the hanging remark and the question I'd been saving.

Jack says: Well it's hard to say. There seems to be a clique of several people in the group. Not everybody. But people in the clique seem to sneer whenever I say anything.

George says: Do these people have anything positive to say about the work? Do they make good contributions or do you feel that they are just freeloaders?

Jack says: They're certainly not freeloaders. Actually they work hard -- they just worked on other things at the first of the semester. They make good contributions. One of the clique is actually very supportive of all of us. And one person is very caught up in herself.

George says: Maybe it all hangs on your original choice task. Sometimes people think that gathering information is a low level job and that synthesising it is what really counts. Maybe they are stuck with the notion that all you can do is the donkey work?

I was happy enough with this - it looked as if "we" were moving away from taking action, but I knew we would soon return to it.

Jack says: You could be right. And if the group is like most groups, they're still synthesising right now. Maybe if I talk with Jody I could see if I can make a contribution.

Here was the statement about action that I had been hoping for - and of course Olin knew that I was looking for it. Finally Olin made Jack say what we both knew I wanted him to say.

George says: That sounds like a good idea to me. You must have been doing some drawing together of the threads yourself - in your head if not on paper. Why don't you have a shot at contributing to the analysis and conclusions; that way you'll get some personal satisfaction at least.

Jack says: O.K. I have a class with Jody next period. I'll see if I can help. Thanks, George!

My experience is that once "resolution" is offered by one of the characters, the other one tends to take it up pretty quickly. We had been "working" for about 45 minutes and were ready to stop.

George thinks: Now we wait to see how he acts in the world.

This is always the real test (of bubble dialogue as well as everything else).

Organizational Learning and Hypermedia: An Action Science Perspective

Emrah Orhun

Bilgisayar Mühendisligi Bölümü Ege Üniversitesi, Bornova, Izmir, Turkey

Abstract. Action science views organizational learning as a process of collaborative inquiry in learning context. Computer support for organizational learning needs to facilitate the representation of the logic embedded in social action in a form that will allow the knowledge to be falsifiable, usable online, and informative of alternatives. This requires integration of mediational means for organizational action and the capacity to observe and discuss action. Hypermedia has a potential to meet these requirements.

Keywords. Organizational learning, computer-supported cooperative work, action science, hypermedia

1. Introduction

Hypermedia technology provides new methods for information representation and management around a network of multi-media information units connected together by links. Organizational learning is a process through which members of an organization act as learning agents for the organization to respond to changes in the internal and external environments by detecting and correcting errors in their actions (Argyris & Schön, 1978). The aim of this paper is to examine the ways in which hypermedia can be used to support the process of organizational learning.

Human action typically employs mediational means such as tools and language, and these mediational means shape the action (Vygotsky, 1978). Consequently, technology is not only the design of artifacts, but it is the design of practices and possibilities to be realized through artifacts (Flores et al., 1988). The introduction of a computer-based information system into an organization is a redesign of work which intervenes with the practices of a community of people (Winograd, 1986).

The notion of cooperation is of vital importance in the design of any technical means to support human interaction (Holand & Danielson, 1991). Computer-supported cooperative work (CSCW) is the field of study that endeavors to understand the nature and characteristics of cooperative work with the objective of designing adequate computer-based support. The CSCW field has not yet produced "a theoretical framework that will help us understand the complex interactions between the technical subsystem, the work organization, and the requirements of the task environment and guide us in designing a usable system to support collaborative work" (Bannon & Schmidt, 1991).

The adoption and usage of CSCW applications have been limited despite the significant improvements and the level of collaboration they would provide in the workplaces. The major explanation for this focusses on the CSCW's reliance on concepts with strong positive connotations such as "cooperation" and "collaboration" while understating the levels of conflict, control, and coercion (Kling, 1991). The impact of CSCW on work depends on preexisting patterns of authority, obligation, and cooperation, and an organization's openness to change, as well as the technology's information processing capabilities.

The first question that needs to be addressed in this paper is "What theory of organizational learning can inform us in designing computer support for organizational learning?".

Such a theoretical framework for organizational learning is shaped by the view of organization, the notion of learning and the way one thinks about intervention.

Argyris and Schön (1978) consider the main approaches to organizational learning in the categories which view organization as a group, agent, structure, system, culture, or politics representing the perspectives of disciplines such as sociology, psychology, anthropology, systems theory and political science.

The mainstream approach to organization theory during the past four decades has emerged from the view of organizations as agents involved in the efforts to achieve objectives by the rational selection of appropriate means (Argyris & Schön, 1978). The focus in this approach is upon the acquisition and application of knowledge useful for effective performance of organizational tasks, and the organizational world is conceived as knowable through scientific method. The prominent elements are decisions, for it is through decision-making that knowledge is applied to the performance of organizational tasks. This approach has led to the development of decision-support systems and expert systems.

Underwritten by an epistemology derived from the logical positivists' analytical distinction between value and fact, the idea of decision has been challenged as a result of the insights from contemporary philosophy of language and ethnomethodological research that point to an epistemology of everyday life in which value and fact are fused (Harmon, 1989).

An alternative to the mainstream approach is action science which derives from a more plausible epistemology of social life. Action science appproach has the potential to better guide the design of computer-based systems for organizational learning due to its focus on the social context in which decision possibilities as well as interests are socially produced. The theory of action (Argyris & Schön, 1978) is one possible approach to action science that focuses on how organizations learn and attempts to integrate cultural, cognitive, social-psychological, and systems perspectives with the notion of organization as agent.

The aim of this paper can now be restated in the question "How can hypermedia be used to support the process of enhancing an organization's capacity for learning from the perspective of the theory of action?".

The paper continues in sect. 2 with an overview of the action science perspective and the theory of action. Section 3 discusses mediational requirements of organizational learning and sect. 4 examines the potential of hypermedia for supporting organizational learning.

2. Action Science and Theory of Action

Argyris et al. (1985) consider Lewin's (1948) early concepts of action research, an activity that involves studying social systems by changing them, as the seeds of action science. The intended change typically involves reeducation, a term that refers to changing patterns of thinking and acting that are presently established in individuals and groups.

Action science is an inquiry into how human beings design and implement action in relation to one another and aims to produce knowledge that can inform action (Argyris et al., 1985, p. 4). Argyris et al. (1985) suggest that action science is an examplar of critical theory as formulated by the Frankfurt School (Habermas, 1971). Critical theory is based on the distinction of three types of sciences (Berstein, 1976, p. 191). "Empirical-analytic sciences" serve "technical" interests by seeking hypothetical-deductive theories to extend the control of human beings over nature. "Historical-hermeneutic sciences" serve "practical" interests and are concerned with communicative action, the interpretation of texts and the understanding of meaning. "Critical social sciences" serve "emancipatory" interests by offering a critique of what is from the perspective of what might be.

Justification of the normative stance of critical sciences is based on internal criticism of the practices of the community to which it is addressed. The methodological framework of critical sciences builds upon self-reflection, by which humans can Schön transform their self-awareness and can act to change their world.

As an example of critical theory, action science advocates that communities of inquiry be enacted in communities of practice. Such inquiry is a form of practical deliberation, one that is guided by the norms of science as well as by norms of practice (Argyris et al., 1985). The practice of action science involves creating conditions in a community of practice in which members can participate in public reflection on matters of concern to them and also on the norms and rules of inquiry they enact.

Action science builds on an epistemology of practice that views practical knowledge as a domain of tacit knowledge that can be made explicit. Reflecting-in-action is a way of making explicit some of the tacit knowledge embedded in action (Schön, 1983).

Action science epistemology focuses on framing or problem setting, as well as on means-ends reasoning or problem solving. The agent facing a complex situation, draws on tacit knowledge to frame the situation and act. The consequences of the action generate information about the situation and about the suitability of the framing and the agent's action.

Argyris et al. (1985) contrast the implications of the action science epistemology and the mainstream epistemology for learning. With the mainstream epistemology which focusses on meands-ends rationality, failure to achieve the intended ends leads to a reexamination of means and a search for more efficient means. With the action science epistemology, failure to achieve intended consequences may lead to reflection on the original frame and the setting of a different problem. Argyris and Schön (1978) refer to the first approach as single-loop learning and to the second as double-loop learning.

Action science intends to promote learning at the level of norms and values and inquiry focusses on double-loop learning.

Argyris and Schön (1978) introduce the concept of Theory of Action as they develop an epistemology of action for the organizational context. Theory of action focuses on how human beings design their action.

Designing action requires that agents construct a simplified representation of the environment and a manageable set of causal theories that prescibe how to achieve the intended consequences. Agents learn a repertoire of concepts, schemas, strategies, and programs for drawing from their repertoire to design representations and action for unique situations. Argyris and Schön (1978) call such design programs theories of action. These theories may be thought of as a very large set of complexly related propositions of the form "In situation s, to achieve consequence c, do action a".

Argyris and Schön (1978) propose two kinds of theories of action. Espoused theories are those that an individual claims to follow. Theories-in-use are those that can be inferred from action.

Argyris and Schön (1978) belive that individual theories of action contribute to limited learning systems. They distinguish three types of organizational learning. Single-loop learning occurs when the detection and correction of organizational error permits the organization to carry on its present policies and achieve its current objectives. When organizational error is detected and corrected in ways that involve the modification of underlying norms, policies, and objectives, then double-loop learning occurs. Deutero-learning consists of inquiry into the learning system by which an organization detects and correct errors.

Argyris and Schön (1978) suggest that individual theories of action often contribute to systems which inhibit double-loop and deutero-learning. Their goal is to help individuals develop theories of action that will contribute to the kinds of inquiry on which double-loop and deutero-learning depend.

To this end they advance a perspective on organizational intervention based on action science in order to facilitate double-loop and deutero-learning. Such an intervention engages action scientists with participants in a collaborative process of critical inquiry into problems of social practice in a learning context. This process of critical inquiry is a form of public reflection and experimentation that follows rules that can themselves be used in action contexts in order to ensure the validity and usefulness of the inquiry.

Action science have devised processes and tools to achieve particular aims in the face of threats and obstacles relevant to them. In analyzing the data on action and in representing the knowledge constructed from them in order to make explicit the propositional logic embedded in social action, action scientists must meet the criteria of their practice. The constructed knowledge must be falsifiable in an action context, usable on-line by agents, and informative of alternatives.

Action science draws on three analytic tools to satisfy these criteria. The first includes the highly Abstract explanatory and normative models that guide the direction of the analytic process. Argyris and Schön (1978) have developed two such models, Model II and Model O-II, which constitute a normative epistemology for the action context. The second tool is the ladder of inference by which participants can make explicit the data they select and the inferential steps they take to negotiate meanings and construct attributions. The third tool is maps that describe the tacit logic which informs social action and theimplications of the action for the behavioral worlds of the agents.

Argyris et al. (1985) specify a set of processes and rules that guide the action scientists and the participants in the process of inquiry using the tools mentioned above.

3. Design of Computer Support for Organizational Learning

Mediational means such as tools and language shape the human action. Mediational means, and hence mediated action, are shaped by cultural, historical, and institutional factors (Vygotsky, 1978). This section addresses some of the issues in designing computer-mediated action for the context of organizational learning.

Theory of action perspective focuses on the organizational inquiry into theories-of-action held by individuals and on the individuals' ability to appreciate and transform the learning systems they form. The design of computer support for organizational learning is driven by the need to make explicit the theories-of-action held by the individuals. Certain features of organizational information such as ambiguity, vagueness, excess, sparsity, and untestability make it difficult to express theories-of-action and to recognize inadequacies in them. Computer support for organizational learning has to provide capacity for discussions about the computer-mediated learning system as well as for observing and discussing action.

The integration of mediational means for organizational action and the capacity to observe and discuss action may enhance the capacity for organizational learning as this can facilitate the connection of abstract knowledge to action.

The rest of this section addresses the issues of design methodology, work models, and representation relevant to the design of computer support for organizational learning.

3.1 Design Methodology

From the action science perspective design of computer support for organizational learning is part of the critical inquiry into how an organization does and does not learn and how it can be transformed. As such it is an intervention into the ongoing practices. A design may be viewed as a proposal describing a desired state of world and should be subject to public experimentation and agreement as any other hypothesis and be guided by the norms of the action science.

Such a design may prove to be productive or counter-productive and should be generated and assessed by the people involved. This leads to a conception of the design process as a cooperative prototyping through which user community envision future work (Bødker & Grønbæk, 1991) and participate actively and creatively in cooperation with designers.

Cooperative prototyping requires that the group of designers and users have access to tools that support rapid development and modification of prototypes (Bødker & Grønbæk, 1991).

3.2 Work Models

Various models of work developed during the last decade have focused on office functions and procedures with a view of developing and automating well-defined functions and procedures (Auromaki et al., 1988). Recent models of work have focused on the social aspects of work and have viewed organizations as systems of communicative action engaged in cooperative work. These models are concerned with the processes involved both at the level of individual turns in an activity (e.g., models based on speech act theory (Äuramaki et al., 1988; Flores et al., 1988), and at the level of describing the activity itself (Hennesy, 1991).

Hennesy (1991) notes that most of these models include some structural aspects of an organization such as departments, roles, people, and workspaces (Smith et al., 1991) as well as the procedural aspects of the activity but neglect the aspects of information sharing. Activity oriented models are based on the concept of role playing within structured activities. These models aim to provide a mechanism, often a language, for describing different patterns of cooperative work (Benford, 1991).

Those models of work which view organizations as structures for the social coordination of action generated in conversations (Flores et al., 1988) require that the computer supports the organization, Schöncoordination, and distribution of action in time and space (Bowers, 1991). There are two issues in providing such support: articulation, and sharing an information space.

Bannon and Schmidt (1991) cite Strauss (1985) who has suggested that cooperative work requires "articulation work" which involves all the tasks needed to coordinate a particular task. Bannon and Schmidt emphasize that computer support for cooperative work should have facilities that allow users to freely negotiate task allocation and articulation by providing multiple alternative channels of interaction. Those alternative channels should allow both ambiguity and clarity that would be required in any non-trivial collective activity.

Irish and Trigg (1989) suggest that the kinds of collaborative activities that should be supported by the computer include representing the work, discussion about the work, and discussion about the shared use of the medium. The authors refer to these activities as substantive, annotative, and procedural. They suggest that supporting of the meta-discussions about the shared use of the medium may allow the discussion to be connected to the work. Another advantage of supporting meta-discussions would be the possibility of recording and preserving the rationale for the decisions taken.

How to achieve mutual intelligibility in a shared information space is a core problem in the CSCW field (Halasz, 1988). As cooperative work involves interaction of people with different conceptual frameworks in a distributed way, computer support for cooperative work faces problems in addition to the technical problems such as concurrency control in multi-user applications (Bannon & Schmidt, 1991). In order to facilitate a sufficient degree of understanding of the actions of a collaborator, a shared information space must provide contextual knowledge of the conceptual framework of the originator in addition to task-related information. The shared information space must be transparent as the identification of the originator is needed for the assessment of the information.

The action science approach to organizational learning focuses on framing or problem setting as well as problem solving. The collaborative inquiry for organizational learning examines both problem-solving means and the framing of the problem. The representation of the alternative perspectives on that framing is essential for the possibility of double-loop learning. As the learning system is also subject to inquiry, the procedures and rules for sharing the medium need to be understood and, if necessary, changed by the collaborators.

3.3 Notations and Representations

The question of formalism and selection of notations for representing work have been central to many contributions to computer-supported cooperative work. Bowers (1991) notes that a persistent assumption in several work is that to represent something explicitly makes people aware and in control of what previously was unconscious and controlled them. He suggests that this is the intention behind the arguments for making speech types explicit in The Coordinator (Flores et al., 1988).

There seems to be general agreement among CSCW researchers on the requirement that the medium should embody constitutive distinctions of human social action which are generative of the complex organizational phenomena (Flores et al., 1988) and that alternative channels of interaction should allow both clarity and ambiguity. The extent of formalism and consequences of the chosen notations however have created disagreements. For example, the Coordinator (Flores et al., 1988) supports conversational actions through formal requests, offers, promises, etc. and a free-form conversation for action. Bannon and Schmidt (1991) quote Robinson (1989) who has noted how some reviewers have criticized the system because it forces people to be explicit about their comments and has objected to the exclusion of the "cultural" dimension of conversation.

Bowers (1991) discusses the incompleteness of representations and the problem of when to stop formalization.

Schön Representations restrict the range of possibilities for action and our interpretations of it by imposing categories. As a result of those restrictions, however, they develop a potential to anticipate and identify breakdowns and to inform action more effectively due to easier retrieval.

Another issue related to representation is the extent of automation. It may be possible to automate some of the tasks related to the organization, coordination, and distribution of action in time and space. The Coordinator automatically generates a new message with a generic text corresponding to the choice of act. For example, if the response is "Promise", the initial message is "I promise to do as you request" (Flores et al., 1988). The user can augment or replace this text using embedded wordprocessing facilities. Kreifelts et al. (1991) describe the design of an automated mediator which coordinates and synchronizes asynchronous message exchange.

Lai et al. (1988) argue that when we understand enough to formalize in a computer system some, but not all, of the knowledge relevant to acting in a given situation, "semiformal" systems are most useful. They define the properties of such systems as: (1) representing and automatically processing certain information in formally specified ways; (2) allowing humans to process the same or other information in ways that are not formally specified; and (3) allowing the boundary between formal processing by computers and informal processing by people to be easily changed.

Object Lens is such a semiformal system which allows users to represent information about people, tasks, products, messages etc in a form that can be processed by both them and the computer.

The creation of a flexible semiformal system requires the knowledge to be represented in a visible and changeable way. These requirements can be met by providing semistructured objects and semiautonomous agents with template-based interfaces that make the knowledge visible and changeable.

This model of semiformal system describes a flexible and general computer system for cooperative work that can be tailored to the needs of a particular domain. The implementation of a semiformal system will benefit from integrating features of object-oriented data-bases, electronic messaging, knowledge-based systems and hypermedia. The next section discusses the potential of hypermedia for supporting cooperative work and organizational learning.

4. Hypermedia in Cooperative Work

4.1 Characteristics of Hypermedia Systems

Hypermedia provides new methods for information representation, management, and presentation, and has been in increasing use in the organization and manipulation of irregularly structured information in applications ranging from software engineering to collaborative authoring.

Hypermedia is a generalization of the hypertext concept and allows non-sequential representation of interrelated information which can be graphics, video or sound as well as text.

Most hypermedia systems can be characterized by the following features (Akscyn et al., 1988):

- Information is represented in a data-base of typed units variously called nodes, frames, notecards, etc.
- Units of information are interconnected by links to form a semantic network. Users navigate in the network from unit to unit by selecting links.
- Information access is accomplished using "browsers" containing a graphical map of the network structure or by simple query-based search.
- A programmer's interface allows the extension and tailoring of the system.

Workstation-based systems that have been applied in many domains include Notecards (Halasz, 1988), Intermedia (Yankelovich et al., 1988), and KMS (Akscyn et al., 1988). Hypercard, Guide, and Hyperties are among the commercially available PC-based systems with more limited functionality.

The potential power and limitations of hypermedia systems have been discussed by several authors. The main source of hypermedia's claimed power is the capability for external representation of human associative memory that would augment the human cognitive performance (Bush, 1945).

Early proponents (Bush, 1945; Engelbart, 1968; Nelson, 1981) of hypermedia viewed the connections in hypermedia as linked ideas and suggested that associations could be easily formed, and that information could be personalized, freely annotated, freely viewed, and readily accessed (Jonassen, 1989).

The limitations of hypermedia systems have been outlined by Conklin(1987) and Halasz(1988). The major problem of hypermedia is disorientation, or "getting lost", in which users forget where they are in the system, cannot discover how to get somewhere else they know, or forget what is in the system (Conklin, 1987).

A second problem arises due to the cognitive overhead needed to create, manage and choose among links (Conklin, 1987).

Other limitations of hypermedia have been discussed by Halasz(1988) in addressing the issues of composite structures, virtual or dynamically-determined structures, computation, versioning, support for collaborative work, and extensibility and tailorability.

Numerous efforts have been undertaken to create more adequate mechanisms for managing the loose connections of unstructured information that are characteristic of hypermedia, including mechanisms for accessing, organizing and filtering it (Neuwirth et al., 1989).

General mechanisms proposed for overcoming the limitations of hypermedia include graphical browsers, query/search mechanisms (Halasz, 1988), and paths

(Zellweger, 1988). Task-specific mechanisms on the other hand focus on information structures and processing mechanisms of the application domain such as writing (Neuwirth et al., 1989) or argumentation (Conklin & Begeman, 1988).

4.2 Hypermedia in Cooperative Work

The notion of supporting cooperative work has been in the agenda of hypermedia research since the beginnings. Bush's (1945) "memex" was intended to be part of a global information network. In Engelbart's (1975) NLS system computer support was provided to enhance communication between "knowledge workers". Nelson's (1981) Xanadu project visions having all the world's literature online in a single unified system to which people can add their own links and annotations on others' work as well as their own text.

The following outline of the research findings on the collaborative use of some hypermedia systems identifies the limitations of hypermedia systems for supporting cooperative work.

Irish and Trigg (1989) report on their experience with using Notecards for a variety of collaborative work including document authoring, legal, scientific, and public policy argumentation, and design and development of instructional materials.

They argue that hypermedia is an appropriate medium for supporting collaborative work, as it can represent the work, the discussion about the work, and the discussion about the shared use of the medium. They claim that in a hypermedia-based medium the discussion can be grounded by linking from the meta-discussion to the work itself. By recording such discussion explicitly in the medium, the rationale for the decisions taken can be preserved for historical purposes. They have observed in some cases that such discussions lead to proposals for tailored tools to support a particular adopted convention.

Notecards was built around the concept of notecards which are stored as a network in a structured file called a notefile. The mechanics of information sharing is limited as simultaneous access by multiple users to notefiles is not supported and cross-referencing between notefiles is restricted.

The absence of specific support for annotation and meta-discussion in an early version of Notecards led users to develop their own conventions and techniques for collaborative use of the system. Irish and Trigg (1989) describe how these conventions and techniques were incorporated later into collaboration-specific tools. One such tool was called history cards and semi-automated the creation of history cards that describe the work done and the links to the relevant cards.

The authors identify a need to provide ways to tie events together, thus forming a more coherent story about the changes made to a network. They propose to integrate history card facilities with the Guided Tours facility so that history cards can be followed as a linear, coherent path. They also argue for providing support for the process of adopting and automating conventions regarding the use of the medium.

Catlin et al. (1989) describe extensions for Intermedia to better support small groups of collaborators, particularly those involved with document review and revision. The Intermedia system povided support for multiple users linking to and from a document simultaneously and for creating links between fine-grained selections in multiple document types. The extensions provided the functionality for annotation creation and management, and revision. Intermedia's navigational linking mechanism was extended so that the links in the system can be used for both navigation and data transfer.

The project showed the need for a number of communications tools to help groups of authors and annotators coordinate their efforts. Catlin et al. (1989) suggest that more than simple email is necessary. They require facilities for synchronously communicating with users currently working on the same task, as well as asynchronously sending messages to users not currently working.

KMS's features for collaboration has been discussed by Akscyn et al. (1988). KMS is a distributed hypermedia system for workstations which has been used to support organization-wide collaboration for a broad range of applications including electronic publishing, software engineering, online documentation, project management and computer-aided instruction.

The unit of information in KMS is screen-sized workspaces called frames. A link is a property of an item in the frame. There are two types of linked items, tree items and annotation items. KMS allows annotations to be placed directly on the frame they refer to. Annotations are ignored by KMS programs that process hierarchies and therefore do not appear in the author's "official" document. Akscyn et al. (1988) claim that those features provide increased incentive to users to comment and encourage authors to give other people early access to their documents. They suggest that meta-discussions can be conducted by sharing frames without the need for other communication systems such as email or bulletin boards.

gIBIS (Conklin & Begeman, 1988) is an application-specific hypertext system designed to facilitate the capture of early design deliberations. It implements a model of argumentation called Issue Based Information Systems (Rittel, 1980) which has been developed for complex design problems. The IBIS model focuses on the articulation of Issues, Positions that resolve Issues, and Arguments for the Positions. An argumentation network is formed by linking together Issues, Positions, and Arguments. There are nine kinds of links in IBIS. For example, Arguments can be linked to their Positions with either Supports or Objects-to links.

gIBIS system facilitates building and browsing IBIS networks by cooperating design team members over a local-area network. gIBIS provides multiuser support including shared, coordinated access to IBIS networks, automatic notification of changes to the networks, and access control and locking.

Conklin and Begeman (1988) report that many users of the gIBIS tool have regarded IBIS "as a powerful method for research thinking and design deliberation". The explicit rhetorical structure of IBIS which makes apparent the general structure of an unfolding discussion has been found useful.

Conklin and Begeman(1988) have identified several shortcomings in the gIBIS implementation. They suggest that further support for issue resolution and linking of design artifacts or traditional documents to the issue-based network are important. They point to the problem of intelligibility in the shared information space. To quote:

"Unless each author was careful to write clearly and completely, the readers found that, while they had a sense of understanding the individual nodes, they could not follow the thread of the writer's thoughts as it wound through several dozen nodes."

They are trying to attack this problem with higher level constructs that aggregate a set of nodes and linearize the network.

Regarding the problem of disorientation in large networks, they suggest to divide the networks into regions that are meaningful in terms of the problem.

One common problem with semiformal systems arises when one is trying to map ideas into discrete structures, particularly when the problem is not well understood and ideas are vague and shifting. gIBIS users have required support for recording ideas before structuring them.

Conklin and Begeman(1988) discuss the difficulty of placing meta-discussions in the IBIS framework. Meta-discussions placed in the network "adds complexity to the browser display without illuminating the substance of the problem being discussed and initiates a discussion that may well lead to a change in the network, after which the meta-discussion will have only historical interest". They are experimenting with a scheme whereby a user may append a "metaline" at the end of the body of any node and may then begin an annotative or procedural discussion in that part of the node.

The two issues of articulation and sharing of an information space in the use of hypermedia for cooperative work have been discussed by Halasz (1988). For supporting articulation work, Halasz refers to the suggestion by Trigg et al. (1986) that existing hypermedia systems focus primarily on supporting substansive and annotative activities.

Regarding the sharing of information space, hypermedia systems need improvement both in the mechanics of sharing and in the support for mutual intelligibility. Supporting the mechanics of simultaneous multi-user access involves incorporating and extending the facilities of technologies for shared databases such as concurrency control and change notification.

Achievement of mutual intelligibility in hypermedia-based collaborations requires the development of a rhetoric of hypermedia (Landow, 1987). This rhetoric will provide conventions for creating hypermedia links that will be understandable by others who share the rhetoric.

Availability of a sufficient versioning mechanism to maintain and manipulate a history of changes to a network becomes more important for collaborative use of hypermedia. Such a mechanism may also support partitioning of the network to provide private contexts through which users interact with the shared network (Halasz, 1988) and may allow simultaneous exploration of several alternate configurations.

5. Conclusions

This paper has attempted to assess the potential of hypermedia for supporting organizational learning. An action science perspective has been taken toward organizational learning, viewing it as a process of collaborative inquiry into the practices, norms and values as well the the learning system by which an organization detects and corrects errors. A distinction of this perspective is making explicit the tacit theories-of-action held by the members. Computer support for organizational learning must facilitate the Schönrepresentation of the logic embedded in social action in a form that will allow the knowledge to be falsifiable in an action context, usable online by agents, and informative of alternatives.

This requires a computer environment that will integrate mediational means for organizational action and the capacity to observe and discuss action. Furthermore, it is desirable to have the capability to automate some of the tasks related to the organization, coordination, and distribution of action in time and space.

These requirements lead to the notion of semiformal systems with which users represent and process information both formally and informally, and partially automatically, while being able to control the extent of formality and automation.

Most hypermedia systems fall short of fully meeting these requirements. Removal of the hypermedia's shortcoming requires the resolution of some of the issues - e.g., computation, virtual structures - discussed by Halasz (1988). The integration of hypermedia and knowledge-based systems would be beneficial to the advancement of hypermedia systems in this respect.

Hypermedia systems have been used to support cooperative work ranging from loosely-coupled and unstructured (e.g., writing separate papers) to tightly-coupled and structured (e.g., software engineering). The wide applicability of hypermedia systems stems from the flexibility of the basic hypermedia model and the support provided for extensibility and tailorability.

The potential power of a hypermedia-mediated work environment is derived through structuring a hypermedia system to fit with the documents, processing mechanisms, tasks, settings, and their interrelationships specific within that domain (Scacchi, 1989). Supporting organizational learning will similarly require structuring a hypermedia system to the context of work as well as to the process of collaborative inquiry into organizational action.

Success in realizing the potential power of a hypermedia-supported organizational learning environment will depend on how effective the system will be in supporting the social interactions among people. A collaborative prototyping approach to the design of the system will be a necessary condition for this. Achievement of the desired degrees of visibility, changibility, intelligibility, and ease-of-use requires more work to address the main issues of hypermedia.

Whether a hypermedia-supported intervention into an organization's capacity for learning will be productive needs to be tested in the collaborative inquiry into the organizational learning system. Such testing can be guided by the norms and rules of action science. A sustained organizational interest in change and learning will be essential to overcome many breakdowns such an intervention will create.

Acknowledgements

I would like to thank Birgül Egeli for presenting at the Workshop a description of our work at Ege University on hypermedia and learning. The present paper reflects on the contents of the Workshop and has benefitted from conversations with colleagues at the Shell Centre for Mathematics Education. This work has received support from Turkish Scientific and Technical Research Council.

References

Akscyn, R.M., McCracken, D.L., & Yoder, E.A. (1988). KMS: A distributed hypermedia system for managing knowledge in organizations. Communications of the ACM, 31(7), 820-835

Argyris, C., & Schön, D.A. (1978). Organizational learning: a theory of action perspective. Reading, MA: Addison-Wesley

Argyris, C., Putnam, R., & Smith, D.M. (1985). Action science. San Francisco, CA: Jossey-Bass

Äuromaki, E., Lehtinen, E., & Lyytinen, K. (1988). A speech-act-based office modeling approach. ACM Transactions on Office Information Systems, 6(2)

Bannon, L.J., & Schmidt, K. (1991). CSCW: four characters in search of a context. In: J.M. Bowers & S.D. Benford (Eds.), Studies in Computer Supported Cooperative Work. Elsevier Science, 3-16

Berstein, R.J. (1976). The restructuring of social and political theory. Philadelphia, PA: University of Pennsylvania Press

Benford, S. (1991). Requirements of activity management. In: J.M. Bowers & S.D. Benford (Eds.), Studies in Computer Supported Cooperative Work. Elsevier Science, 285-298

Bødker, S., & Grønbæk (1991). Cooperative prototyping studies -users and designers envision a dental case record system. In: J.M. Bowers & S.D. Benford (Eds.), Studies in Computer Supported Cooperative Work. Elsevier Science, 315-322

Bowers, J.M. (1991). The Janus faces of design: some critical questions for CSCW. In: J.M. Bowers & S.D. Benford (Eds.), Studies in Computer Supported Cooperative Work. Elsevier Science, 335-350

Bush, V. (1945). As we may think. Atlantic Monthly, 176(1), 101-108

Catlin, T., Bush, P., & Yankelovich, N. (1989). Internote: extending a hypermedia framework to support annotative collaboration. Proceedings, Hypertext '89

Conklin, J. (1987). Hypertext: an introduction and survey. Computer 20(9), 17-41

Conklin, J., & Begeman, M.L. (1988). gIBIS: a hypertext tool for exploratory policy discussion. ACM Transactions on Office Information Systems 6(4), 303-331

Engelbart, D.C., & English, W.K. (1968). A research center for augmenting human intellect. Proceedings of 1968 Fall Joint Computer Conference, 33 Part 1, Montvale, NJ: AFIPS Press, 395-410

Engelbart, D.C. (1975). NLS teleconferencing features: The journal, and shared-screen telephoning. Digest of papers, IEEE Computer Society Conference, 173-176

Flores, F., Graves, M., Hartfield, B., & Winograd, T. (1988). Computer systems and the design of organizational interaction. ACM Transactions on Office Information Systems, 6(2), 153-172

Habermas, J. (1971). Knowledge and human interests. Boston: Beacon Press

Halasz, F.G. (1988). Reflections on Notecards: seven issues for the next generation of hypermedia systems. Communications of the ACM 31(7)

Harmon, M.M. (1989). "Decision" and "Action" as contrasting perspectives in organization theory, Public Administration Review, March/April, 144-150

Hennesy, P. (1991). Information domains in CSCW. In: J.M.Bowers & S.D.Benford (Eds.), Studies in Computer Supported Cooperative Work. Elsevier Science, 299-312

Holand, U., & Danielson, T. (1991). Describing cooperation. In: J.M. Bowers & S.D. Benford (Eds.), Studies in Computer Supported Cooperative Work. Elsevier Science, 17-27

Irish, P.M., & Trigg, R.H. (1989). Supporting collaboration in hypermedia: issues and experiences. Journal of the American Society for Information Science, 40(3), 192-199

Jonassen, D.H. (1989). Hypertext/hypermedia. Englewood Cliffs, NJ: Educational Technology Publications

Kling, R. (1991). Cooperation, coordination and control in computer-supported work. Communications of the ACM, 34(12), 83-88

Kreifelts, T., Victor, F., Woetzel, G., & Woitass, M. (1991). A design tool for autonomous group agents. In: J.M. Bowers & S.D. Benford (Eds.), Studies in Computer Supported Cooperative Work. Elsevier Science, 131-144

Lai, K-Y, Malone, T.W., & Yu, K-C. (1989). Object lens: a "spreadsheet" for cooperative work. ACM Transactions on Office Information Systems, 6(4), 332-353

Landow, G.P. (1987). Relationally encoded links and the rhetoric of hypertext. Proceedings, Hypertext '87 Workshop, University of North Carolina at Chapell Hill, November. ACM, New York, 331-343

Lewin, K. (1948). Action research and minority problems. In G. Lewin (Ed.), Resolving Social Conflicts. New York: Harper & Row

Nelson, T. (1981). Literary machines. Swartmore, PA: T. Nelson

Neuwirth, C.M., & Kaufer, D.S. (1989). The role of external representations in the writing process: implications for the design of hypertext-based writing tools. Proceedings, Hypertext '89, 319-339

Robinson, M. (1989). Double level languages and co-operative working. Support, Society and Culture. Mutual Issues of Cybernetics and Science. Proceedings, March 27–April 1, 1989. Amsterdam, 79-114

Rittel, H. (1980). APIS: a concept for an argumentative planning information system. Working Paper No. 324, Institute of Urban and Regional Development, University of California, Berkeley, CA

Scacchi, W. (1989). On the power of domain-specific hypertext environments. Journal of the American Society for Information Science, 40(3), pp. 183-191

Schön, D.A. (1983). The reflective practioner. New York: Basic Books

Smith, H., Hennessy, P., & Lunt,G. (1991). An object-oriented framework for modelling organisational communication. In: J.M. Bowers & S.D. Benford (Eds.), Studies in Computer Supported Cooperative Work. Elsevier Science, 145-158

Strauss, A. (1985). Work and the division of labor. The Sociological Quarterly 26(1), 1-19

Trigg, R., Suchman, L., & Halasz, F. (1986). Supporting collaboration in Notecards. Proceedings, Conference on Computer-Supported Cooperative Work, December 3-5, Austin, TX, 147-153

Vygotsky, L.S. (1978). Mind in society. M. Cole, V. John-Steiner, S. Scribner, & E. Souberman (Eds.). Cambridge, MA: Harvard University Press

Winograd, T. (1986). A language/action perspective on the design of cooperative work. Proceedings, Conference on Computer-Supported Cooperative Work. December 3-5. Austin, TX, 203-220

Yankelovich, N., Haan, B., Meyrowitz, N., & Drucker, S. (1988). Intermedia: the concept and construction of a seamless information environment. IEEE Computer 21

Zellweger, P. (1988). Active paths through multimedia documents. In: J.C.van Vliet (Ed.), Document Manipulation and Typography. New York: Cambridge University Press, 19-34

Supporting Cooperative Work with Workflow Management Technology

Thomas Schäl

HDZ/KDI, RWTH Aachen, Dennewartstraße 27, D-52068 Aachen, Germany
and Istituto RSO, Via Leopardi 1, I-20123 Milano, Italy

Abstract. There is a growing school of thought that traditional functional management is no longer appropriate in an era when quality, innovation and services are of paramount importance. As a response to the required change a model for process working in organizations is developed. The implications for design of computer systems for cooperative work in fairly stable as well as dynamic organizations are outlined. The proposed design approach sees customer satisfaction as the key-issue for system design in the changing organizational context. New information systems promise to be feasible for turning this change and organizational restructuring into reality. Many of these new systems are coming from the field of Computer Supported Cooperative Work (CSCW). Workflow Management Technology is proposed as the most appropriate solution to support process organizations. It integrates aspects of human work traditionally separated by information system design. Support functionalities for workflow systems are outlined.

Keywords. Computer supported cooperative work, groupware, office automation, office procedure, business process automation, workflow management technology, customer satisfaction

1. Introduction

Industrialized societies find themselves in a time of pace and deep changes (Piore & Sabel, 1989). Companies can only survive and develop in this context, if they adopt quickly and sufficiently to ongoing change (Hinterhuber, 1984). Changes in the environment influence at the same time more and more the companies' internal (Beckhard & Harris, 1987). The increased pace of innovation and resulting level of complexity requires that organizations become more adaptable to internal and external conditions, e.g., by continuing education of the employees the organization becomes flexible and capable to respond to new environmental conditions. This paper is concerned with the kind of less hierarchical more flat and networked organization which is able to cope with changing tasks, technologies and environments - what is called the learning organization (Argyris & Schön, 1978; Shrivastava, 1983; Senge, 1990; Huber, 1991; Simon, 1991).

1.1 From Functional to Process Organization

Today, a growing number of organizations is managing important parts of the business in a new and different way. It is called process management, and it differs sharply from the traditional functional approach that has been used by all but the smallest organizations for more than half a century. *Process management* is the way to approach the *transformation of traditional bureaucratic organizations* into *market-oriented companies*. Process working gives organizations the possibility to dramatically improve response times, service and quality by cutting across departmental demarcations.Processes have two important characteristics: they have *customers* who are internal or external recipients of the process' output, and they *cross organizational boundaries* which separate the functional units in the hierarchical organizations. As a consequence, process management implies team working.

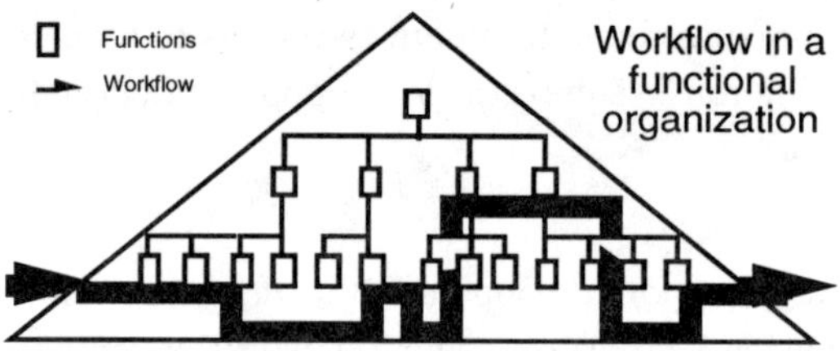

Fig. 1.1 Workflow step by step in a hierarchical organization (Source Butler & Cox, 1991)

In a conventional functional organization, a procedure typically involves multiple functions. The procedure progresses step by step as it passes between the functional units according to their assigned competences. Queries and exceptions are handled at higher levels in the hierarchy due to their declared level of responsibility, competence and decision authorization (these are the typical examples of escalation in case of a breakdown in a hierarchical organization; everybody shifts responsibility in case of trouble to the next higher level). As a consequence the procedures are highly time and cost consuming; and the quality of work is low.

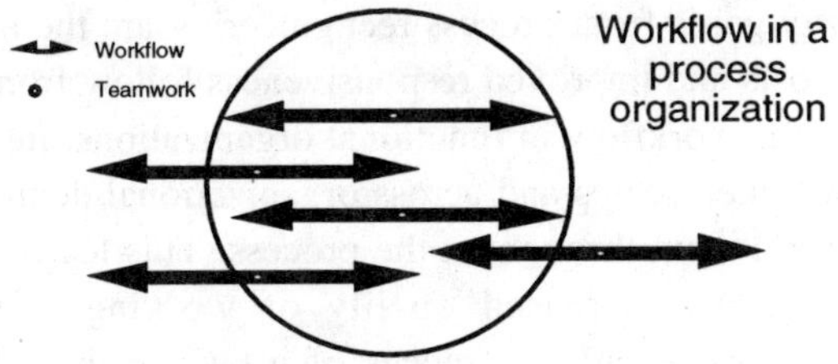

Fig. 1.2 Workflow in a process organization performed by teams (modified from Butler & Cox, 1991)

In a process organization, the procedure may involve the same set of work activities, but all these activities are managed as a single operation which is done either by an individual or by a group of people. The procedure, that is the pre-defined set of activities, becomes part of a process which has to be managed in order to achieve a specific goal. In process management the work does not follow a mono-direction stream. It frequently follows a multiple circuit path, which includes check points and feed-back to assure quality and efficiency of work.

1.2 Process Reengineering

Along with the *process organization* goes the concept of *process re-engineering*: Instead of using computers and information systems to automate the way a business has always run, the concept calls for rethinking the operations first; one reengineers the process in first place, before applying computing power to the newly defined process. This approach shifts work-analysis directly to design, as Schmidt (1990) argues:

> 'Since the objective of work analysis is to improve a given situation in some way - for instance, by designing and implementing information systems, by redesigning work [...], - the basic approach is reformist or therapeutical: the work analyst investigates a particular social system of work to change it for the better. Accordingly, the analysis cannot take the current behaviour of the social system of work for granted. To the contrary, the analyst must uncover the hidden rationale of current practices as well as the accidental choices of the past, the procedures turned rituals, the formalized mistakes, etc. [...] In a sense, then, work analysis can be linked with 'reverse engineering' - in the sense, namely, that the analyst approaches the given system as a product of a design effort and 'takes it apart' with the purpose of putting it together again, perhaps differently.' (Schmidt, 1990)

Business companies are becoming interested in process reengineering, because their competitiveness cannot be guaranteed any more only by cutting costs. The resulting productivity gains from process reengineering are the best known at the moment. Reduced costs and improved responsiveness follow from the elimination of multiple steps in the workflow in functional organizations. Instead of the work passing up and down hierarchies and across organizational demarcations, it stays with an individual or a team throughout the process. This leads to a third benefit of process working: the increased quality of working life. In functional organizations, people work only on pieces of a task in the whole process. In process working, by contrast, individuals are encouraged to carry out entire tasks from the beginning to the end. Moreover, the people who do the work undertake their own monitoring and control, and handle exceptions themselves instead of making it an escalation in the hierarchy or contributing to a centralized coordination and control function. These middle-level functions can be overcome with process working; as a consequence a lot of overhead costs are cut. Furthermore, process working provides a platform for continuing innovation. It promises competitive advantage which will be sustainable over time.

1.3 The Customer Satisfaction Paradigm for Work Process Reengineering

Attempts are being made to augment service and quality in business organizations as well as public administration. Improving service and quality means identifying clients and designing the necessary work processes to produce the services in a more customer-oriented organization. Quality is defined by matching predefined standards. However, quality standards have to be measurable. Therefore they mostly consider physical dimensions, time and other tangible attributes for which a numerical value can be defined and later compared to the delivered product or service. But sometimes the customer can be dissatisfied by high quality products, while good management of poor services is often more appreciated. Airlines report on quality (for example intended as flying on schedule and not having overbooked planes) that customer satisfaction increases when they make an error but handle it well, compared with not having made an error in the first place (Keen, 1991). Improving quality in public services involves the same steps as in the private sector. But, quality can have vastly different meanings to various public sector constituents. To real estate developers it might mean fast approvals for projects. To a local neighbourhood group it might mean a public agency which gives weight to their concerns about the developer's plans. Quality standards in this example are, e.g., efficiency for the estate developer (time) and effectiveness for the neighbourhood group (exhaustive consideration).

The examples show that quality does not guarantee the perception of good service. Instead, customer satisfaction expressed by the customer has to be *the* issue for organizational redesign.

The problem of customer satisfaction is that it cannot be measured like quality in terms of predefined standards. Customer satisfaction is a result of a good communicative relation which serves to manage commitments and breakdowns within the client/supplier chain. The supplier has to understand the service requested by the clients (in the case of public administration, those of all citizens in their huge variety). The client's satisfaction can only be understood after the delivery of services in gathering the clients' satisfaction through communicative interactions.

A new process management model has to be defined starting from the fundamental change of perspectives required to cope with the growing complexity of customer demand. The work process model has to change from a production centred (transformation) model to a market centred (language) model. Re-designing working processes, we need to look at every element of processes, work, and the information technologies that support or enable them in terms of customer satisfaction. But how to do this? Peter G.W. Keen states that:

> 'any methodology for analysing business processes must begin and end with the customer. Too many traditional approaches to business analysis, especially in the area of information technology planning, are poorly fitted to this new thinking. Many begin with the firm and work forward to the customer, instead of starting with the customer.' (Keen, 1991)

Therefore, we need powerful tools and methodologies which make customer satisfaction an integral part of the planning and design specifications of organizational design through information technology.

2. Workflow Management Technology for Process Organizations

2.1 Perspectives in Process Representation and Modelling

Many aspects of a process have to be described by its modelling, e.g., what is going to be done, who is going to do it, how and why will it be done, who is dependent on its being done, etc.. Process modelling languages differ in the extend to which they highlight the answers to these questions. Process modelling languages and representations usually present one or more different perspectives related to these questions. Traditionally, the modelling of information systems has focused on analysing data flows and transformations. This modelling accounted only for the organization's data and that portion of its processes which interacted with data. Newer uses of information technology extend computer use beyond transaction processing into communication and coordination. Successfully integrating these systems into the enterprise often requires modelling even the manual organizational processes into which these systems intervene (Curtis et al., 1992).

Medina-Mora et al. (1992) distinguish three different process types in an organization:

Material processes relate to human activities which are rooted in the *physical world*. An observer notes the moving of materials to specific places where physical things are changed, transformed, manipulated, consumed or combined into new items. This was the relevant domain in the tradition of factory automation. Material process redesign has been used to increase efficiency in production and industrial engineering.

Information processes overcome the failure of material processes to capture what is important about every-day activity in the 20th century. Theorists and information technology providers have developed sophisticated ways to analyse and facilitate the flow of information. Current techniques of data flow analysis, data-base storage and retrieval, transaction processing, network communication, etc. have provided a structure for effective information processing. What is lost in the information process perspective is the recognition that:

> 'information in itself is uninteresting. Information is only useful because someone can do something with it [...].' (Medina-Mora et al., 1992)

This perception leads to the domain of *business processes*, where people communicate in order to get something done, agree about what will be done, who is responsible for doing it, when will it be done, and whether the customer is satisfied with what has been done. Following Medina-Mora et al. (1992), business processes are implemented in information processes, just as information processes are implemented in material processes. However, in moving the focus on the language/action structure of working processes, rather than on forms or data-base transactions, one reveales a higher level of organization (Medina-Mora et al., 1992).

Process models can have different goals and objectives, depending on the intended use. Two possible categories of process modelling (from Curtis et al., 1992) reflect the objectives to:

• *facilitate human understanding and communication* which requires that a group is able to share a common representational format,

• *support process improvement* which requires a basis for defining and analysing processes

Table 2.1 Process modelling objectives and goals

Facilitate human understanding and communication
• Represent process in form understandable for humans • Enable communication about and agreement on processes • Support commitment negotiation • Allow for informal communication • Formalize the process so that people can work together more effectively • Provide sufficient information to allow an individual or team to perform the process • Provide escape hatches to handle exceptions and emergencies • Provide heuristic plans for situated action • Form a basis for training the intended process
Support process improvement
• Identify all necessary components of a process • Identify customers and conditions of satisfaction • Reuse well-defined and effective processes for future processes • Compare alternative processes • Estimate impacts of potential changes to a process without first putting them into actual practice • Assist in the selection and incorporation of technology (tools) into a process • Help people maintaining ownership of the process and its outcomes • Facilitate organizational learning regarding effective processes • Support managed evolution of a process

2.2 Workflow Definition

The fundamental unit for getting things done (products or services) in a work process is a *workflow*. A workflow is a unit of work that happens repeatedly in an organization of work. It has a beginning, a middle, and an end. And, most importantly, in a workflow things get completed that are related to, or result in, producing satisfaction for customers. In fact, every workflow has a customer, who may be a customer or client of the organization, or may be another work group or an individual in the organization (an internal customer).

Most workflows have sequential and parallel steps. They involve the movement and tracking of people, documents, products and information. Workflows reflect the recurrent organization of work, but their organizational and communication patterns are based on both standard and *ad hoc* actions taken by specific persons in order to fulfil a particular condition of satisfaction requested by someone. Some workflows are formally defined. Others are inherited from experience, without any explicit design.

Differences on how persons manage the flow of work within their cooperative network form the basis for defining three types of workflows: *case-based*, *ad-hoc* or *general purpose* workflows (Gable, 1992):

• *Case-based workflows* are central to the organization's business. Examples are insurance claim processing, bank loan application processing, client order production processing, etc. These workflows might occur in a specific department where individuals in a workgroup try to accomplish the department's objective, satisfying the case's customer.

• *Ad-hoc workflows* arise from specific, temporary needs of unique project teams. A group has a specific objective, but once the task is accomplished and the project objective is achieved, the team will disband and its members will go on to other assignments. In turn, group members often initiate other ad-hoc workflows than the initial one during the project's duration.

• *General purpose workflows* are designed to perform day-to-day administrative operations in an organization. Examples are processing purchase orders, expense reports, time sheets, etc. General purpose workflows are generally *secondary workflows* which are recurrent in case-based or ad-hoc workflows.

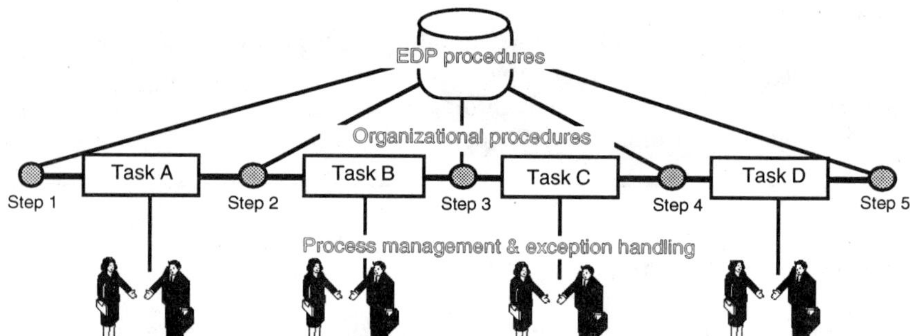

Fig. 2.1 Different aspects of a working process

Workflows can be seen as the sequence and interrelation of information, activities and communications within the process. Within this new context, organizational and EDP procedures reduce their possibility to manage the work processes; exception handling becomes a wide and ever growing area of working behaviour. In this sense workflows combine EDP procedures and organizational procedures with process management and exception handling (see Fig. 2.1).

2.3 Limitations of Procedure Processing Applications

Procedure processing applications try to formalize the work process. This is done by applying models for information flow and procedural working.

The information flow model represents office work in terms of units of information (forms, memos, etc.) and is used to define operations performed on each information unit. It describes the relation of information units in terms of source, operation, integration, destination and relation to other flows.

The procedural model is based upon the view that office work is basically procedural in nature, involving the execution by office works of pre-defined sequences of steps. It emphasizes the task-oriented nature of work and operands (units of information). Inter-procedural and inter-departimental flows are less identified than in the information flow model (Ellis & Naffah, 1987). Therefore office procedure systems are:

> 'suitable for describing office work that is structured around actions (e.g., sending a message, approving, filing) where the sequence of activities is the same except for minor variations and few exceptions. [...] These systems do not deal well with unanticipated conditions.' (Barber et al., 1983)

While the information system may be designed to match the current social structure around the identified information flow and procedural working practice, the change of technology engenders a change of the social structure. An office is a social environment to which any introduction of procedural change, goal changes or automated equipment causes perturbations. This has been a bitter experience of office automation projects and installations, designed to match the traditional allocation of tasks in the office. The office automation experience has demonstrated that the potentials of information technology in the office in terms of productivity, flexibility, product quality, etc. cannot be realized without a corresponding change in the allocation of tasks among staff (Hammer, 1984; Schmidt, 1990). Many technologically successful systems have failed due to ignorance of human and social factors. For example, the mass movement of secretaries away from individual managers to word processing pools violated social maxims. In addition, if people who needed to exchange information are moved so that they are no longer close to the coffee machine, then information which was transferred in this informal way, may not longer be exchanged (Ellis & Naffah, 1987).

Procedure automation has always concentrated on the predictable part of social work. It incorporates a model for task allocation, defined levels of responsibility, formalized patterns of communication, etc.. In all these systems, information is treated as something on which office actions operate producing information which is passed on for further actions or is stored in data-bases for later retrieval. There is nothing wrong with this modelling approach in itself, but computer systems have to be more than the rigid implementation of formal structure and control in the work process.

> 'The problem with incorporating models of plans (established procedures, organizational structures, or conceptual schemes) in computer systems is not that plans are fictitious. Rather, plans serve a heuristic function in action by identifying constraints, pitfalls and strategic positions in the field of work. [...] In fact, 'plans' may serve different functions. Consider organizational procedures, for example: Procedures may of course codify 'good practice', recipes, proven methods, efficient ways of doing things, work routines. In flexible work organizations such procedures are of little value and may actually impede flexibility. [...] Whatever the function, however, organizational procedures are not executable code but rather heuristic and vague statements to be interpreted and instantained, maybe even by means of intelligent improvisation.' (Schmidt, 1991)

Organizational procedures may, of course, codify good practice, recipes, proven methods, efficient ways of doing things, work routines, etc. (Bannon & Schmidt, 1992). However, organizations require to be able to cope with unanticipated contingencies in procedures. In this case, the perceived plan becomes a resource. Suchman (1987) has analysed the role plans play in situated actions and states that:

> 'plans are inherently vague [... and] are resources for situated action, but do not in any strong sense determine its course. While plans prepopose the embodied practices and changing circumstances of situated action, the efficiency of plans as representations comes precisely from the fact that they do not represent those practices and circumstances in all of their concrete detail.' (Suchman, 1987)

In this ambiguous relation between pan and procedure, the organization may still have the necessity to introduce some formal or statutory conditions, where non compliance of the constraints may evoke severe organizational sanctions. In any case, organizational procedures in computer systems cannot be programmed code, but rather heuristics and organizational statements that have to be interpreted to get the job done. Computer support of cooperative work should aim at supporting self-organization of cooperative settings as opposed to desrupting cooperative work by computerizing formal procedures (Bannon & Schmidt, 1992). The value of a heuristic vision of procedures lies in two main issues. The first point is generated by the nature of social processes. The second concerns dynamic changes and resulting complexity in the working environment.

The first issue relates to social processes in working together. In all social life, in societies as well as in groups and organizations, there are two basic social processes. One is the pressure to and striving for uniformity and conformity. The other process is social differentiation. Specialization in skills and divergence in opinions and values are necessary for the functioning of most groups and organizations. The resulting conflict is beneficial for both, the social group and the individuals involved. Therefore, organizations can be seen as a framework for cooperation as well as for conflicts (Kensing & Winograd, 1991). These social processes occur via communication which should be supported by information systems, e.g., by functionalities for negotiation.

Concerning the second issue, change is frequent and expected in many domains. An employee's vacation day, e.g., forces others to change their routines accordingly. Change also results from promotions, employee turnover, sickness, laws, etc.. Some of these changes are short-termed and local, turning after some time into the initial state, while others are radical (e.g., a change in law might necessitate a new procedure with different personnel).

The occurrence of unforeseen events and the notion of breakdown and its implications is a fundamental concept in cooperative work (Winograd & Flores, 1986). Its implications are generally not taken into consideration by traditional office automation procedures. They rationalize the working process only by foreseen steps and do not support unpredictable events. The general meaning of breakdown implies that something which is missing in order to proceed, is needed. These missing items can be pre-defined as in the model of a work breakdown structure in project management. This notion of breakdown implies a certain recurrence and possibility of planning for missing items (breakdowns) and resolution.

A second type of breakdown is an unforeseen event where something is missing for continuing a task, process, etc.. Running out of ink while writing a report is an example of this second type of breakdown. Whenever we have this type of breakdown which we cannot overcome by ourselves, we look for help. We communicate with others requesting them to do something for us.

This final point concerning exception handling in a working process has especially to be supported by information systems which want to succeed in supporting cooperative work. This point is crucial for process support systems, because:

> 'it is impossible, both in practice and in theory, to anticipate and provide for every contingency which might arise in carrying out, a series of tasks. No formal description of a system (or plan for its work) can thus be complete. Moreover, there is no way of guaranteeing that some contingency arising in the world will not be inconsistent with a formal description or plan for the system. [...] *Every real world system thus requires articulation to deal with the unanticipated contingencies that arise*. Articulation resolves these inconsistencies by packaging a compromise that *gets the job done.*' (Gerson & Star, 1986)

As a consequence, systems have inherently to be 'open-ended' with escape hatches to handle unanticipated exceptions and emergencies. This is much more tenable than aiming toward notions of total automation of procedures, total removal of paper from the office or peopleless offices (Ellis & Naffah, 1987). The whole issue for CSCW applications discussed in this chapter is summarized in the statement that:

> 'organizational models in CSCW applications should be conceived as *resources* for competent and responsible workers. That is, the system should make the underlying model accessible to users and, indeed, support users in interpreting the procedure, evaluate its rationale and implications. It should support the users in applying and adapting the model to the situation at hand. It should allow the users to tamper with the way it is instantiated in the current situation, execute it or circumvent it, etc.. The system should even support users in modifying the underlying model and creating new models in accordance with the changing organizational realities and needs. [...] In all this, the system should support the process of negotiating the interpretation of the underlying model, annotate the model or aspects of it, etc.' (Bannon & Schmidt, 1992)

2.4 Workflow Management Technology

On the basis of past experience of procedure automation systems, new software products and enhanced office information systems' functionalities are developed which fall under the new domain of workflow management technology. It is not a new technology for procedure automation, but takes the formal part of procedure processing applications and combines it with general communication and information sharing facilities. This added value to traditional office automation results from support requirements for cooperative work which have not been considered before (Schäl & Zeller, 1991).

Workflow management technology can be divided into the three general categories of document routing, ad-hoc workgroup tools and process automation.

Document routing at its simplest establishes the flow of information and routes documents accordingly. It represents, however, only a small portion of what is required to manage and track the flow of work in a process. Products in this arena are, e.g., Staffware, Workhorse and FileNet.

Ad-hoc workgroup tools, such as Lotus Notes or The Coordinator, provide some measures of workflow management technology through standard message based templates or forms. The flow component of such systems is generally a dedicated server which can either route messages addressed to general destinations (such as 'administrator'), or to a specific person (such as 'Mr. Smith'), or to automated agents that take action on behalf of people (such as processing information).

These tools show different views of the state of work depending on who does the inquiry. Although these capabilities can provide effective solutions to some users, they are not adequate for organization-wide workflow support. These products have to be modified for integration with existing information systems, especially if a mainfraime is involved to track transaction across the enterprise networks.

Process automation refers to the systems designed to define business processes and to implement them in software. At its best, workflow management technology encompasses both a methodology for redesigning work and the means to automate and manage the resulting new workflows. In any case, without understanding business processes, it is impossible to build software to automate and manage workflows in a reasonable way. Systems in this arena are, e.g., ActionWorkflow, X-Workflow, Workman and ProcessIT.

Emerging workflow technology in the area of CSCW addresses the different aspects of cooperative work in their temporal distribution in the process. In particular workflow management technology allows the reintegration of previously separate communication, information and data flows in a working process at any time and in its unpredictable complexity. At the same time, workflow technology is an environment which supports the integration of previously separate Information Systems (Office Information Systems, Electronic Data Processing, Telecommunication Networks, etc) (Schäl, 1992). The main point is that these new developments allow for a combined support and re-integration of data, information and communication flows with a high degree of flexibility for process management and exception handling. The combined modelling of information and conversation flows allows to trigger, e.g., the information about a specific document in the information-base, the status of the process in the conversation-base.

The combined support of information and communication flows allows also for *event-based information retrieval* (De Michelis et al., 1992). Problems concerning information management are relevant in cooperative processes. This is true in general for the production, revision, filing and distribution of documents. This is also true when cooperation is not directly related with creating and modifying documents, because people collaborating on a common task need to share the information characterizing that task. When we access an information source for the first time, we adopt a 'logical' criteria of selection. However, when we access an information we have created together with others or we accessed before, the natural method is to reconstruct the process or context of its creation, or when we used it for the last time. Let us make an example: When an office worker needs a document which he/she remembers related to a specific person during a meeting and a following discussion about its content on e-mail, the person has two possibilities for his/her research. The search in structured information storages; structured retrieval methods are efficient, but not always successful. In this case he/she has to trigger back to the event. Consulting again his/her e-mail messages in the period around the meeting he/she might find out about the person who wrote the document. he/she might also find the message with which the document was sent to him/her as a file.

All activities can be recorded in workflow applications. This means that organizations develop a knowledge base of their 'being' in the world. Each action of the organization (as an ensemble of individual actions) can be scanned by event-based information retrieval mechanisms to see the results in the specific context of action at time, i.e., the learning process is reflected in an always up-dated history of the organization.

2.5 Functional Requirements for Workflow Management Technology

Available workflow products have different functional modules for different users and purposes. These are e.g., the authoring environment, the workflow user interface, the workflow motor, the workflow maintenance tools, and the workflow auditing functions. The actual state of the art is still far away from the vast number of needed functionalities. This chapter gives an overview of functionalities which might only be developed and/or integrated in the next generation of workflow systems (Marshak, 1992).

Workflow systems offer some general tools in these modules which allow to adopt the software to the specific requirements. In fact, effective workflow systems must be customizable by end users and application programmers. This ensures, that the people who know the business best can effect full-blown applications. Application developers can transcend individual application development efforts. This means that developers can build a baseline application which can then be leveraged by multiple user-driven application efforts. For end-users, however, a graphical environment is a requirement.

In order to develop the workflow software script, the user organization needs first to model the business process to be managed, and input this then to a workflow procedure set-up facility. Most workflow systems allow the programmer to display the process in a flow diagram form as the model for the process is built up. It is only now, that we enter into a stage of process automation which has flexible workflow scripting languages.

The *authoring environment* has to be a graphical front end for workflow definition. Successful products have to support the definition of workflow steps (actions), sequences (the order of actions), routing (who does the action) and conditions (rules for routing and doing actions). The graphical tool should include a flow chart generator that assists the user in visualizing the flow. A more sophisticated tool will also write the script from the flow chart (actions, roles, rules, rooting, etc.) Advanced auditing environments might also include simulation functionalities. e.g., to see the consequences of a certain step for the entire flow. Workflow authoring tools should allow to re-use steps and rules from other workflows (library of primitives and already developed workflows).

Once a workflow has been developed, it should be able to be *copied or modified*, or modified directly be the user with sufficient permission rights. It would also be useful, if the user could modify the workflow within the already started process, rather than having to go back to the authoring environment. This functionality could also be used to *debug a blocking situation* in an already running workflow. A *simulation facility* would be very helpful for this kind of modifications (e.g., delete a step or change a rule and see what changes in the process).

Actions have to be accomplished by persons. Workflow products should allow to assign an action either to a specific person or to a role, such as a sales manager. Most products allow only for the one or the other, but not for a mixed assignment between physical persons and alias roles. When defining roles, the local name service should be invoked. Email names should be coupled to workflow roles and possibly stored in X.500 directories.

When a step is routed to a role which corresponds to more than one person (e.g., salesman), there has to be some mechanism which notes that an individual has taken the action. The *assignment of actions* within a role to individuals could also be determined by rules, e.g., work-load balancing.

Each action should have a *deadline*. This deadline is usually expressed as a period for the task accomplishment. Deadlines might also be specified by specific or recurrent dates in the flow. Some workflow products support only the entire period for the process and do not break the flow down into single action deadlines. This makes it difficult to handle deadlines at the end of the process due to missing control for previous steps. Another point is that deadlines cannot be established for each action by the initiator, but have to be negotiated among the workflow users for every specific process. These negotiated deadlines constitute a network of commitments among the participants. The process of deadline negotiation is not supported for the time being by any of the existing products.

Users should also be *alerted if actions are overdue or pending* for a certain period. This might also activate automatically or by manual choice an alternative workflow path for time recovering.

The workflow application should *notify the user* that he or she has a step or action to do. These new actions to perform should be handled in a *mail inbox* (e.g,. in a special folder or action items listed separately from other email messages), but not isolated from the email application. Most system currently handle a to-do list, but the integration with general purpose electronic mail is still missing (one has to quit the workflow application in order to read or to send an email message, also if this is directly related to the action due).

The user should be able to view his or her *mailbox and/or to-do list* in a variety of manners: by date, deadline, process, type of step, sender, workflow initiator, etc.. These *views should be tailorable* for the users.

Workflow applications must provide *electronic routing* of information, images, documents, files, etc., in a process to be supported. This routing can be simply *sequential* (predefined steps) or *rule-based* (dependent upon static and dynamic criteria). Furthermore, the flow might have a *ramification into parallel tracks* which converge again at a certain step.

The workflow systems available today seem to focus on tracking and managing the handling of tasks rather than facilitating and managing the tasks themselves (as well as the flow between them). What is missing is the work itself. A well-designed workflow application should do more than route the information associated with a task and keep track of progress. It should also facilitate the process itself.

As workflow applications should support the work done in a flow, they should provide the users with the *proper tool(s)* to complete each action or step. This means that people need for the accomplishment of a task the right or prefered tool, like spreadsheets, form handlers, word processors, etc. New generation workflow applications can launch any application, like a spreadsheet or a procedure template, at a designed point in the routing of a workflow. These workflow applications allow also to *connect data from different sources* (data-base, file, template, etc.) to the actual application (e.g., paste data or text from different sources into the application file). This connection is insured by *running background applications* which are activated by scripts in the workflow application.

By designing a workflow application, one has to take into consideration also *individual preferences* for standard office tools. The foreseen use of a spreadsheet at a certain point could be supported by all existing spreadsheets if they fit into the task. Therefore the workflow designer should leave users their prefered office applications and concentrate on the connection of data to these applications (spreadsheets, word processors, graphical applications, etc.).

Certain steps will require creating forms for the workflow application. These forms must be *easy to create and to modify*, also by the workflow users.

However, not all data and information is generated or taken from company data-bases as the workflow progresses, but off-line paper documentation (e.g., from external sources at customers). The transcription of the information treated in these documents into workflow documents might be expensive or eaven impossible. Therefore workflow applications have to be able to *route a variety of data, documents, forms, mail messages*, as well as *scanned images* from external sources. The *image management* will be done in background by Document Management Systems and is not a functionality for workflow applications.

The same logic counts for non-electronical actions in a process. Not every step in a workflow might be computer-based, e.g., check references with a client via phone. The workflow should allow the person assigned to indicate that the action was done, and to report on results if necessary.

Workflows should also provide functionalities to *link the workflow to other processes or sub-flows* which are recurrent in the organization (e.g., administrative processes). Therefore workflows and workflow users should be able to launch other workflows or agents, as well as agents should be able to launch workflows.

In order to allow this integration of different applications and data into a workflow application, it will be *developed jointly by the workflow users and information systems people*. The user (possibly aided by consultants) will prototype the application, determine the flow and define the forms and applications to be used. The computing specialists will write the back-end operations, e.g., sophisticated data-base access. This is an advantage for for both, users and information systems people: the user has direct control on the application definition and development; the developer concentrates on technical problems and is not in the old dilemma to try to understand what the users intend for the process application.

The *auditing module* provides functionalities for monitoring the status of objects, single flows or groups of flows for the same process. In addition to simple tracking reports (flow X is in step Y which is worked on by Z), systems should provide more sophisticated reports (e.g. on productivity, quality, etc).

A workflow system is a combination of centralized control and local interactive processing. Hence, it is ideally suited to a *client-server architecture*. The development of workflow technology with the indicated functions and integration of data, information and communication flows implies a heavy use of a software layer which takes care of message exchange and routing. This software layer for *message control* is a key component of a technological architecture which is the most appropriate to cooperative information services (Madnick, 1991].

3. Conclusion

Workflow management technology is a new issue for information systems people and managers. By the end of 1993 nearly all IT suppliers will have their workflow product for standard office systems on the market. Growth is currently hindered by a lack of knowledge amongst senior management about the concepts and benefits of workflow management technology. The technical aspects of current workflow products will be improved within one or two years. Developments will include graphical user interfaces for process scripting and the integration of multimedia (e.g., voice messages). New systems might also cross corporate boundaries and link up with EDI systems.

Although the concept of process working and the use of workflow technology seem simple, it is not easy to introduce it in the real world. The first problem for change are the vested interests in organizations. Process working runs counter to conventional thinking: It overcomes the idea that career advancement means scrambling up the traditional hierarchical ladder in the bureaucracy. Instead, process working offers people the opportunity of carrying out challenging and broader tasks and of continually learning new skills that enable them to work better. Process working therefore changes totally the personnel system, and how people are motivated and encouraged in their professional career.

The second problem lies in the grouping of people in a process. They need a wide range of skills, covering all actual functional expertise involved in procedures and in less structured or rule-based working processes. Process working implies the coupling of multidisciplinary skilled teams. There is still little experience of group working in many sectors and there are many unsolved questions (e.g., how to select people, how to train them, etc.).

The third problem lies in the capability of people in information systems to change their perspective concerning technology. They have to develop the existing systems into workflow management technology which supports the interpersonal relations in the working processes and provides means for communication, information and data management.

A fourth point is, that in realty the problem is not technology, but the organizational model for the business which is supported by workflow management technology. Denning (1992) puts this problem as follows:

> 'By automating business processes that are not explicitly oriented toward customer satisfaction, organizations have accelerated the production of dissatisfied customers. It does not matter whether the organization is a business firm or a research institute. The problem is the design of processes for conducting business.' (Denning, 1992)

Therefore, the crucial point for success of workflow management technology is the proper process analysis before implementation. The simple mechanization of steps in an existing procedure does not reach the potential benefits. Processes have to be analysed and redesigned by proper methods and skilled persons or companies. The best implementations might be developed with the use of outside consultants or experienced system integrators. Workflow solutions cannot generally be bought from the package supplier.

References

Argyris, C., & Schön, D.A. (1978). Organizational Learning: A Theory of Action Perspective. Reading, MA: Addison-Wesley

Bannon, L., & Schmidt, K. (1992). Taking CSCW Seriously: Supporting Articulation Work. Computer Supported Cooperative Work (CSCW): An International Journal, 1-1/2, 7-40

Barber, G.R., de Joung, P., & Hewitt, C. (1983). Semantic Support for Work in Organizations. In: R.E.A. Mason (Ed.), Proceedings of the IFIP 9th World Computer Congress. Paris, 19-23 September. Amsterdam: North-Holland, 561-566

Beckhard, R., & Harris, R.T. (1987) Organizational Transition: Managing Complex Change. Reading, MA: Addison-Wesley

Butler Cox (1991). The Role of Information Technology in Transforming the Business. Butler Cox Foundation, Research Report 79, London

Curtis, B., Kellner, M.I., & Over, J. (1992). Process Modelling. Communications of the ACM, September, 35-39, 75-90

De Michelis, G., Donzelli, P., Schäl, T., & Zeller, B. (1992). Computer Support for Cooperative Work in Space Science Activities. In: G. van der Veer et al. (Eds.), Human Computer Interaction: Tasks and Organization - Proceedings of the Sixth European Conference on Cognitive Ergonomics, CUD, Rome, 71-85

Denning, P.J. (1992) Work Is a Closed-Loop Process. American Scientist, July/August

Ellis, C.E., & Naffah, N. (1987). Design of Office Information Systems. Berlin: Springer-Verlag

Gable, J. (1992). Workflow Processing Software. Workgroup Computing Series: Strategies & LAN Services, DATAPRO, Information Services Group, Delran, NJ, July 1992, 1-11

Gerson, E.M., & Star, S.L. (1986). Analysing Due Processes in the Workplace. ACM Transactions on Office Information Systems 4(3), 257-270

Hammer, M. (1984). The OA Mirage. Datamation 30(2), 36-46

Hinterhuber, H.H. (1984). Strategische Unternehmensführung. Berlin: De Gruyter

Huber, G.P. (1991). Organizational Learning: The Contributing Processes and the Literatures. Organizational Science, 2(91), 88-115

Keen, P.G.W. (1991). Shaping the Future: Business Design through Information Technology. Boston, MA: Harvard Business School Press

Kensing, F., & Winograd, T. (1991). The LanguageAction Approach to Design of Computer-Support for Cooperative Work: A Preliminary Study in Work Mapping. In: R. Stamper et al. (Eds.), Collaborative Work, Social Communication and Information Systems. Amsterdam: North-Holland, 311-331

Madnick, S.E. (1991). The Information Technology Platform. In: M.S. Scott Morton (Ed.), The Corporation of the 1990s – Information Technology and Organizational Transformation. New York: Oxford University Press

Marshak, R.T. (1992). Requirements for Workflow. Office Computing Report 15(3), 3-8

Medina-Mora, R., Winograd, T., Flores, R., & Flores, F. (1992). The Action Workflow Approach to Workflow Management Technology. In: J. Turner & R. Kraut (Eds.), Proceedings of the 4th Conference on Computer-Supported Cooperartive Work. ACM, 31 October–4 November, Toronto, Canada, 281-288

Piore, M.J., & Sabel, C.F. (1989). Das Ende der Massenproduktion: Studie über die Requlifizierung der Arbeit und die Rückkehr der Ökonomie in die Gesellschaft. Frankfurt am Main: Fischer

Schäl, T. (1992). Information Systems in Public Administration: From Transaction Processing to Computer Supported Cooperative Work. Proceedings of the IFIP WG 8.5 Working Conference on Computer Supported Cooperative Work and Public Administration, Linz, Austria, February 25-27

Schäl, T., & Zeller, B. (1991). Design Principles for Cooperative Office Support Systems in Distributed Process Management. In: A. Verrijn-Stuart et al. (Eds.), Support Functionality in the Office Environment. Amsterdam: North-Holland, 85-101

Schmidt, K. (1990). Analysis of Cooperative Work. A Conceptual Framework. Risø-M-2890, Risø National Laboratory, Roskilde, Denmark

Schmidt, K. (1991). Riding a Tiger, or Computer Supported Cooperative Work. In: L. Bannon, M. Robinson, & K. Schmidt (Eds.), Proceedings of the Second European Conference on Computer-Supported Cooperative Work. Dordrecht: Kluwer Acadamic Publishers, 1-16

Senge, P.M. (1990). The Fifth Discipline: the Art and Practice of the Learning Organization. New York: Currency

Shrivastava, P. (1983). A Typology of Organizational Learning Systems. Journal of Management Studies 20(1), 7-20

Simon, H.A. (1991). Bounded Rationality and Organizational Learning. Organization Science 2(91), 125-139

Suchman, L. (1987). Plans and Situated Actions. The Problem of Human-Machine Communication. Cambridge, UK: Cambridge University Press

Winograd, T., & Flores, C.F. (1986). Understanding computers and cognition - a new foundation for design. Norwood, NJ: Ablex Publishing Corporation

Computer Support for Cooperative Work: Computers Between Users and Social Complexity

Giorgio De Michelis

Laboratory for Cooperation Technologies, Department of Information Sciences
University of Milano, Italy

Abstract In this paper I consider work processes as complex customer-performer relations. From this point of view I identify the cooperation network carrying on a work process and analyse the cooperation forms of its members. I show how cooperative work is primarily characterized by the complexity of the inter-actions constituting it, and how it is possible to measure the transaction costs of a work process generated by its complexity.

Moreover, I analyze the factors affecting the transaction costs of a work process, and I discuss how one of them, the threshold of sustainable complexity of its participants has a particular relevance, as it defines a limit that cannot be overcome through organizational means.

CSCW systems are finally discussed as computer-based tools that can enhance the threshold of sustainable complexity of a cooperative network. This view allows us to evaluate CSCW systems, as well as any computer based system, within a well defined categorization.

1. Introduction

Computer Supported Cooperative Work (CSCW) designates a new class of applications of computer-based tools within work processes. It is not possible, today, to propose a precise definition of what CSCW is: the debate on this point is still ongoing among researchers in the field, and it accompanies their research (AAVV 1986, 1988, 1990, 1991, 1992; Bowers & Benford, 1991). It is possible anyhow to say that CSCW systems propose a radical paradigm shift with respect to traditional office automation (Ellis & Naffah, 1987), as they aim to support informal, non-routinized, work processes instead of (or besides) office procedures.

While we have some papers discussing why and how CSCW applications fail, e.g., (Grudin, 1988), very little has been said about why and how CSCW applications would succeed.

It is my opinion that this is due to the fact that we know too little about work processes, and that the little we know is hard to use in designing tools to support them.

In this paper I will try to overcome these limits, indicating a new perspective to analyse, evaluate and design work processes.

While my work takes inspiration from many authors, (let me quote here at least Fernando Flores, to whom I pay a decisive debt for his work on commitments and conversations and on business work-flows, and Oliver Williamson, for his development of the transaction costs theory) the proposed framework is new, as it reinterprets and redefines most of the basic concepts of those contributions as well as other categories of organization theory and management sciences, and I am the only person responsible for it.

It has to be underlined at this point that my approach to transaction-costs theory is rather different from the way it has been applied to understanding Information Systems development within firms by many authors, e.g., (Ciborra, 1987; Ciborra & Olson, 1988; Malone et al., 1987; Malone & Crowston, 1991), as my objective is to define the factors contributing to the transaction costs within business processes.

The crucial point of the theoretical framework I have developed is that I consider complexity as the basic category for understanding work processes and for designing their change.

More precisely, in my view the transaction costs of a business process depend upon its complexity: therefore I pay attention to the way complexity affects transaction costs, and to the means of absorbing them. Within this discourse I show how CSCW applications can be considered the computer-based applications explicitly dedicated to enhancing the ability of human beings to manage the complexity of work processes without augmenting their transaction costs.

2. Work Processes are Complex

It is not my intention in these pages to discuss in depth what work is as a human activity: I will assume a very general and to some extent fuzzy image of work, considering it the pragmatical dimension of human life.

With this work image in mind, if we look at organizations such as firms and public institutions, we can completely characterize their behaviour (analysing their structure, evaluating their performances,...) through the work processes their members are involved in.

From this point of view, anyhow, a work process is something more than the simple flow of actions giving rise to an expected outcome: to understand the behaviour of an organization it is important to embed any flow of action within the relation giving it a sense. The meaning of any action is strictly dependent on the context in which it is performed, i.e., on the customer-performer contract defining it.

In accordance with Fernando Flores and his coworkers (Keen 1991; Medina-Mora et al., 1992), I consider a work process as a business process starting with a customer generating the request for an action, continuing with a performer agreeing with the customer to do the required action and then completing it, and finishing when the customer declares his/her satisfaction (the Action Work Flow®; see Fig. 1).

Any organization, as a firm or a public institution, can be characterized by the work processes its members are able to initiate, and its behaviour can be described in terms of those work processes.

The image of a work process given in Fig. 1 has many interesting facets: let us recall some of them.

A work process is a communicative relation between a customer and a performer (both can be persons or juridical entities represented by persons) within which both, the performer as well as the customer, assume responsibilities and consume time and resources.

A work process is a cycle, as the relation between a customer and a performer does not restart at any new request for action the former proposes to the latter: i.e., a work process has memory of the requests of the customer the performer has already fulfilled (as well as of those he/she has not fulfilled).

Fig. 1. The Action Work Flow® (Source: ATI)

It can be interesting to display when a customer performer cycle is simplified, through the contraction of two of its phases (an example of a contracted cycle is the business process of mail marketing, where the requests the Buyer makes in the catalogue are by definitions accepted as contracts by the Seller; see Fig. 2). Simplified customer performer cycles, whenever the contractions do not cause too frequently some breakdowns, allow a sensible reduction of transaction costs, as they do not allow some cost effective interactions (in the case of mail marketing no negotiation about prices and/or delivery times is allowed).

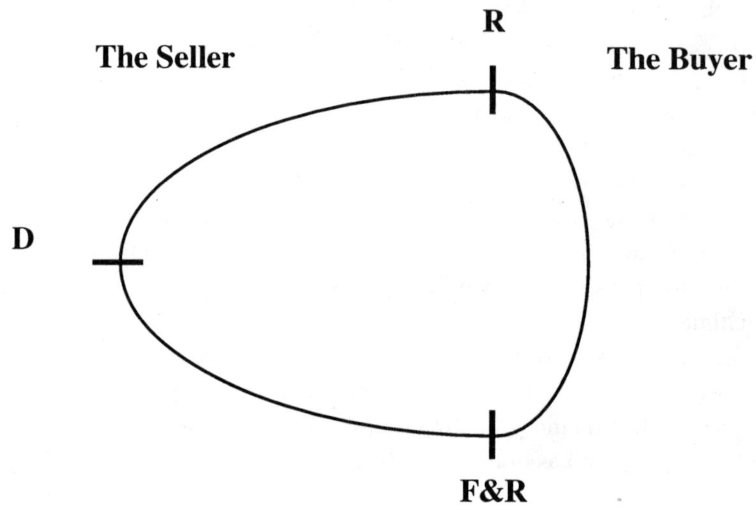

Fig. 2. Example of a contracted cycle

When a work process is not simple, the performer, as well as the customer, can open within it other work (sub-)processes with other performers, becoming their customer, to get the help (in terms of human, informational, technical or material resources) necessary for the completion of any of its phases. The performer together with all the persons with whom he/she has opened a work sub-process constitute the *cooperation network* of the work process itself. The cooperation network notion we use in this paper shares many facets with the notion of *community of practices* ethnographers have recently developed at the Institute for Research on Learning (Lave & Wenger, 1991; Jordan, 1992) within the situatedness paradigm (Suchman, 1987). I underline this fact because it can be the basis for a better understanding of different views on work practices.

The opening of a work sub-process within any work process can either be the response to an unforeseen breakdown, as in the case in which it is opened to solve a problem, or a necessary step, as when, for example, it is due to the necessity of an approval by any responsibility center in order to continue the process.

In the latter case the work process is proceduralized and can be represented as a cycle of cycles (see in Fig. 3 the cycle of cycles of a credit procedure within a Bank (Agostini et al., 1993); also in this case I pay tribute to the work of Fernando Flores and coworkers, who developed a well supported methodology to design chains of action workflows (Medina-Mora et al., 1992), and to the experiences of Buni Zeller and others at RSO in supporting work processes with workflow management systems (Schael & Zeller, 1991)).

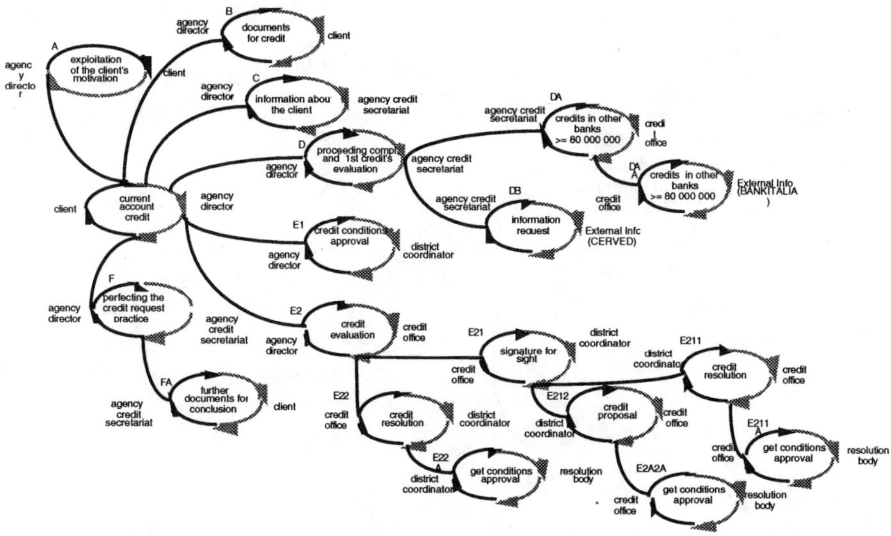

Fig. 3. Cycle of cycles of a credit procedure

The cycle of cycles image of a work process emphasizes the communicative relations embedded within a procedure, showing, for example, if a customer-performer relation is closed or not, i.e., if the relation opened by a request of the customer, is completed with his/her declaration of satisfaction, or not. Only closed customer performer relations fully exploit the responsibility (if any) of the role played by their actors, avoiding the breakdowns incomplete responsibility definitions may generate.

The cycle of cycles model has, in my opinion, some weaknesses when the procedure embedded within a work process has a high degree of independent sub-processes and/or when it is embedded within an highly structured organization. In both cases, in fact, some problems appear that cannot be reduced to customer performer relations but require, to be represented and understood a richer set of basic categories.

Let us propose here some other ways to represent work processes as means of designing them so that they can fulfill specific performance needs (what follows can be considered as an integration proposal of the methods developed at BDA and ATI by Fernando Flores and co-workers).

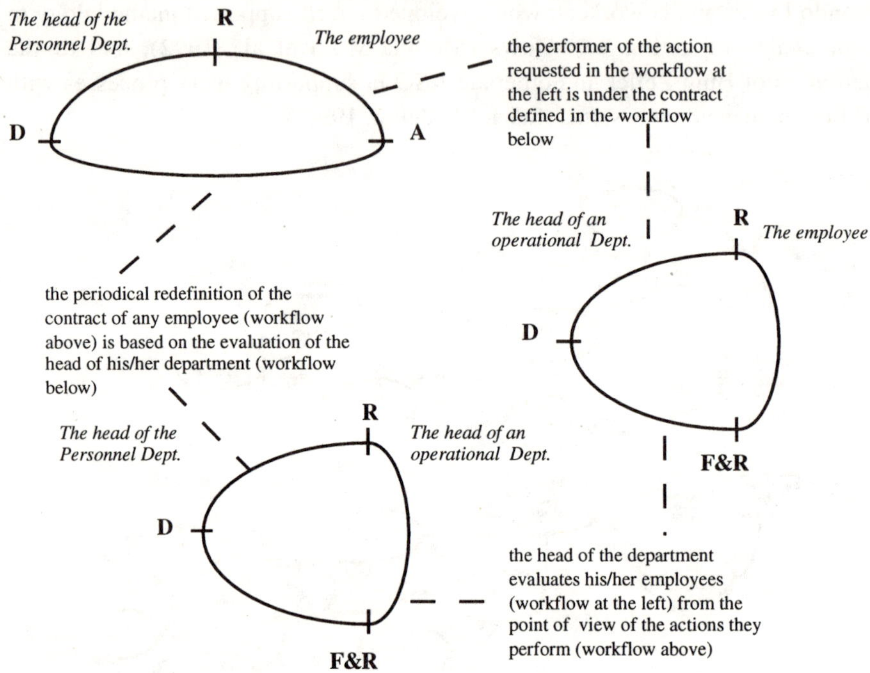

Fig. 4. Customer-performer cycles

Initially, it is frequent that a customer-performer cycle (generally a family of similar customer-performer cycles) is substituted by a set of (eventually simplified) customer performer cycles, eliminating some unnecessary inter-actions (or some unnecessary steps within interactions) in it.

As an example (Fig. 4), it is very frequent that, within a company, hiring a group of designers for a new project, is not performed through the four phases customer-performer cycle, but through a couple of cycles, the first one of which is dedicated to defining and maintaining contractual regulations between the company (its Personnel Department) and the designers (where the third and fourth phases are contracted to one, because there is no specific fulfillment in it), while the second one is dedicated to the attribution of roles within the new project to the designers by the Project Manager (where the second and third phases are contracted to one because no negotiation generally follows the request).

The two cycles are related through a third cycle (also having the second and third phase contracted to one), through which the Project Manager periodically gives his/her evaluation of the designers to the Personnel Department. It is evident that, as well as the above contractions of the customer-performer cycles are generally not generating breakdowns, the two cycles are more effective than the simple cycle, whenever within the duration of a contract many projects are developed.

It is also frequent that the action negotiated within a customer-performer cycle, is defined as a procedure, in order to standardize the roles various performers play within it. In this case the work process is designed as a single cycle, with more than one performer, i.e., all the performers playing a role within the procedure. Within the procedure all the client performer cycles are contracted since the standardized nature of the actions constituting it allows us to consider any negotiations either before or after their completion as exceptional (breakdowns). The radical contraction of customer performer cycles within proceduralized work processes improves their effectiveness, if breakdowns are rare and allow easy recovery. A representation of a procedure embedded within a customer performer cycle is displayed in Fig. 5.

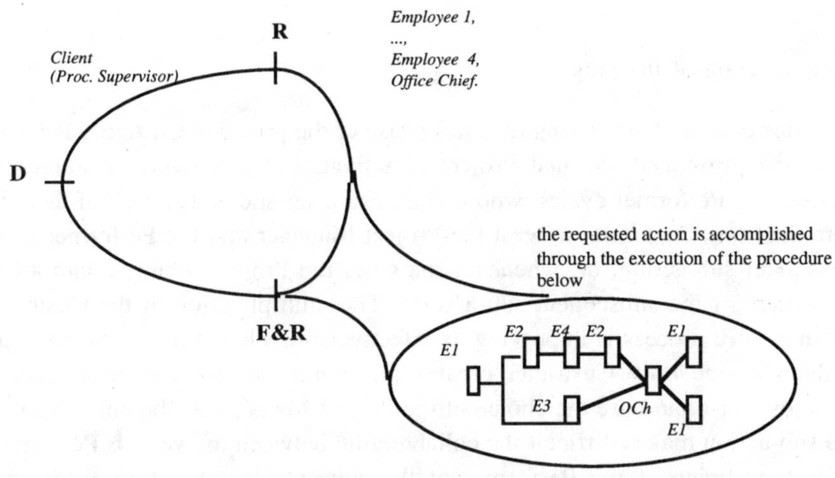

Fig. 5. Procedure embedded within a customer performer cycle

Also if the action requested within a work process has not already been standardized into a procedure, it is possible to anticipate its splitting into sub-actions assigned to different performers defining, during its second phase (going from the Request of the Customer to the Agreement between the Customer and all the involved Perfomers), the structure of the project necessary to complete it.

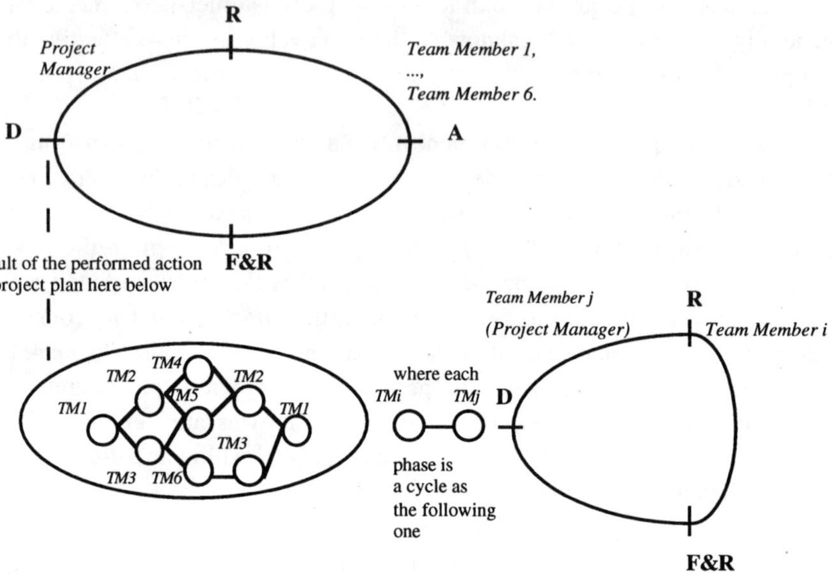

Fig. 6 Structure of the project

In this case (Fig. 6), during the third phase of the principal Customer Perfomer cycle, the previously defined Project is activated as a network of contracted Customer(s) Performer cycles, whose main characteristic is that each of them has more than one Customer: at least the Project Manager and the Performer of the subsequent sub-action, or, whenever the case, the Project Manager and all the Performers of the subsequent sub-actions. The multiplication of the Customers within a work process is improving its effectiveness, whenever the choice of any of them as the only Customer creates the conditions for the occurrence of breakdowns (in our case the choice of the Project Manager as the only Customer of a sub-action makes difficult the collaboration between the various Performers, while the choice of the Performer of the subsequent sub-action as the only Customer of a sub-action does not allow the Project Manager to grant the consistency of the ongoing sub-actions with the expected outcome of the whole project).

A complex work process may be considered as the result of a sequence of changes from a simple Customer Perfomer cycle into a more complex network of (whenever possible, contracted) Customer Performer cycles, induced by the continuously emerging necessity of overcoming its too high transaction costs. Redesigning a work process means creating and (re)organizing the firm (or the institution) performing it, as it affects the roles of its members, its (macro- and micro-)organizational structures, its behavioural norms.

Any person participating to the cooperation network of a work process, plays within it a role, characterized by the sub-processes to which he/she can participate, by the tasks he has to accomplish within the procedures that are activated within it, and/or by the actions he has to complete within the projects in which he participates, characterizing its recognized abilities and responsibilities within that work process. The role of a member of a cooperation network changes during the development of a work process, as an effect of the occurrence of changes within the sub-processes he/she participates in.

On the one hand, we can abstract from a work process its action flow, taking into account only the performed input-output transformation. Action flows can either be designed during the negotiation phase (and modified during the execution phase, when necessary), as in a project, or be predefined in a standard way as procedures. Frequently the actions constituting a project are standardized as procedures, while those constituting a procedure are discretionary.

On the other, we can abstract from a work process its communication flow, i.e,. its network of conversations for action, using the well known Flores and Winograd terminology (Flores & Ludlow, 1981; Winograd & Flores, 1986). The communication flow of a work process characterizes the mutual relations between the members of its cooperation network, their sharing of knowledge created within the work process, in short, their cooperation.

There are four main forms of cooperation two members of a cooperation network may enter in: collaboration, coordination, codecision with equal roles and co-decision with distinct roles. The four forms of cooperation listed above can have a double characterization, the first one of which depends on the nature of their cooperation (a discussion of the nature of the cooperation forms has already been given in (DeMichelis, 1990) and in (DeMichelis et al., 1992)), while the second one depends on the role played by them within the customer performer cycles.

Collaboration occurs when two or more people work together in order to achieve a single common goal: no distinction can be done between what is done by any single participant to the collaboration. Collaboration is carried on by customers and performers within a customer performer cycle.

Co-ordination occurs when two or more people work together to achieve, each one of them, different sub-tasks that are part of a more general task: each single participant has therefore his/her own task, but all tasks are mutually related. Coordination is carried on by people being customers and performers in cycles at different levels: e g., a customer must co-ordinate him/herself with the performers of the subcycles opened by his/her performer.

Co-decision with equal roles occurs when two or more people participate in a decision process, where all of them share a full responsibility of the decision to be taken. Codecision with equal roles is carried on by persons being customers of the same performer (with one single satisfaction condition).

Co-decision with distinct roles occurs when two or more persons participate in a decision process, where each one of them has his/her own responsibility on a particular issue related to the whole decision to be taken. The managers of the different functions of a firm defining its annual budget participate in a codecision with distinct roles. Codecision with distinct roles is carried on by persons being all performers for the same customer (having one condition of satisfaction, embodying all the actions all the performers perform).

The way customers and performers are distributed within a work process allow us to evaluate all the interactions constituting the communication flow of a work process. It is important to stress that cooperation is not restricted to customer performer cycles, but that the analysis of customer performer cycles allows us to capture all the cooperation occurring within a work process.

The interleaving of acting and conversing within a work process cannot be modeled: whenever a breakdown occurrs within a work process, a new work sub-process is opened, but breakdowns are generally inpredictable. On the one hand, if we try to take into account all the possibilities of breakdown occurrences, we have a combinatorial explosion of the dimensions of our model (and still we are constrained by the level of detail of our model, losing the possibility of describing the occurrence of breakdowns within actions); on the other, if we select only the most meaningful or probable potential breakdowns, we oversimplify the model.

The cooperation network of a work process is not (generally) a unit of the organization to which the performer belongs: also if, very frequently, firms and public institutions try to shape their organizational units on their work processes in order to unify the control on the organizational structures and the control on the work processes, it is impossible to constrain the cooperation network of a work process within the boundaries of a predefined organization without reducing its effectiveness with respect to possible breakdowns or duplicating the resources needed to perform them.

The relation between work processes and organizational structures is more subtle than a simple embedding: at the macro-organizational level, any work process needs a supervising authority controlling and maintaining it, while at the micro-organizational level the rational distribution of human resources among work processes can be accomplished by embodying them into organizational units. The performers of a work process generally do not belong to the structure headed by the person having responsibility over it, who therefore acts as customer for their performances, while the members of an organizational unit, generally, perform actions within different work processes, and therefore the head of that organizational unit is a performer within a customer performer relation with all the people responsible for the work processes on which his/her collaborators play a role.

The above recalled observations on work processes can be synthetized in the following claim: work processes are complex, as their distinctive features (customer-performer relation, cooperation network, action flow, communication flow, cooperation forms, organizational structure, etc.) cannot be captured by a single point of view. In other words we can say that the complexity of a work process is the image of the combination of the multiplicity (the number of persons in its cooperation network, the number of persons with which each of them communicates, the number of conversations between any couple of persons,....) and of the autonomy (of the persons in its cooperation network, of the persons with which any person of its cooperation network interacts, of the persons participating in other work processes,...) it exhibits.

Why is the complexity of a work process relevant? Because it can be the cause either of the impossibility of performing it to the satisfaction of the customer (the requested action is not performed within the fixed time constraints, with adequate quality attributes,...) or of the distance between the value of the resources spent to perform it and the value the customer attributes to it.

The complexity of its work processes is therefore the main cause of the transaction costs (Williamson, 1975, 1985) a firm must afford to do its business, and analysing the former can be a good starting point to evaluate the latter ones. The characterization of transactions costs (of both the ex ante and the ex post transaction costs in the terminology of Williamson: see (Williamson, 1985)) we can derive from the analysis of the work process complexity gives finally meaningful guidelines for redesigning it when its cost-benefit ratio is too great.

3. Transaction Costs and the Complexity of Work Processes

The transaction costs of a business process constitute, with its trasformation (production) costs, the main element of its economic value. In the theory of Williamson, the attention is focused on understanding the conditions under which different contractual models emerge within an industry. He characterizes in a synthetic way the actors and the object of a transaction, and analyses how they affect its cost. In this paper we have a different aim: we want to discuss the factors characterizing transaction costs as a mean of understanding the different business process models emerging within a firm. Our investigation is not at the micro-economic level: we are interested in understanding the behaviour of firms and not of industries.

If we take into account the work process model we have discussed in the previous section, we can try to give a more detailed characterization of its transaction costs, defining the factors affecting them and, if necessary, their weight. Let us discuss here below some of these factors. The list is not exhaustive, but it has been compiled in order to give a rich account of their nature.

1. The transaction costs of a work process depend on the number of members of its cooperation network: a larger number of participants implies in fact a higher cost of any assessment of the actions to do.
2. The transaction costs of a work process depend on the number of the interactions (collaborations within sub-processes, coordinations, codecisions with equal and distinct roles) it embodies: a larger number of interactions implies in fact a higher cost of the management of the whole process.
3. The transaction costs of a work process depend on the number of participants of any of its sub-processes: if any sub-process has a number of participants comparable with that of the work process itself, then any sub-process has the same complexity as the latter, and we have no a divide-et-impera effect.
4. The transaction costs of a work process depend on the number of other work processes to which any of its participants participates also: if a large number of participants has other commitments in other work processes, then the complexity of those work processes reflects on the work process under consideration.
5. The transaction costs of a work process depend on the number of sub-processes running in the same period of time: any breakdown in any sub-process impacts all the concurrently running sub-processes.
6. The transaction costs of a work process depend on the number of newcomers to it: newcomers require some effort to become participants of the co-operation network.
7. The transaction costs of a work process depend on the number of participants leaving it: a participant of a cooperation network is an intelligent source of knowledge within it, and if he/she leaves it the cooperation network loses that source.
8. The transaction costs of a work process depend on the threshold of sustainable complexity of its participants. Any person in any moment of his/her life has the capability to manage in a smooth way a certain number of inter-actions of different type or, in other words, to participate effectively in a certain number of work-processes: if and when he/she tries to increase that number, he becomes always more ineffective not only in the new processes into which he/she enters but also in all the work processes he/she was already participating in. Ineffectiveness grows exponentially when the maximal number of processes is passed. For any person, the above recalled maximal number depends on the threshold of complexity he/she is able to sustain.

It is evident that in order to understand the impact that the above listed factors of transaction costs have on the (economic) performance of a work process, we must relate them to some characteristics of the process itself: Williamson mentions the specificity/genericity factor (Williamson, 1985); we can reclassify it through the complexity concept in terms of variance. The relevance of the above listed transaction cost factors depends on the degree of variance the process exhibits (i.e., on the number of the unforeseen sub-processes that may occur).

While the others factors of the transaction costs of a work process depend on the design of the work process itself, and can be afforded only modifying it (either on the client side, modifying the services offered, or on the performer side standardizing or destandardizing some parts of it - on this issue I will dedicate a different paper), the last factor plays a different role: on the one hand, the threshold of sustainable complexity is relevant only when a significant number of members of a cooperative network have reached it; on the other, in this case it behaves as an inverse multiplicative factor of the transaction costs: the higher the threshold, the lower the costs. Enhancing the threshold of sustainable complexity of the members of a cooperation network means, therefore, reducing globally the transaction costs of the related work process. It is in fact evident that if a person is able to cope effectively with a larger number of interactions without generating breakdowns within them (either forgetting to do some required action, or forgetting the matter of an interaction, or confusing two different interactions, etc.), the transaction costs of the work process in which he/she is involved will be smaller, as he/she will not impact it with the breakdowns he/she is provoking. Let me illustrate further the above point: until a person is able to cope with his/her interactions within a work process without needing help, he/she will not produce any new cost: whereas, when he/she reaches the point of no longer being able to handle them without the help either of his customers or some other persons, he/she will be individually a cost factor of the process itself.

On what does the threshold of sustainable complexity of a participant to a work process depend? Is it possible to recognize its components?

The answer to the above questions seems quite evident if we recall some common-sense reasoning.

On the one hand, the threshold of sustainable complexity of any person heavily depends on his/her culture: if he/she has a good education, if he/she is literate, then he/she will be able to cope with a large number of interactions of different type without losing in terms of performance; instead, a person with a low cultural degree, unable to express him/her self with exactness, will generally be unable to handle complex interactions without generating some breakdowns.

On the other, the threshold depends also on the tools he/she can use to interact with his/her partners: fast and reliable transportation means efficient communication media, good addressing mechanisms, efficient storages for the information and for the documents generated within the work process are some examples of tools that are needed to manage interactions with other people efficiently.

The two above factors are not independent: the ability to use tools is part of the culture of a person (using sophisticated communication media is a skill we attain through learning and pratice), whereas with good tools we can become capable of doing things for which we have not any special skill (with a word processor I am able to create high-quality documents that I was not able to create only with a typewriter).

The impact of tools on the threshold of sustainable complexity depends finally also on the way we dispose of them: if they are always at our disposition, whenever we need them and wherever we are, then they are maximally effective in enhancing our threshold of sustainable complexity, while, if to use them we need to move ourselves, if they require a modification of our work setting to be used, etc., then they balance the complexity they absorbe with the one they create, weakening their positive impact on the threshold of sustainable complexity.

Therefore, what is relevant with respect to enhancing the threshold of sustainable complexity is not only the set of tools we are provided with, but also how the latter are integrated within our work space. This is the reason why computer-based tools are good candidates for enhancing the threshold of sustainable complexity of a person: they in fact are the only tools the user is provided with that can be embedded into a single working environment (today into a work-station, in the future in a computerized room (Moran & Anderson, 1990)) presenting themselves as integrated components of a multifuctional system, that whenever we want can be virtually recreated in any location.

In concluding this section, we must observe it is not casual that we didn't mention any organizational mean to enhance the threshold of sustainable complexity of a person. Also if it can seem that a good secretary is great for enhancing the threshold of sustainable complexity of a person, it has to be argued that the secretary is absorbing a part of the complexity of the interactions of his/her manager, and not enhancing the threshold of sustainable complexity of the latter. This point may appear as a subtlety without relevance, but if we imagine a cooperation network where any participant is supported by a secretary, then we can figure that we have doubled the number of persons needed to perform the work process, and therefore we have augmented its transaction costs, and not reduced them, as when we enhance the threshold of sustainable complexity of the participants.

4. CSCW and the Threshold of Sustainable Complexity

As we have claimed in the previous section, the threshold of sustainable complexity of a person participating in a cooperation network depends upon the tools he/she can use within it. CSCW systems (called also Groupware (Ellis et al., 1991) or Workgroup Computing Systems) are, by definition, the systems aiming to support work processes and, therefore, our attention can now focus on them. It has to be recalled that the benefits CSCW systems can offer work processes are not only those deriving from the enhancement of the threshold of sustainable complexity. CSCW systems have also the ability to reduce complex procedures to simple operations, as when they automatically search a needed information into a data-base and merge it into a document.

Let us analyse in this section the main classes of problems generated by the complexity of work processes, and in the next one, how CSCW systems can help reduce the transaction costs they generate.

The framework for our classification will be based on the analysis of cooperation we have carried on in the previous sections. We will therefore use categories that are based on cooperation forms (we concentrate our attention on coordination and collaboration, as we think we need more theoretical work to fully understand codecision forms) and on the ways the context of a work process can impact a cooperation network.

The following constitute a first list of problems that have some relevance within work processes.

1- Breakdowns

The first one does not require many comments. Whenever a breakdown occurs within a procedure the performer needs help to (re-)create the conditions for doing his/her task. Finding this help can be difficult, and the expected performances of the procedure can no longer be maintained, requiring therefore a redesign of the process in order to avoid that customer satisfaction is affected. Moreover, solving the breakdown can represent the occasion for exploring and opening new possibilities.

2- Coordination problems

The definiton of the agreements among all the participants who must perform some activity within the process can be very complex, requiring many inter-related negotiations. Moreover, the coordination among various interrelated actions is such that the quality of a previous action can affect the outcomes of any subsequent one, requiring some new work that had not been foreseen in advance. Finally, once a coordination process has been defined, any breakdown occurring within any of the coordinated actions can require a redefinition of the whole process.

3- Collaboration problems

In a cooperation process there are some activities and decisions that involve more than one participant. Those collaborating in performing these activities and making these decisions need to share their knowledge and understanding in order to offer their best contribution. This can be very difficult if the participants have different experiences and/or if they are not supported by very effective communication channels.

Very often a performer is not able to accomplish his/her task alone. He/she requires help, and the participant requested for help needs to share with the requestor the knowledge about the task to be performed and the language to express it, in order to be able to give the requested help. This knowledge is difficult to access, as it is partly formalized in some documents, partly only in the minds of the requestor.

4- Multiplicity of tasks

Many persons working in large organizations and having some responsibilites within are involved contemporaneously in many cooperation processes, some of which are highly relevant to them, while others are less interesting. The distribution of their time among the activities they are involved in is frequently a very complex problem, since it requires both good selection procedures and the ability to create the work environment within which any of those activity can be performed.

5- Organizational noise and organizational opacity

Organizational units are, by definition, structures created to support their members in performing their tasks. Organizational units can be very ineffective in this respect if they become too large.

On the one hand, their members can have difficulties in finding the help they need, in that unfamiliar with the role distribution (in terms of competence, experience and responsibility) within an organizational unit (we can say that there is organizational opacity).

On the other, they can have difficulties in finding the help they need as the communication protocols are very inefficient, because it is difficult to access directly the participant able to give help due to bureaucratic (too much and too rigid hierarchy) and/or technological (low efficiency of the communication media) reasons (we can say that there is organizational noise).

Point 5, above, offers new arguments to show that organizational means (such as rules, structures and roles) cannot be sufficient in coping with the complexity of cooperation processes. Very frequently, trying to give a solution to one of them with an organizational means can generate new complexity without solving it.

Computer-based tools, on the contrary, have limited capabilities (as they have not the creative autonomy of human beings!) but can avoid their users a lot of unnecessary activities, allowing them to concentrate more attention on their main problem.

5. Requirements for a CSCW Environment

The five above-mentioned categories of problems generated by the complexity of cooperation processes give some hints on the requirements that a CSCW Environment should satisfy.

5.1 Breakdowns Within Procedures

As mentioned before, the most frequent problem any office worker can encounter within his/her work is a breakdown occurring while he/she is accomplishing his/her task. When a performer lacks the information he/she needs to fill a form, or when he/she does have enough time to respect the fixed deadline for the task, then he/she needs to inter-act with other participants (his/her manager, the person familiar with the missing information, the expert who can solve his/her problem, etc.), in order to recreate the conditions for performing his/her task.

Generally, the computer-based system where a performer executes his/her routinized task (also when it provides some exception handling mechanisms; see (Kreifeltz & Woetzel, 1987; Kreifeltz et al., 1991)) is separated from the system supporting his/her conversations with other participants, and therefore when a breakdown occurs he/she has to leave the first system and enter the second one.

A CSCW environment shoud support, without losing effectiveness at the procedural level, a natural and immediate switch from the first system into the other one, and viceversa, to accompany the user while his/her attention switches from the routine to the solution of the occurred breakdown.

There are two ways to satisfy this requirement which have been proposed in the CSCW field.

1- Considering work-flow management systems

Fernando Flores and his collaborators at Action Technology Inc. have designed Business Design Technology, which allows users to design action work-flows based on the cycle presented in Fig. 1 (Medina-Mora et al., 1992).

The work-flow of any procedure is characterized in Business Design Technology integrating the action synchronization with the conversations occurring between the requestor and the performer, so that the possible breakdowns can be taken into account while designing the procedure, supporting the conversations necessary to receive the required help.

2- Supporting the switch from routines to conversations

In the Office Environment that has been developed within the ESPRIT-2 Ithaca project, the human communication' support (UTUCS) is built upon a Work Flow manager (COP) in such a way that, on the one hand, conversations are implemented as distributed procedures, while, on the other, at any moment the user can transform an action to perform within a procedure into the object of a conversation with the other participants in the procedure (Agostini et al., 1992, 1993; De Michelis & Grasso, 1993). Procedures and conversations about them appear to the user as two different views on the same object, effortlessly supporting him/her while his/her attention is switched from the task to the occurred breakdown and back.

5.2 Coordination Problems

Most of the proposed CSCW tools (whether working products or still academic prototypes) are bound to a particular communication medium: either to asynchronous electronic mail (e.g., The Coordinator (Winograd & Flores, 1986), Information- and Object Lens (Lai & Malone, 1988), as well as the CHAOS System (De Cindio et al., 1986, 1988) under development at our Department), or to synchronous meetings (e.g., the system John Whiteside is developing at DEC Laboratories, (Whiteside & Wixon, 1988)), etc.

In any case, human experience cannot be partitioned on the basis of the media being used: human beings characterize their affairs on the basis of their contents, not on the basis of the communication media they are using.

When a person (a professional, a manager, a researcher, etc.) is managing his/her commitments, coordinating his/her with his/her colleagues', he/she is generally involved in complex conversations carried on through different communication media: I ask a colleague by phone to prepare for me the document I need; after that we meet at my office for discussing the critical aspects of the document; later we both participate in a meeting where the deadline for the document is changed; during the preparation phase I frequently exchange messages with my colleague through electronic mail for discussing the first releases of the document, and so on. S. Reder and R. G. Schwab, analysing a real work setting, have observed that more than 80% of the conversations occurring during the work time contain at least one channel switch (Reder & Schwab, 1989).

On the basis of the above considerations, I believe that a cooperation support becomes increasingly more effective the more it helps its users no matter the communication medium they are using.

The synchronization needs of a participant in managing his/her coordination therefore require a communication support capable, on the one hand, of supporting various communication channels (such as electronic mail, electronic conferences, but also face-to-face as well as group meetings, etc.) in an efficient and effective way, and, on the other, of supporting a conversation, no matter what channel it is carried on. The already mentioned (see Sect. 3.1 above) UTUCS System is, as far as I know, the first CSCW system with these capabilities. It has to be underlined that also C. V. Bullen and J. L. Bennett in their research on Groupware in Practice have observed that supporting conversations, instead of simple message exchanges, is a major quality of coordination supports (Bullen & Bennett, 1990, 1990b).

5.3 Collaboration Problems

Collaboration occurs when the task to be accomplished is in itself complex, since it cannot be accomplished by one person alone. If we do not consider expert systems or other A.I. technologies, it is rather difficult to design a computer-based support able to reduce its complexity.

What can be offered to the participants of a collaboration process is a support allowing them to focus on the task to be accomplished, without caring about the many problems that a collaborative process creates. In particular, I think that a computer-based support can be very effective in improving the way in which information is shared among the participants.

The information basis of a collaboration process (collecting all the documents created in it) can be retrieved either on the basis of its logical structure, captured in the data-base containing it; or by its pragmatics, i.e., through the events in which its items were created and used; or, lastly, by conceptual search methods, allowing the selection of a document for the concepts occurring within it. Different participants in different situations may get benefits from any of the three above retrieval methods: while the actor who participates in the collaboration process from the very beginning, for example, generally uses pragmatics-based search methods, the newcomer to the collaboration process may need conceptual search methods in order to navigate the already created documents.

The information basis of a collaboration process can be supported, therefore, by a well-defined data-base containing the documents created in it, whose items are linked to the conversations within which they were created, modified and used, so that the latter allow access to them, enriched by a conceptual search tool.

5.4 Multiplicity of Tasks

Our personal diary contains all the commitments of various types that we have with others. Many of us are at the same time involved in many different commitments, and managing them is one of the most complex tasks we accomplish every day.

Having many different commitments implies having many different tasks, each of which requires different tools, different information, different communication channels linking us with others. Switching from one task to another requires changing the tools, the information and the communication channels we have at hand. The material desk at which we spend our work time does not offer any support to this problem: whether we maintain it in order, eliminating all that is not needed before initiating an activity, or we keep it full of everything that might be useful once in the day, we have not solved the problem, because in the first case we spend a lot of time preparing ourselves to do something and in the second case the confusion makes any search complex and tiring.

The computer-based work environment must therefore allow its user to organize his/her electronic desk so that whenever he/she begins to do something it transforms itself into the appropriate desk for doing that thing. To the extent that it offers an easy way to build, exchange, and access various different work centers (see (Holt et al., 1983; Holt, 1988)), it becomes capable of making the virtual (logical) environment for performing an activity real (usable).

The same approach proposed here to design the individual computer-based work environment can be extended to designing the work space of cooperating groups. Moving from work centers to work spaces, the whole network of work-stations supporting the members of a group becomes capable of trasforming itself in accordance with the activities they are performing. Designing an elegant and effective solution for computer-based work spaces is in any case difficult, since it requires us to synchronize the work-centers of each member of the group.

A partial but effective implementation of a group work space can be built by creating a window on which the members of the group can interact synchronously, communicating in written or spoken form.

5.5 Organizational Noise and Organizational Opacity

The structure of an organizational unit (rules, role distribution, and communication protocols) is the outcome of its history. It can be rather obscure both for the newcomer, who does not know anything about it, and for the experienced member of the group, since he/she may not have registered its changes. But while the newcomer needs to see the structure of the organizational unit he has become a member of in order to learn it, the experienced member generally acts as if the structure of the organizational unit he/she is a member of were transparent to him/her.

A computer-based work environment must therefore be able to support both characteristics, transparency and visibility, of the organizational structures.

As the structure of an organizational unit may undergo continuous changes, updating the computer-based work-environment can be a rather difficult task. Moreover, if the environment is not updated at the right time while the visibility support becomes ineffective but does not create wrong behaviours, the transparency support can become an obstacle to performing the right actions.

In order to overcome this problem, the computer-based work environment should be able to learn the changes of the structure of the organizational unit it is supporting, in regard to the events of the group happening through it. A learning environment requires a deep understanding of the rules on the basis of which the structure changes; it must also restrict the tool to supporting only the constitutive rules of the organizational unit, disregarding any prescription the members of the organizational unit must obey (the CHAOS system is a prototype with such learning capabilities (De Cindio et al., 1986, 1988b)).

The requirements discussed above have two important features: they are realistic, as each one of them is satisfied by already existing prototypes and products, and they are integrable, since they characterize services that don't have contradictory aspects. It is therefore possible to imagine a user environment exhibiting all or most of them.

Acknowledgments

This paper is a written and improved version of the lecture the author gave at the NATO Workshop on "Organizational Learning and Technological Change", in Siena in September 1992.

The author previously discussed the relations between the complexity of work processes and Computer Supported Cooperative Work at a seminar organized in Bologna by Achille Ardigo' and Graziella Mazzoli at the end of 1991. His contribution at that seminar is in press in Italian. A discussion with Antonio Morawetz of that paper provoked a deeper study of the work of Willliamson, allowing a better understanding of transaction costs theory and an improvement of the analysis of the transaction costs of work processes.

After the NATO Workshop, this paper was written while the author had some interesting discussions with members of the ESPRIT-BRA COMIC (#6225), and in particular with Liam Bannon, Kari Kuulti and Kjeld Schmidt.

Many of the ideas proposed in this paper have been deeply influenced by the coopertion the author had in RSO with Federico Butera,Thomas Schael, Danilo Viviani and Buni Zeller.

This research has been conducted under the financial support of the ESPRIT-BRA COMIC (#6225).

References

AAVV (1986). Proceedings of the Computer Supported Cooperative Work Conference 1986. Austin, TX: MCC

AAVV (1988). Proceedings of the 2nd Computer Supported Cooperative Work Conference 1988. New York: ACM

AAVV (1990). Proceedings of the 3rd Computer Supported Cooperative Work Conference 1990. New York: ACM

AAVV (1991). Proceedings of ECSCW'91. Dordrecht: Kluwer

AAVV (1992). Proceedings of the 4th Computer Supported Cooperative Work Conference 1992. New York: ACM

Agostini, A., De Michelis, G., Patriarca, S., & Tinini, R. (1992). A prototype of an integrated coordination support. UM-DSI Technical Report, Milan

Agostini, A., De Michelis, G., Grasso, M.A., & Patriarca, S. (1993). Reengineering a business process with an innovative workflow management system: a case study. UM-DSI Technical Report, Milan

Bagnara, S., Rizzo, A., & Stablum, F. (1991). Partners' credibility development and assessment in distributed decision making. Journal of Human Factors 2

Bignoli, C., De Michelis, G., & Tinini, R. (1991). UTUCS: a support for synchronous and asynchronous communication. In: K. Gorling & C. Dattlers (Eds.), Proc. International Workshop on CSCW, Informatik Informationen Reporte, IIR Berlin

Bowers, J., & Benford, S. (Eds.) (1991). Studies in Computer Supported Cooperative Work. Amsterdam: North-Holland

Bullen, C.V., & Bennett, J.L. (1990). Groupware in practice: An interpretation of work experiences. Cambridge, MA: MIT Center for Information Systems Research

Bullen, C.V., & Bennett, J.L. (1990). Learning from user experience with groupware. In: (AAVV, 1990)

Ciborra, C.U. (1987). Reframing the role of computers in organizations: The transactions cost approach. Office: Technology and People 3

Ciborra, C.U., & Olson, M.H. (1988). Encountering electronic work groups: A transaction costs perspective. In: (AAVV, 1988)

Coase, R.H. (1937). The nature of the firm. Economica 4

De Cindio, F., De Michelis, G., Simone, C., Vassallo, R., & Zanaboni, A. (1986). CHAOS as a coordination technology. In: (AAVV, 1986)

De Cindio, F., De Michelis, G., & Simone, C. (1988). Computer based tools in the language-action perspective. In: Proc. Euteco'88. Amsterdam: North-Holland

De Cindio, F., Simone, C., Vassallo, R., & Zanaboni, A. (1988). CHAOS: a knowledge-based system for conversing within offices. In: H. Lamersdorf (Ed.), Office knowledge: Representation, management and utilization. Amsterdam: North-Holland

De Michelis, G. (1990). Computer support for cooperative work, London: Butler Cox Foundation

De Michelis, G. (1992). A CSCW Environment - Some Requirements. In: S. Scrivener (Ed.), Computer supported cooperative work. Uxbridge, UK: UNICOM Publications

De Michelis, G., De Cindio, F., Simone, C. (1989). Groups in the language/action perspective. In: Proceedings 2nd European Meeting on Cognitive Science Approaches to Process Control, Siena

De Michelis, G., Donzelli, P., Schael, T., & Zeller, B. (1992). Computer support for cooperative work in Space Science Activities. In: G. Vanderveer, M. J. Tauber, S. Bagnara, & M. Antalovits (Eds.), Human-computer interaction: Tasks and organization. Rome: CUD

De Michelis, G., & Grasso, M.A. (1993). Conversations and routines. Structured Programming, 1993

Ellis, C.E., Gibbs, S.J., & Rein, G.L. (1991). Groupware: some issues and experiences. Comm. ACM 34(1)

Ellis, C.E., & Naffah, N. (1987). Design of office information systems. Berlin: Springer-Verlag

Flores, F. (1982). Management and communication in the office of the future. San Francisco, CA: Hermenet

Flores, F., & Ludlow, J.J. (1981). Doing and speaking in the office. In: F. Fick & R. Sprague (Eds.), DSS: Issues and challenges. New York: Pergamon

Flores, F., Graves, M., Hartfield, B., & Winograd, T. (1988). Computer systems and the ontology of organization interaction. ACM TOIS 6.2

Holt, A.W. (1988). Diplans: A new language for the study and implementantion of coordination. ACM TOIS 6.2

Holt, A.W., Ramsey, H.R., & Grimes, J.D. (1983). Coordination system technology as the basis for a programming environment. Electrical Communication 57(4)

Jordan, B. (1992). New research methods for looking at productivity in knowledge-intensive organizations. In: Proc. of the Conference on Productivity in Knowledge-Intensive Organizations, Grand Rapids, MI

Keen, P.G.W. (1991). Shaping the future: Business design through information technology. Boston, MA: Harvard Business School Press

Kreifeltz, T., & Woetzel, G. (1987). Distribution and exception handling in an office procedure system. In: G. Bracchi & D. Tzichritzis (Eds.), Office systems: Methods and tools. Amsterdam: North-Holland

Kreifeltz, T., Hinrichs, E., Klein, K.H., Seuffert, P., & Woetzel, G. (1991). Experiences with the DOMINO office procedure system. In: (AAVV, 1991)

Lai, K., & Malone, T.W. (1988). Object Lens: a "spreadsheet" for the cooperative work. In: (AAVV, 1988)

Lave, J., & Wenger, E. (1991). Situated learning: Legitimate peripheral participation. Cambridge, UK: Cambridge University Press

Malone, T.W., Benjamin, R.I., & Yates, J. (1987). Electronic markets and electronic hierarchies. Comm. ACM 30(6)

Malone, T.W., & Crowston, K. (1991). Toward an interdisciplinary theory of coordination. CCS TR 120, MIT

Malone, T.W., Grant, K R., Lai, K., Rao, R., & Rosenblitt, D. (1986). Semi-structured messages are surprisingly useful for computer-supported coordination. In: (AAVV, 1986)

Malone, T.W., Grant, K.R., & Turbak, F. (1986). The Information LENS: an intelligent system for information sharing in organizations. In: Proc.of the CHI 86 Conference on Human Factors in Computing Systems, Boston

Medina-Mora, R., Winograd, T., Flores, F., & Flores, R. (1992). Workflow management technology: examples, implementations, and new directions. In: (AAVV, 1992)

Moran, T.P., & Anderson, R.J. (1990). The workaday world as a paradigm for CSCW design. In: (AAVV, 1990)

Reder, S., & Schwab, R.G. (1989). The communicative economy of the work group: Multi-channel genres of communication. Office: Technology & People 4(3)

Schael, T., & Zeller, B. (1991). Cooperative office support systems for distributed process management. In: Proc. MEOW 91. Amsterdam: North-Holland

Suchman, L. A. (1987). Plans and situated actions. New York: Cambridge University Press

Whiteside, J., & Wixon, D. (1988). Contextualism as a world view for the reformation of meetings. In: (AAVV, 1988)

Williamson, O.E. (1975). Markets and hierarchies. New York: Free Press

Williamson, O.E. (1985). The economic institutions of capitalism. New York: Free Press

Winograd, T. (1988). A language/action perspective on the design of cooperative work. Human Computer Interaction 3(1)

Winograd, T., & Flores, F. (1986). Understanding computers and cognition. Norwood, NJ: Ablex

Communicating Through Pictures

Masoud Yazdani and Stuart Mealing

Department of Computer Science, Exeter University, UK

Abstract. In this article we present design consideration for a visual language which may allow users with different linguistic backgrounds to communicate freely. This cross language communication uses icons which are graphically clear, semantically unambiguous, adaptable and above all simple.

Keywords HyperCard, icon, conceptual dependency, cross language communication, grammar, spelling, language

1. Introduction

How could two people who do not know each other's language communicate with each other? Koji Kobayashi chairman of the Japan's Nippon Electric Co. (NEC) foresees "a situation that would make it possible for any person in the world to communicate with any other person at any place and any time".

Kobayashi (1986) aims to exploit future developments in machine translation, speech synthesis and recognition. An English speaker may pick up the telephone and speak with a Japanese person in English. It would be the task of the telecommunication system to recognise the sounds and translate them into Japanese and later translate the reply into English. This proposal still is far from becoming a real possibility.

As an intermediate solution Yazdani (1987) proposed to the EC's DELTA project an environment which could assist users of electronic mail systems to compose and understand messages in major European languages.

Email systems are currently passive, they do not provide tools needed for those who would like to use them as a multilingual facility. The proposed LE-MAIL (Language Education - Monitoring And Instructional Learning) is intended to be active and provide support (including a spelling checker and a grammar checker etc). It may, through a suite of linguistic systems, monitor the text being created, and offer help and specific remedial advice in language use.

As the result of funding shortages the Le-Mail proposal was incorporated into the DELTA project for the European Open Learning Service (EPOS). The EPOS

Project envisages the possibility of a network of students all over Europe, able to contact each other via a telecommunications network and having access to a range of educational products and databases. The aim is for the system to provide support for communication between course-ware designers, tutors, students and others, working in as a team in different locations using computers which themselves may be spread over a large geographical area.

Originally, we saw our undertaking at the University of Exeter as the feasibility study of a multi-lingual interface to EPOS, the idea being that a student about to send a message to a tutor (or another student) who does not speak his language should be able to use the program to preprocess the note before sending it.

In order to do this task we began by building (Jones, 1990) a suite of linguistic systems for Spanish, French and English (to be extended to other European languages) which could monitor the text being created. The system would detect spelling and grammatical errors, offer specific remedial advice as well as offering more general help with the grammar and spelling of the language.

In this project we intend to encourage users to communicate with speakers of other languages electronically. The growth of the use of various email systems indicates that our service would be attractive to many potential users. The difference here is that we shall support the user with various linguistic tools which allow the user to communicate with others on the network who may not know his/her language.

The user composes a message through a special educational wordprocessor which assists him/her to compose messages in a foreign language. The writing tools would:

a) detect spelling and grammatical errors,

b) offer specific remedial advice,

c) offer more general help with the grammar and spelling of the language.

The wordprocessor would also incorporate help facilities. The advice will be provided using the knowledge of both the mother tongue and the target language.

The correspondent also requires tools to understand the message. Online dictionaries and grammar analysers which could be used as translation aids already exist and as the user will be in charge of the process the task would not be as complex as machine translation. Our system would not be capable of dealing automatically with the semantics of the messages and the receiver of the message would need to ask for clarification (from the pen pal) if the meaning of any message was not clear.

The work on LE-MAIL was stopped in 1991 in favour of a more integrated approach with the rest of the EPOS project. We were asked if we could use the remaining time on the project to investigate the possibility of producing a multilingual "Welcome Pack" which could describe EPOS to the users as they saw the system for the first time. As a result we produced a rapid prototype of a "Welcome Pack" to show the concept of an iconic presentation.

2. An Iconic Language

Mealing and Yazdani (1991) present design consideration for a visual language. Our criteria for icons are that they should be:

* graphically clear,
* semantically unambiguous,
* without linguistic bias (culture, race?),
* adaptable (open to modification to express shades of meaning),
* simple (perhaps created within a 32 x 32 matrix).

We have tried to work in a simple computational environment in order that the ideas may have a wider application. However, we anticipate that a number of additional design aids may be used at a later date, including:

* colour,
* movement (micons),
* background (picons),
* 3-dimensional (CAD style space of 3D icons).

This iconic communication system uses icons which represent units of meaning greater than single concepts. In return the icons can explain themselves if need in order to clarify the meaning and provide the context.

An icon should be able to explain its meaning:

1) in terms of more fundamental icons,
2) trace its development from photographic reference,
3) maybe even through natural language.

Such "self explaining icons" use simple animations to help the user understand the meaning of the message clearly and thus avoid the problem of ambiguity associated with static icons.

Our work can be seen as an attempt to show the potential of computer animation to the work of the Isotype movement who have attempted to avoid ambiguity and complexity of natural language. The Isotype rules for creating new icons can be incorporated into our language and extended. New icons therefore may be produced by:

* superimposition
* conjunction
* concatenation
* transformation

* inheritance
* duplication
* etc.

More recently Mealing (1992) presents a prototype hotel booking system developed in HyperCard which uses icons to allow a potential guest and hotel manager to communicate. The system is an initial attempt to create an interactive, iconic dialogue using hotel booking as the theme. This prototype system would allow a user to compose his complete booking requirement ironically and send the message to the hotel manager to reply to (again ironically). The design of the system follows the general structure of the LE-MAIL system described above where the process has been divided into two independent stages of composition and comprehension.

3. Hotel Booking

Hotel booking offers us the opportunity to apply iconic language in a simple dialogue - between a potential guest and a hotel manager. We have assumed no interaction between hotel and customer during the compilation of the message and its reply. A typical scenario could find a stranger in a foreign city operating a touch screen in the window of a tourist office, or a traveller contacting a foreign town's accommodation bureau through his home computer terminal.

The compilation of the booking message is accomplished in stages and at each stage the current domain is cued by a picture resident in the background. In sequence these are: a 'typical' hotel front (Fig. 1), a 'typical' hotel reception area (Fig. 2), and a 'typical' hotel bedroom (Fig. 3), each new screen holding the background picture for a second before the information is faded in over it. Therefore, when dealing with the required room type(s) the background picture on screen would be of a room. Sub-domains might later require a typical bathroom, dining area etc. and in the future such images could even be live video.

The first screen (Fig. 1) shows a hotel overlaid by an appropriate caption, and clicking anywhere on the image starts the booking sequence. (This prototype is created in HyperCard™ and is operated with a standard mouse, but conversion to other input devices, such as a touch screen, is anticipated by the design). The screen then invites input of destination, cued by a map, and selection of hotel type, by selecting from cyclable 'star' ratings. Movement to the next screen is initiated by clicking on the 'tick' icon, a convention observed throughout the package.

The second screen (Fig. 2) shows a hotel reception area and invites selection of the dates and times of arrival and departure. The number of nights that have thus

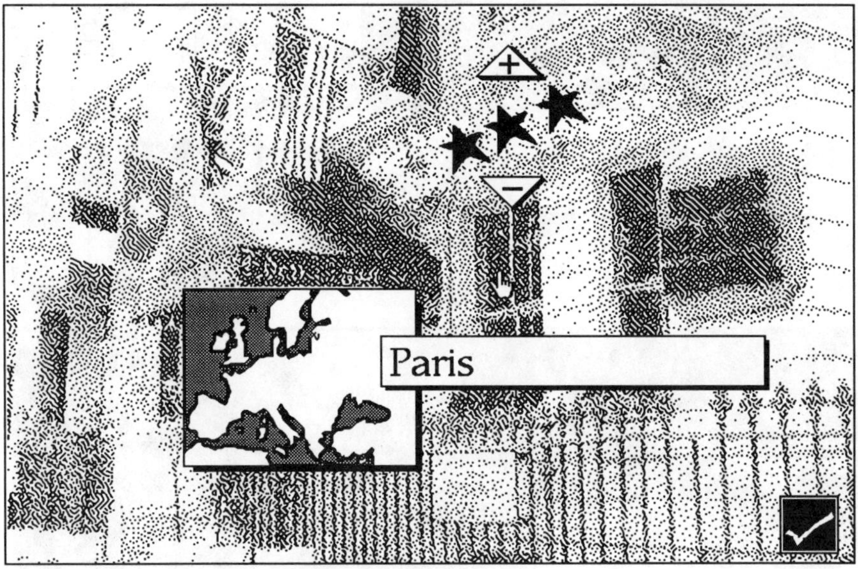

Fig. 1

Fig. 2

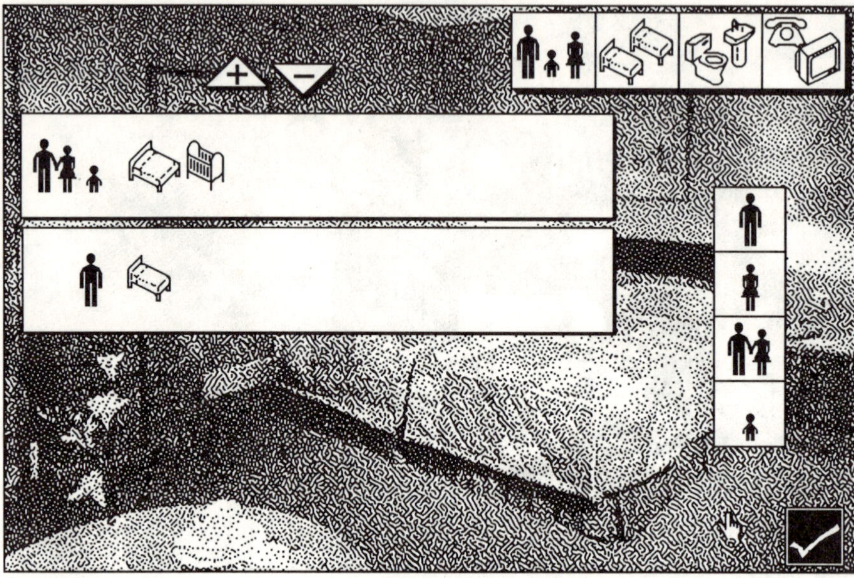

Fig. 3

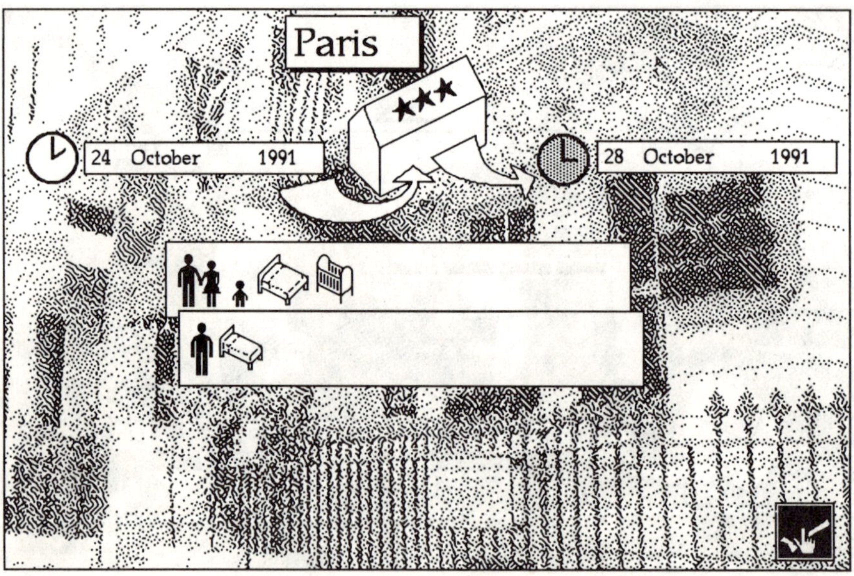

Fig. 4

been booked is indicated by black bars which appear (and disappear) as each night is added (or removed), but whilst this added display is useful its meaning is notimmediately clear in its present form. The 'tick' icon moves the user to the next screen.

The third screen (Fig. 3) shows a room overlaid with icons permitting the selection of room type(s). Four icons each 'unlock' further related icons to enable (a) the number and type of occupant to be shown (b) the number and type of beds required (c) the type of bathroom facilities required and (d) a range of other available facilities (such as TV) which can be selected. Throughout the application, the user is presented with a limited range of choices at any one time. The features are selected by clicking on the relevant icon, which produces a clone beside it, and then dragging that clone into the room which is shown as a rectangle, these mobile icons can also be deleted. More rooms can be requested by clicking on the '+' and '-' icons and the rooms' occupants and contents can be rearranged to suit. This method of presentation was felt to allow the user more flexibility in organization than if each room had to be defined separately, and whilst no controlled tests have yet been conducted, the interface has been found to be intuitively obvious. In order to cue the user's understanding of the principles of organising a room, the screen opens showing one room with occupants. When a satisfactory arrangement of rooms, occupants and facilities has been achieved, the 'tick' item moves the user to screen four (Fig. 4) which displays the complete booking requirement. If this is satisfactory, a further 'tick' sends the message to the hotel.

The message is revealed to the hotel in stages (Fig. 5). Confirmation of the acceptability of each part of the message (by selecting a 'tick') moves on to the next part of the message, unavailability (indicated by selecting a 'cross') brings up the range of available alternatives.

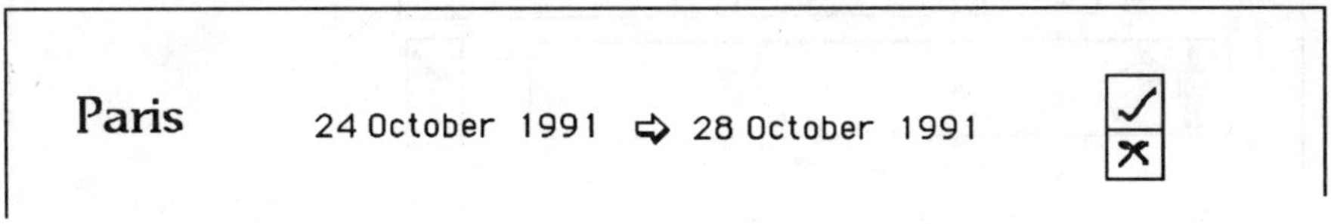

Fig. 5

In Fig. 6 the choice of 'star' rating was unavailable and the alternatives are presented so that an alternative can be offered.

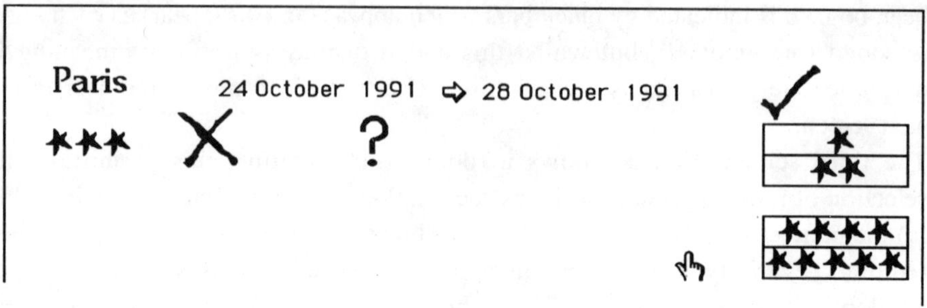

Fig. 6

In Fig. 7 it can be seen that the chosen bedding arrangement for the first room is not available, and two single beds are offered instead of the double bed requested. The large ticks and crosses indicate to the customer what is, and is not, available and the question mark precedes the alternative which is being offered ("Is this OK?"). The layout of the rooms and alternatives is intended to enhance their understandability. The cursor is about to select a 'tick' to indicate that the second room is available as requested.

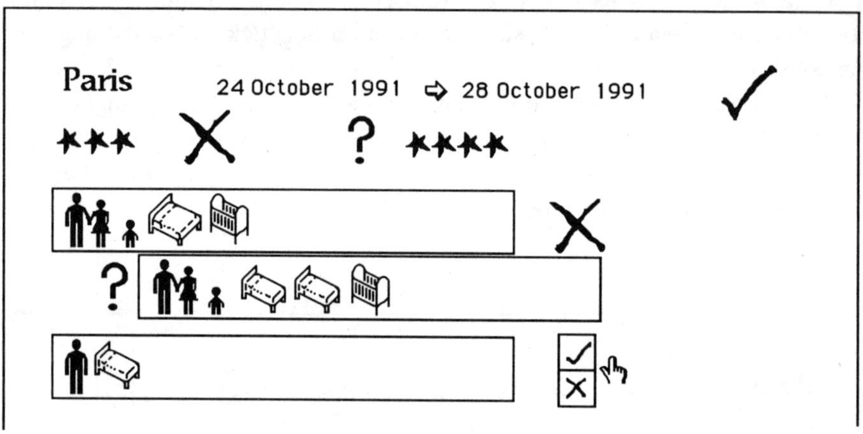

Fig. 7

The final message (Fig. 8) is sent back to the customer who will then be able to accept or reject the alternatives offered, continue the dialogue, and confirm a booking. The application does not pretend to be comprehensive or the most practical in real terms, but is an initial attempt to create a simple, interactive, iconic dialogue using hotel booking as the theme. It does, however, offer much that could be used in a real system, and serves its purpose in starting to explore the possibility of communicating with icons.

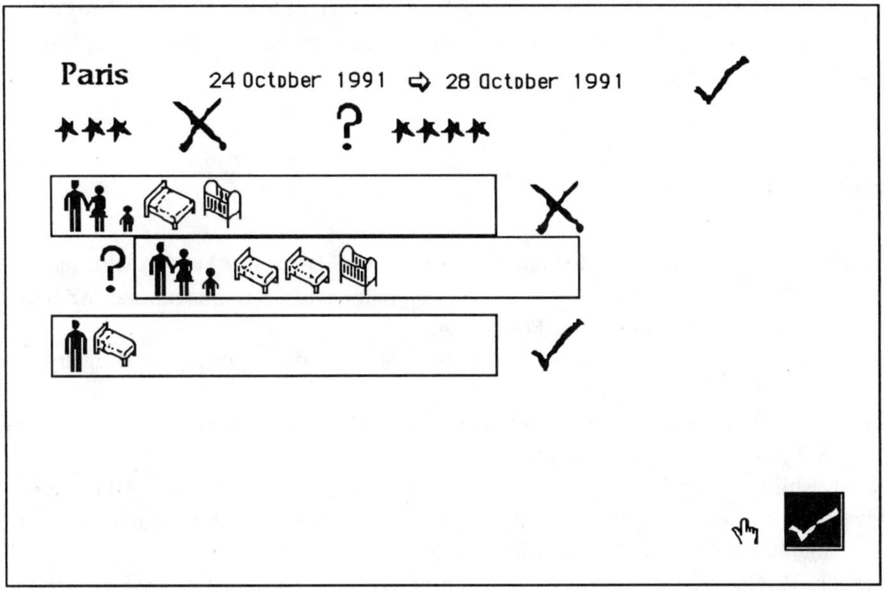

Fig. 8

4. Conclusion

Our work has found support from a small group of researchers elsewhere (Beardon, Dormann, Mealing, & Yazdani, 1992) who are interested in cross language communication using icons. Unlike previous computer based icon systems, the purpose of our work is to explore unrestricted person to person communication. In this work we wish to go beyond the role of pictures as adjuncts to the written word but as replacements to it.

Beardon's (1992) work in particular is worthy of serious consideration as he has used Schank's (1975) Conceptual Dependency formalism, with the ultimate referential objects of the message being self-explaining icons. Messages are composed by selecting options from a small number of carefully designed screens which can be accessed recursively. The recipient receives an animated version of the message using actors which are represented by simple icons.

Beardon's system differs from Mealing and Yazdani's in that there is a small element of syntactic structure which the user needs to learn. However, it shows that it is feasible to build a non-textual language for communication with few learned conventions. How much learning the user of an iconic language needs is a matter of debate between various members of our small research community. What is unanimously agreed is that it is a worthwhile research topic to explore the various options in more detail.

References

Beardon, C., Dormann, C., Mealing, S., & Yazdani, M. (1992). Talking with pictures: Exploring the possibilities of iconic communication. Computers and Writing V Conference, Sheffield City Polytechnic

Beardon, C. (1992). CD-Icon: an iconic language based on Conceptual Dependency. Intelligent Tutoring Media 3(4)

Jones, S. (1990). A multilingual grammar analysing system. Department of Computer Science, University of Exeter

Kobayashi, K. (1986). Computers and communications. Cambridge, MA: MIT Press

Mealing, S., & Yazdani, M. (1991). Computer-based iconic communication. Intelligent Tutoring Media Vol.1 No. 3

Schank, R. (1975). Conceptual information processing. Amsterdam: North-Holland

Yazdani, M. (1987). Artificial Intelligence for Tutoring in Tutoring and Monitoring Facilities for European Open Learning

Yazdani, M., & Goring, D. (1990). Iconic Communication. Department of Computer Science, University of Exeter

Complex Organizations, Knowledge-Intensive Applications, Software and Systems Development: A Case Study

Anna Giannetti and Francesco Gerbino*

SOGEI (Società Generale di Informatica) SpA, Technical Management
Via M. Carucci 99, I-00143 Rome, Italy

Abstract. In this paper we present some theoretical and experimental issues related to organizational and cognitive models of the design and use of new information technologies in an industrial software development/maintenance environment. The approach attempts to reconcile, from a unified viewpoint, some results obtained through CEC and company funded research projects. It is our opinion that a large amount of "enterprise knowledge" should be captured both in terms of human expertise and skills, and in terms of system and domain knowledge acquired or produced during software system development. Varied and distributed expertise as well as shared workspaces and resources will pose further constraints on the way service organizations will evolve in the near future. An "enterprise knowledge communication network" can well constitute a learning infrastructure embodying organizational principles as well as decision support tools which are important for engaging and sustaining successful learning. Understanding and thus modelling organizational learning as a continuously interactive process which influences the decision-making behaviour as well as the service-oriented behaviour and providing it with intelligent support software tools, can result in the workplace of the future, where specialisation is valued and innovation encouraged.

Keywords. Organizational learning, knowledge acquisition methods and tools, software development and maintenance, knowledge based systems, social communication

---

* Views and conclusions are those of the authors and should not be interpreted as representing the official opinion or policy of FINSIEL SpA, STET, IRI SpA, and Commission of European Communities (CEC) or any person or agency connected with them.

1. Introduction

The major aim of this paper is that of providing a conceptual and methodological framework for designing better technological artifacts and for evaluating their impact on existing organizations. This represents a first attempt at coordinating research results, carried out in the CEC ESPRIT II framework[1] and partially supported by the company[2]. To begin, we wish to situate our contribution to the topic of the workshop and book in terms of our interpretation of the workshop's title "Organizational Learning and Technological Change".

We shall consider learning as a fundamentally *knowledge-based process*, i.e., that of acquiring, manipulating, consolidating, refining and thus modelling knowledge. In fact to acquire knowledge from any (human or computerised) source is tantamount to generate an individual (but potentially socializable) model of that knowledge. This assumption may be rather rationalistic but we do not intend to follow the Artificial Intelligence paradigm blindly. In fact we shall consider learning as a fundamentally *social process*, i.e., that of communicating knowledge under different formats and forms and in different contexts. Moreover, all communication activities develop over many stages which include amongst others that of mutual sharing of knowledge, assessment, judgement and evaluation of what it is communicated. learned. We have thus decided to couple the terms *learning&communication* as if they were a single concept. In fact there is no learning without communication, which in its turn implies positive or negative feedback or reactions and even mutual assessment of what has been learned or communicated. In this respect we distinguish the *process* of learning, which is highly social and "situated", from the actual products or *results* of learning which usually reside in single-minded agents and can only be assessed and evaluated individually through communication. This, maybe rather hyperindividualistic, view reflects the idea that any introduction of new technological artifacts which does not deal with cognitive processes and behavioural reactions looses track of the individual components of the organization. The notion of Learning appears to be necessarily related to the notion of Assessment and Evaluation as well as the notion of Communication is always related to the notion of Response or Uptake (Atkinson & Heritage, 1984). Both Learning and Communication need a shared procedure for interpreting the meaning of what is being said or presented (Sperber

[1] ESPRIT II project 5330 NOMOS "Knowledge Acquisition for Normative Reasoning Systems", coordinated by SOGEI (I)-Partners: AXON (P), CNR-IDG (I), HELLASLEX (Gr), INESC (P), IRETIJ (F), ORION (Gr), STEP-INFORMATIQUE (F), TECSIEL(I) and ESPRIT II project 5111 DOCKET "Document and Code Elicitation Toolset", coordinated by UMIST(UK)-Partners: Computer Logic (Gr), CRIAI (I), Software Engineering Services (D), SOGEI (I), Universidade Portucalense (P).

[2] The research has been carried out at the Research and Innovation Dept. of SOGEI SpA.

& Wilson, 1986). The response can be acceptance or refusal for a certain communicative act. It is common opinion (Failla & Bagnara, 1989) that one of the most difficult barriers to the full development of a professional in a company is the incapability of self-assessing her own performance. A distributed learning network would provide immediate feedback to every professional accessing the network. Another important aspect is the group-assessment. In fact a group can act as a self-regulating and self-organising sub-system. Learning is also an ubiquitous process: that is, it happens everywhere and at any time in the organization.

Organizational Learning is both individual and group learning which happens in organizations/societies/cultures following organizational/social/cultural models, values and principles. As well as individuals that learn, there are organizations that learn (so called Learning Organizations), where learning is no longer an *extra*, something which is carried out outside the usual activities, roles or responsibilities, but the backbone of the whole organization. At the same time, individuals that learn become integral part of Learning Organizations.

Besides, Technological Changes, whilst being somewhat "external" to the basic learning process, may be considered as enhancing and supporting factors to the processes of knowledge acquisition, manipulation and communication throughout any complex organization. Technological supports can be both intelligent, knowledge-based systems and intelligent, distributed networks supporting hundreds of different communication services which allow the worker to carry out and test the results of his/her activities in a simulated world, where real-time responses are guaranteed and where novices and experts interact daily even and, especially, outside the traditional educational sites, such as classrooms. In this paper, we shall consider those technological changes which mainly result in software tools having an impact on the traditional way of carrying out tasks and activities in a real workplace.

We shall also consider Organizational Changes and Technological Learning as the complementary parts and side-effects of what has been claimed before.

Organizational Changes happen because of external or internal factors, such as new businesses or market crises or new production/service technologies (Stamper et al., 1991). Technology and its applications is one of these factors and one of the most dramatic. In fact there are at least two inter-dependent processes which have an impact on complex organizations: the gradual replacement of routinary skills produce "de-skilling" with a consequent demotivation of old-fashionally trained personnel and changes in the consideration and position of some jobs, whilst the development of intelligent assistant tools produces what can be called a sort of "re-skilling" that is the need for more sophisticated professional skills and continuous and permanent training (Negrotti, 1990). Once we consider organizational structures as a set of interacting (business) means, such as knowledge, expertise, skills, communication codes, rhetorical visions and tools for reaching ever-changing dynamic (business) goals, we can also assert that changes in the means (from humans to human-machine systems or, in the near future, human-machine-human systems) can well generate changes in the means as well as in the goals, which will thus be reflected in organizational changes.

Technological Learning is the central issue of the global diffusion of technology. Very often, the cost of introducing a new technological artifact into the workplace is highly under-estimated and its social and cognitive impact only superficially considered (Bagnara & Failla, 1990). We should try to make these hidden costs more apparent so as to justify the technological choices and investments, from an economical or quality improvement point of view.

Technological learning is also considered in the course of the paper as *technology-based learning,* i.e., computer or human learning supported by automated tools. New emerging subfields of Artificial Intelligence are Machine Learning and (automated) Knowledge Acquisition. Machine Learning (Carbonell & Langley, 1987) is an area where computational techniques and methods for acquiring new knowledge, new skills and new ways of organising existing knowledge are investigated. From a practical point of view, it is concerned with the design and development of algorithmic strategies and tools able to *automatically* conceptualise, classify, categorise and discover inner relations among different aspects of observable phenomena, such as perceptions or empirical data. (Automated) Knowledge Acquisition (Gaines & Shaw, 1991) is a more traditional sub-field of Expert Systems development mainly concerned with the acquisition of knowledge from human informants, by means of intelligent support tools. The boundaries between the two sub-fields are becoming fuzzier as they become more mature fields of research. In both cases learning is both the *modality* by which knowledge is acquired and the *central feature* of the knowledge to be acquired and processed. This viewpoint on knowledge not as a static entity fixed once for all, but as a dynamic entity always subject to refinement and individual subjective judgement is influencing the way by which human expertise and knowledge can be acquired, accumulated and exploited.

Technology and learning thus become crucial leverages for organizational changes.

In all these cases the issue of natural human computer interface is crucial. In fact in order to make this (acquired) knowledge *effectively* and *directly* usable by the actual personnel, these systems should be provided with an intelligent cognitive interface which is able to support the user or user group in their interaction and navigation through all different aspects of the system.

It is our opinion that a large amount of "company knowledge" should be captured both in terms of human expertise and skills, and in terms of system and domain knowledge acquired or produced through software system development.

Varied and distributed expertise as well as shared workspaces and resources will pose further constraints on the way service organizations will evolve in the near future and a major value will be that of participating, at all levels of hierarchical structure, in the dynamic, feedback-oriented construction of tasks, activities, artifacts, models, goals and values.

A "company knowledge communication network" can well constitute a *learning infrastructure* emboding organizational principles which are important for engaging and sustaining successful learning.

2 Modelling Organizational Learning and Communication

The scope of this section is that of providing an interpretation model and a methodological framework of the design, development and use of technological artifacts in complex organizations. First of all, technology is two-faceted: It can provide support to already existing activities by "sharing the burden" with users or it can generate new activities which dramatically modify the way by which the actual work is perceived and carried out (Mark, 1986; Chignell & Hancock, 1988). In both cases organizations can be the goal of information systems as well as being generated by new information systems. Usually a company gets its identity through its assets and core businesses. Usually companies tend to see their core business as a technical core. But what it is really its core competence (Prahalad & Hamel, 1991)? What a company, specifically a service company, is really able to offer its customers so that they will choose their services amongst others? This is *knowledge*, knowledge of the user needs, processes and knowledge of the offered products/services. Knowledge is also a pre-condition for improving efficiency and rationalisation (Handy, 1990). Another aspect to be considered concerns the boundaries between living organizations and their environments. For a company which capitalises on its products/services in long-life terms and which tends to perceive the customer as an active participant of the design of its services, boundaries will suddenly tend to disappear. New value-added networks are already interconnecting internal "company" world to external "client" world. Such knowledge-intensive companies are also called "Imaginary Organizations" (Hedberg, 1991).

The two basic features of modern organizations are thus *knowledge*, either consumed or created and *communication*, either internal to the company and external with the clients.

There are occupational categories, like managers and administrators, who can be grouped under the label of "knowledge workers" (Ellis, 1987). These are the people usually engaged in *open tasks,* i.e., creative solution to complex problems, and so they would need office information systems as "intelligent assistants" (Boy, 1991) or "augmenters of their intellect" (Engelbart, 1988). A new generation of these tools, which embody artificial intelligence routines, goes under the name of *decision support systems* which provide rapid gathering and access to data and support for the analysis of the gathered data so as to improve the capability of making decisions. As we can clearly observe, the differentation among open and closed tasks and between knowledge and routine workers, corresponds, to a certain extent, to the traditional organizational hierarchy of a typical (Western) industrial or administrative bureaucracy (Zara, 1992). In recent times it has been realized that there is the greatest need for faster access to various sources of information, stored both electronically and in human memory, both inside and outside the workplace, as well as new sources for more powerful problem solving aids *at each level of the hierarchy* (Malone, 1987). The office of the future is going to become a place where information as well as responsibility will be shared and distributed. Through the introduction of electronic media for communication, the control of the process will become more and more distributed. Such changes call for a new generation of systems which will increase the efficiency of the office in its globality, not solely of the individuals.

In the last few years a new, interdisciplinary area, called *Computer Supported Cooperative Work (CSCW)* (Grief, 1988) has grown around questions such as: How should people cooperate so as to improve the global efficiency of an organizational structure? How can the computer support the cooperative as well as the decision processes? Which software tools are needed? These are difficult questions to answer and multidisciplinary research in different areas such as artificial intelligence, psychology, linguistics, sociology, anthropology, office automation and organizational theory will be needed. The major aim of this new discipline is to apply the more advanced communications technologies to coordinated problem solving and decision making; computers are now sufficiently sophisticated media not only to set up physical communication links but also to help to coordinate activities and support joint problem solving, e.g., by means of role-based access to information (Schal & Zeller, 1990).

We propose the following step-wise method to understand and model an organization:

a) identify core business/competence,
b) identify roles in organization,
c) identify goals for each role,
d) identify goal-oriented learning strategy,
e) identify communication strategy

In a company, which is a services provider, the core business is the service provided to the client and the whole organization should be centred on that issue.

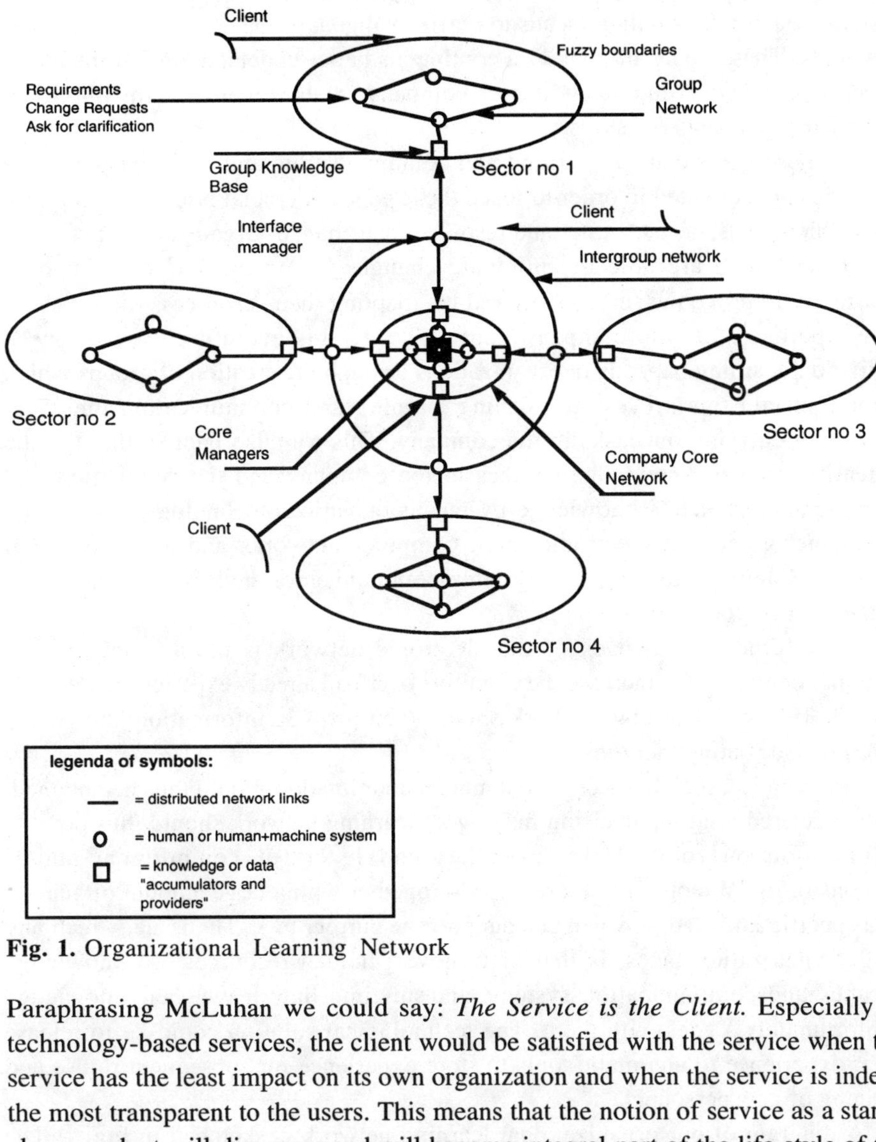

Fig. 1. Organizational Learning Network

Paraphrasing McLuhan we could say: *The Service is the Client.* Especially in technology-based services, the client would be satisfied with the service when the service has the least impact on its own organization and when the service is indeed the most transparent to the users. This means that the notion of service as a stand-alone product will disappear and will become integral part of the life-style of the client.

The functional notion of role substitutes the traditional notion of hierarchical level. Roles in an organization are usually assigned in order to guarantee the fastest response to problems which may arise so as to later make clear responsibilities and claims: roles are functional (Abstract) entities which can be performed by either humans (individual or groups), human-machine systems with decision support software or human-machine-human communication support systems. Each role is only a means to reach a subgoal of the company and it itself has goals. These goals are mainly a continuous better understanding of the client needs and a better understanding of company goals, together with improved technical and managerial skills.

All these goals call for learning and communication, and different plans and strategies are activated in order to reach these goals. A crucial issue is the rapidity with which goals, needs, values and resources can change. Organizations that learn are those which are able to anticipate changes in the outside world and to constantly improve the services offered by adapting their resources, values, skills and expertises, i.e. the company knowledge, to this evolution (Senge 1992). Individual learning happens in organizations through the creation of groups which should sustain this process by making learning and communication the usual means for carrying out tasks in the company. This can also happen through the indentification of knowledge sources in the company and the acquisition and communication of this knowledge by means of various technological supports: telephone, voice and electronic mail, computer networks and group decision support systems according to a flexible communication code which should be known to everybody in the company.

A well-known drawback of the electronic network is the amount of junk information which can increase the cognitive overload already experienced (Hiltz & Turoff, 1985). A learning network should then provide information/knowledge filtering and routing facilities.

Learning is a social process and a mutual coordination should thus be obtained. This is related to group decision making. A learning network should thus provide information on roles, tasks, activities and levels of commitment and/or responsibility (Winograd & Flores, 1986) together with a large amount of general and specific knowledge which can support the current task. The issue, which any large organization faces, is that of turnover: human resources are knowledge sources and communication systems and anytime they leave their knowledge (unfortunately!) goes with them. The technological solution could be to release their experience to automatic tools to store experience for subsequent re-use and training of new personnel.

A diagram of our organizational learning network is sketched in Fig. 1. The main idea is that each sector of the company should be directly sorrounded by the corresponding sector of the client organization, if existing. This has two purposes: first, the client could be interconnected through computer communication links so as to be able to broadcast specific requirements, requests of change or generic needs for new services throughout the network, in a word being able to focus and refocus on the specific product so as to lead to an improvement/change; second, each sector will have the highest visibility and it will focus on a subset of the client organization so as to serve it at its best. During the process of acquiring new knowledge from the external world as well as during the process of developing new software or maintaining old one, new knowledge is generated which should be captured. Each node can be either a human or a human-machine system so that at any time annotations can be taken and stored for subsequent re-use.

All the knowledge produced is stored in so-called *knowledge accumulators and providers* so as to have huge but maintainable knowledge bases containing all domain and application knowledge as well as their intermediate results. The distributed network links will contain cooperative support software as well as project management software and filtering packages which will make the knowledge available at any stage when it is more convenient and to the company role which is going to make the most out of it.

The most important role is played by the so-called "interface management", i.e., middle management as "interface" management, mediating, filtering and dissolving in the network directional inputs and assessing via system responsiveness to each call for responsability and commitment. They put in contact different sectors and different groups in each sector, they collect and gather impulses coming from teams and filter them through the direction pole, where the so-called "core managers" reside. Even the direction pole is sourrounded by the client environment, this is because new services, which will go beyond the traditional software service offers, call for a better characterisation of the client's needs at all levels and from several perspectives. The central knowledge accumulator will generate and visualise knowledge at various degrees of complexities so as to make core management constantly aware of the real assets of the company, by supporting the creation of application portfolios and by assessing the availability and suitability of each single worker for each activity. The client organization in fact sorrounds the service organization: the client organization dissolves its inputs at each site of the distributed network onto the service company information system. These inputs are processed and analysed as much rapidly as possible and mediated by the sectors responsibles and interface managers.

3. Technology and Learning

The way to modelling organizational learning and communication inevitably passes through the issues of technology and models of its design and use. During our analysis we recognise different sites for learning: learning to identify the knowledge sources which are most appropriate for the task in hand (i.e., learning to recognise a need for information), learning how to acquire knowledge from the source (i.e., by interacting with a knowledge acquisition tool), learning how to communicate with the other people in the same group or in different groups, learning how to interpret a new requirement or a new legal variation,etc.. There are several opportunities for learning which encompass different types of technological knowledge, as also shown by (Carrière, 1992). Basically, one can stimulate and hence support learning both at the *technology building phase* and at the *technology exploitation phase.*

In order to design and build a piece of intelligent technology one has to study very carefully how intelligence works so as to engineer it properly. In the effort of building technology many issues are involved and a kind of learning attitude is stimulated, i.e., *reconstructing* and *modelling* the implicit and explicit goals for which the piece of technology can act as a useful/usable tool. Instead, when people, who did not partecipate to the design of that piece of technology, are committed to its usage, a different complementary kind of learning takes place: it is oriented to the actual potential of that piece of technology for supporting that particular task (in the relevant context, at the appropriate time), i.e., *producing* and *practicing* the implicit and explicit goals (i.e., actually executing situated plans and actions) for which the piece of technology should act as a medium. An important aspect of the use of such intelligent technology, as pointed out by several surveys (Gallino, 1988), is its simulative character where world, domain and task knowledge can support the learning process by reinforcing the understanding of the problem and its solution in terms of constraints and resources.

Modern design theory (Simon, 1981) has repeatedly stressed the relevance of a design theory which reconciles the designer and the user in a sort of participatory and evolutionary design. But does it mean that the two streams of learning, above identified, should eventually converge? In fact there is a balance between the technological support given to the actual implementation of organizational (or alternatively individual) roles and the human achievement of those organizational goals, which their roles descend from.

Following the observations about "artifacts in use" (Bannon & Bodker, 1989), we could provocatively assert that the tool *is* the interface. The design of the tool in fact comes from an understanding of how this tool has to be used. Therefore, the design of an interface as a particular module of the system should not focus on the low-level, physical details, but on a complex dynamic cognitive model of the task and of the users, so as to guarantee task adequacy and usability of the tool.

These artifacts are social tools, in the sense that they have to be incorporated as "ways of sharing and dividing work". The suitability of an interface design cannot be decided if not in the *praxis*, i.e., in the practice of use of tools for knowledge sharing and communication support.

As regards the role that technology can play in learning organizations we propose the following step-wise method:

a) identify needs,
b) discover potentials and costs,
c) evaluate organizational impact,
d) identify solutions: methods and tools.

The first phase of the identification of needs is related to the type of learning and communication strategy of each role in the company. It is crucial for several reasons: first of all technological responses should in principle be tuned to social and/or individual needs. In many cases technology, with its design models and practices, anticipates or discovers dormant needs and, in this very case, not only it changes the way of working (i.e., the means used in/by an organization) but even and above all the goals and thus the overall structure of an organization. As we have seen in previous sections, roles tend to vary and to diversify along with the variation and modification of organizational goals.

Usually needs can be *primary* or *secondary* whether they have been generated directly by people or by people interacting with other pre-existing artifacts.

The second phase is the identification of potentials and drawbacks in the introduction of a certain technological artifact in the workplace. Anytime we introduce a new tool to support a task we should take into account that the task can change dramatically. All changes inevitably become opportunities for learning and communication. Training for new software tools should be weighted against its actual improvement of the task they want to support (Giannetti, 1991). We should not forget that the potentials and drawbacks can only be evaluated if at least a partial model of the solution is already present and can be demonstrated.

Technological solutions can then be mapped onto the model of needs and tasks and prototypes can be developed with the intention of testing the solution in a small, controlled environment. Technological solutions are not only tools: they are also new methods for a better execution of classes of task and for the identification of adequate support tools.

The tools and methods presented in the paper should be considered as a technological response to the great loss of knowledge which arises during the software understanding and generation process. At each stage of software developement and at each level of responsibility, knowledge is produced, consumed, manipulated and then possibly lost, even before it is communicated. This occurs largely because there are no suitable intelligent tools which can support the capturing process. For example, let us imagine a computer mediated tool which support the phase of requirements capture. As asserted elsewhere (Davis, 1990) the requirement phase of a large complex software system is the most error-prone and these errors are also the most expensive to correct. On the other end, this phase is the most uncertain phase of software development in the sense that requirements are emergent from the interactions between analysts and the client organization, in a *situated* way (Suchman, 1987). Furthermore they are continouosly evolving, as does the organization. Most of these activities are carried out in (both written and spoken) natural language, and in a particular style which can be assimilated to a sub-language. The knowledge which can be acquired through the tool support can be later re-used so as to refocus on specific portions of the application. The need of integration of several tools/processors (each of them dedicated to support a single aspect of the task) brings about the issue of interconnection of those tools in the traditional mainstream of an organization.

So far, most knowledge-based systems (KBS) have been small, stand-alone systems with little or no connection to the mainstream data processing of an organization (Worden, 1988). Beyond technical reasons (different software and hardware environments) there are also cultural reasons and above all practical reasons. In fact, the main problem is that of managing uncertainty: software engineering practice is designed to minimise the uncertainty and unmanageability of software development whilst the only approach which seems to consider the extreme variability and context-dependency of the knowledge itself is the so-called iterative prototyping approach. KBS have failed as badly as traditional Data processing (DP) systems have, because of their inadequate modelling of user needs and/or because of acceptance issues.

Most organizations are based on a very large number of similar transactions handled by a large number of (possibly unskilled) clerical staff. Information systems have therefore provided support to the most routinary and "automatable" aspects of transactions. This is equally true and applicable to the software development environment where intelligent support tools can be difficult to imagine not only to design. In fact until we have a method which contributes to clarify the steps of the software creation process we will not be able to know in advance what level of performance is adequate for a specific role or what amount of knowledge is needed for its best performance. This information is critical for an adequate integration of KBS into traditional information systems.

Distributed Artificial Intelligence (DAI) is concerned with coordinated intelligent behaviour that allows intelligent agents to coordinate their knowledge, skills and plans to act or solve problems, working toward a single goal or toward separate individual goals that interact (Bond & Gasser, 1988). A distributed artificial system consists of a collection of agents, connected via a communication network. These agents are either humans or machines or human and machines. For the sake of our contribution we only intend to set the scene for our tools and not to develop a theory of coordinated behaviour. We focus here not so much on the coordination activity but on the communication activity which can be either direct (face-to-face) or mediated by a computer. We focus on how these technological infrastructures can support the distributed learning process throughout a complex organization. There are two basic problems with the DAI systems: communication delays and continouos knowledge updating. Cooperative distributed artificial intelligence systems are distributed networks of (semi-) autonomous processing nodes that work together to solve a single problem. In this case each task is decomposed into activities and for each activity the relevant/necessary skill and expertise is identified, at which point the problem is routed to the identified node. Central control is always required for those activities and tasks which have the greatest impact on the company's services offer.

As already stated elsewhere (Durfee, 1988) knowledge based reasoning at each site of the network is crucial to decision making processes. In a distributed problem solving approach a set of distributed resources is allocated in order to solve a problem. The relevant parameters are cost of communication and the type and number of resources. The cost of communication can be reduced by means of a local clustering to reduce the need for non local communication. Another way to cut down communication cost (but also misinterpretation costs) is to model presentational tools which are able to present the same information content under several formats and forms according to the needs and skills of the person/role.

The purpose of any Computer Mediated Communication (CMC) system (Bannon, 1986) is in fact to support and possibly enhance human-to-human communication by augmenting the ease and robustness of processing information. A deeper approach is characterised by an explicit modelling of the communicative actors and their *situated actions* (35] together with their dynamic changes. Since software development people would consider communication not as an end in itself but as a means of carrying out their tasks, they will probably accept a trade-off between the time-consuming activity of communication and the overall textual quality of the message. In this study the cognitive means of structuring and representing information are modelled. Whereas these theoretical observations are embodied in the system design and implementation, we hope to obtain, as an important side-effect of using such a tool, a dramatic improvement of the communicative competence of the user. If users could identify what are the the constraints on the complex communicative act at each stage of the process, they will probably be able to anticipate, at a certain extent, possible breakdowns in communication and then use contextually dependent means of recovering from those breakdowns.

4 Networked Software Development Support Tools

Knowledge is produced / consumed / communicated and updated at every stage of the software development/maintenance process (IEEE, 1992). This knowledge should thus be first identified, then acquired and later communicated, presented and/or visualised. Methodological support is needed for identification of knowledge sources (both humans, texts and machines) and for their inventory. There are several different sources, such as the external sources (client, book, manuals, knowledge based systems) and internal sources (colleagues, managers...). Once acquired, the knowledge shall be communicated by means of distributed knowledge networks.

Tool support is needed for the acquisition of this knowledge and for subsequent constitution and maintenance of knowledge bases. Once the knowledge has been identified and captured, i.e., represented in an adequate formalism which supports some retrieval and reasoning, it can well be distributed across the network so as to maximise exchange and learning. Once the knowledge is acquired, it can be communicated, that is verbally presented or visualised at all sites of the network by means of user-oriented and goal-oriented notations so as to maximise the chances for communication and learning. There are basic issues, which in our opinion, have not yet been sufficiently addressed in the field, such as the overhead of updating knowledge and a goal-oriented visualisation of the information.

Visualization/Presentation of knowledge is indeed a crucial issue in the workplaces of the future: different people's backgrounds and attitudes as well as professional skills and expertise represent a tremendouos gap for any communication system. The generation of information from data-bases or knowledge bases should thus take into account the end-user so as to tailor responses to its own needs and make them more relevant to the task at hand. In some sense, learning cannot happen if not incrementally: after any learning session or cycle the learning context change and so does the system which support this learning.

The issue of updating calls for a better characterisation of the knowledge acquisition process (Woods, 1987). This process is iterative and presents a refinement cycle: when acquiring knowledge there has always to be taken into account the final exploitation tasks. This means that additional knowledge should always be acquired if, during the exploitation phase, something relevant is missing.

Distributed expertise and globally available skills will be the core of any learning organization, where knowledge can be easily accumulated and exploited. In a complex organization each role can reinforce learning by means of valuing specialisations and communication. To deliver knowledge and expertise to a machine can be seen as comprising two basic aspects:

a) It is a *learning* opportunity, because of the cognitive challenges, posed by the machine to the interacting human; we should not forget that in order to support acquisition, the machine has to have been programmed with some sort of (reduced) domain theory methods (and practices) and that knowledge modelling happens interactively and in an highly contextualised way.

b) It is a *communication* opportunity where specialisation, expertise and hence knowledge is valued as the major asset of the company. Once transferred to computer storage, expertise and knowledge is available for distribution and exploitation. At the exploitation phase, the knowledge acquired is re-generated and tuned to the current user needs.

In this section we mainly concentrate on clarifying our perspective for a better integration of two knowledge acquisition and integration toolsets into the company workplace. We focus on the definition of the actual support given by the tools to the various stages of system development. In recent years, there has been an increasing awareness that the software industry is entering a new era (Belady, 1991) where, in addition to commercially available reverse engineering tools, new intelligent support tools, addressing semantic issues related to source code appear to be on the market. For a global system understanding to be achieved, knowledge at different levels of Abstraction and at different stages of transformation and rich of links, should become available for consultation and re-use. Two major issues have been addressed: the *domain specificity* of the most complex information systems and the *integration* of different types of applications to meet requirements at different stages of development. Large and complex software systems are domain-specific as well as enterprise-specific, not bought *off-the-shelf* but carefully and expertly tailored, through the collaboration of designers and users, to the needs of those working at the particular endeavour they have been designed to support.

The life-cycle of a software system is usually seen as consisting of a number of "phases". The following may be distinguished (Gougen, 1991):

needs, in which the desirability of a certain kind of system is identified at an executive level;

requirements, in which properties that the system must satisfy in order to succeed are determined;

design, in which a rough architecture of the system is determined, e.g. as a block diagram of its major components;

specification, in which the behaviour of the components is described;

construction, in which the components are actually built and then assembled to form the system;

validation, in which the resulting system is tested against its specifications;

deployment, in which the system is actually installed in its target environment;

maintenance, in which the system is continually modified, upgraded and debugged.

Although we do not want to commit ourselves, for the sake of this research, to any particular software life-cycle or "waterfall" versus "iterative" development method, we want to stress here the benefits of a massive and coherent use of the methodology currently available in our company (DAFNE, 1993). Our intention here is only to roughly subdivide the software creation process so as to make easier the integration of tool support at each phase.

The notions of software understanding and generation is based on the idea that knowledge about system and domain is generated and hence captured at any stage. Knowledge Acquisition can provide us with techniques, methods and tools for carrying out this capturing. Knowledge is acquired, that means, in our terms, learned and communicated. First, we wish to identify the potential sources for this knowledge acquisition process. As far as requirements engineering is concerned, the major knowledge source is the client and, in the case of administrative systems, such as those developed at SOGEI, the whole set of regulations, laws, decrees and circulars will contain the basic enunciation of constraints to be coded into the information system. A striking feature of the "software knowledge" is that it is subject to a continouos process of *coding*, *de-coding* and *re-coding* into various formalisms, from (spoken or written) natural language (or a domain-specific sublanguage), to structural or functional diagrams, programming languages and chichè and operational system routines. A variety of knowledge sources are involved, both *active* knowledge sources, such as intentional agents in the process and *passive* knowledge sources, such as code and documentation. Cognitive agents communicate internally and with the external world by means of several modalities (text-speech-diagram-code) as more appropriate to the context. Knowledge gaps naturally occur in the process: first, there is a "communication gap" among different types of background knowledge, characterising different people or groups of people, such as the terminological divergence between the users' description of new requirements and its mapping onto existing software systems objects. We also distinguish between intra-group and inter-groups communication. The former is facilitated by the homogeneity of knowledge because of the functional equivalence of the role performed, whilst the latter can be more difficult because of different viewpoints on the knowledge needed at each stage and different communication codes and practices. Most intra-group communication results in the production of a large amount of informal documentation, such as rough notes and spoken conversations, usually unrecorded. This group knowledge is just very difficult to capture and re-use.

All different documents and annotations and dialogues between people will then be encoded (and in some sense at least temporarily chrystalised) in the code. This stresses the compactness and completeness and but also the *crypticity* of the source code with respect to other knowledge sources. All knowledge sources, Abstracting from their origin and type, can be classified according to common characteristics such as *modality* and *time of occurrence.* Different modalities are text, speech, diagram and code. This feature is very important when it comes to the acquisition process. The time issue is related both to the particular point in time when a knowledge source is actually produced or when it becomes available. Some Knowledge Sources can exist at a certain time but not being available, or being available but not sufficiently updated as in the case of functional design specifications and operational code. Some knowledge source can either be generated upon request or be permanently stored (in the head of an individual or scattered throughout a group and hopefully recorded permanently in a *knowledge accumulator and provider* (cf. Fig. 1). During software maintenance and evolution, the desired *goal state* is the new released and operational software system whilst the *triggering state,* i.e., the event which triggers the execution of the task, is an external change request (new user requirements and legal variations). The global task requires communication among agents performing different roles (Kedzierski, 1988). In fact, the overall software creation process can be seen as relying on decisional as well as information/knowledge flows. The decision which flows into the model reflects the decisional power assigned to each company profile. The decision flow works mainly as governing and controlling the information/ knowledge flow among the different roles of the organization and it is filtered in accordance with the skills and expertise of the personnel. It can influence the type, modality and temporisation of the information/knowledge flow.

In the diagram in Fig. 2 we have tried to set logical and physical connections among various tools and their users so as to clarify their mutual interconnections and their relevance to the tasks carried out in the company. The Ministry of Finance (the Client) will produce laws, regulations, decrees and circulars, i.e., all normative texts, which are first stored in a text databank, where they are managed by a Legal Data-Base Management Administrator. This Legal Data-Base will be accessed by the Normative Knowledge Extraction Tool, which extracts. the normative knowledge, in forms of concepts and relations, rules and constraints and will provide normative reasoning tools which are stored in a Normative Knowledge Base. At the requirements analysis phase a Legal Requirements tool will support analysts in their analysis and modelling of these requirements. This tool feeds a Legal Requirements Knowledge Bases where all links between legal texts, legal entities and rules and system entities and rules/controls are stored for subsequent consultation. This Knowledge Base provides input to the software production cycle, where skilled programmers have to translate these models into working systems. At the end of this we have tested code and aligned documentation. Although it has not been visualised in the diagram, at this stage code and documentation become autonomous and can live independently outside of the development environment as operational code, operational guides and user manuals. But the released code and documentation never stops existing in the company (the software provider) so that a continuous change and monitoring should be guaranteed. The Reverse Engineering Tools are applied to the code so as to re-discover inner entities and functional relations. Their output also guarantee the methodological separation between technical, programming knowledge, and the application and domain knowledge. At the same time, Document Analysis Tools (partly re-using linguistic tools already developed for the Normative Knowledge Extraction Tool) can extract relevant and useful knowledge to be fed to the Domain and Application Model Base. Multiple expertise and different problem solving skills, developed in years of practice and training in the application and domain area can be elicited from different expert (eg. legal experts, expert analysts and expert programmers) through the Multiple Experts Knowledge Elicitation Tool so as to populate the Domain and Application Model Base. This Model base will thus contain relevant and updated knowledge which can be re-used both for Maintenance Support and Intelligent Tutoring by both Analysts and Programmers.

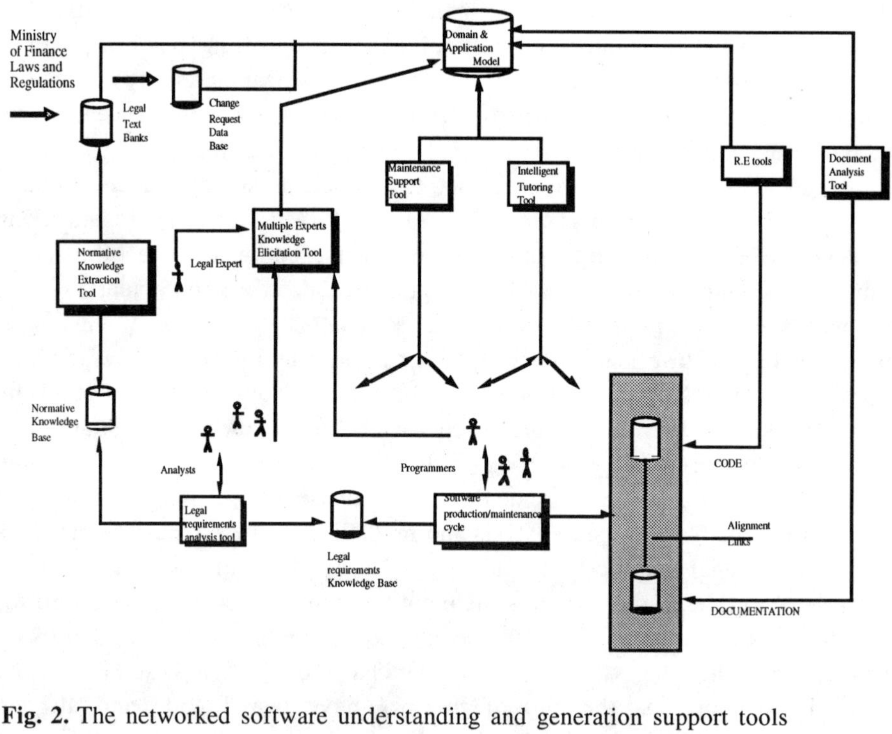

Fig. 2. The networked software understanding and generation support tools

The NOMOS knowledge acquisition approach (CEC ESPRIT) is based on the idea of incorporating sublanguage invariants and of a fully understanding the meaning of a law by modelling the basic legal concepts and their conceptual interdependencies. This tool can perhaps be used during the requirements engineering phase as well as during the needs exploration phase, providing a support for a better understanding of legal requirements and change requests for evolutive maintenance. This tool can even be used for terminological reasons so as to harmonise the client "legal domain sub-language" with the service provider "technical and system sub-language" that is so as to bridge the communication gap between them. The main goal of the tools developed in the NOMOS project (Giannetti et al., 1992) is that of extracting and representing knowledge from normative texts. As already pointed out in the chapter devoted to the background of the research, most software developed by the company concerns fiscal laws and regulations. Strictly speaking, a subset of the whole set of software requirements comes from Ministry of Finance laws, decrees, regulations and circulars. Moreover, the knowledge contained in these texts can be exploited by different roles of the company and at different stages.

The normative domain is very peculiar: the knowledge to be acquired is strongly influenced by the actual appearance of the domain texts and by their interpretation of domain experts. Broadly speaking, what should be acquired from those texts is what has to be used by legal experts for interpreting the norms.

Usually these normative texts, for instance fiscal laws and regulations have the following features:

a) they introduce definitions, obligations and sanctions concerning a citizen's property or actions,
b) they are usually interpreted even for jurisprudential or administration tasks,
c) they are used by experts as a repository of knowledge, and much of their expertise resides in knowing about navigating these texts according to the current case or context.

A further type of knowledge which the system should be able to extract from texts is the *procedural knowledge*, that is the means by which legal experts "interpret" the texts.

The law texts may contain basic definitional rules which, by means of testing the existence of basic attributes, are able to establish whether a particular real world instance belongs or not to the implicitly defined legal concept. This rule can be extracted from the text in the form of a recurrent frame and can be used when required. In particular this rule appears to be crucial for case-based reasoning (Sycara, 1992) where a new case should be mapped in the already (vaguely) pre-defined case. Once the system identifies an instance as belonging to a certain concept, it must derive all the consequences in terms of permits, allowances, obligations, rights, duties, etc. Our approach (Graziadio et al., 1992) is characterised by a multi-layered architecture where each layer corresponds to a software module able to handle every aspect (typographical, lexical, syntactic, stylistic, semantic, textual) of the normative texts.

An example of the use of such tool in the software development mainstream is the following: once the portion of legal text is analysed and acquired by the tool a representation of the knowledge extracted is generated in an adequate formalism. The major characteristic of the law is its definitory or deontic character. Fiscal law is a type of law which deals with individuals or societies and actions carried out by these entities which can be subject to taxes.

For instance, if we take a look to the Italian income taxation law we will find:

Art.2:

All individuals, both residents and non-residents of the State, *shall be subject to the tax, except* those whose income is attributed to others under art. 4. I*n addition to* the persons registered in the Offices of Records of the Resident Population, those persons which have the principle seat of their affairs and interests within the territory of the State or who reside therein for more than six months in the year, as well as citizens residing abroad while serving the interests of the State or other public entities, s*hall be considered* as resident. Non-residents s*hall be subject to the tax only with respect to* income produced within the territory of the State.

This article is of a *definitory* type and it specifies the boundaries of the legal concepts: to be subject to the tax, resident and non-resident. This definition relies on a previous definition of Territory of the State and operates through the use of enumeration of exceptions (or negated instances) or added instances. There are few operators such as *except, in addition to, only with respect* which fix those boundaries and ascribe an operational meaning to these concepts. Once the operators are modelled, these can be stored in a knowledge base where they can be updated and where consistency and coherence checking can be carried out. The tax information system developed at SOGEI already implicitly models these relations as algorithmic conditions without an explicit modelling of how these legal concepts are used in practice. The tax returns forms, as well as the interface objects, i.e., screen of the tax information system, will embody these descriptions and controls. Without many details about the actual functioning of the tools, it should be clear how NOMOS can guarantee a smooth acquisition of knowledge which can be of practical relevance for requirements engineering or even supporting system design. Dependencies between legal entities and legal relations (Cross, 1986) can be modelled explicitly and stored permanently in the system domain and application knowledge base so as to become a reference point for future system evolution and law changes. Service organizations thus learn (and earn(!)) through learning how to model client requirements and through helping the client to shape better their requirements. The NOMOS knowledge base can serve as a support for modelling entities and relationships in the fiscal application domain and related subdomains.

In the software engineering field, there is also a general agreement that over 50% of software costs relate to maintenance and support. Evolutive maintenance can be highly improved by a wider system understanding and by sophisticated techniques of design recovery. DOCKET (Layzell & Champion, 1991) represents an approach to understanding software systems and their supporting components sustained by tools, which can be embedded in a traditional software development cycle, from initial requirements capture to maintenance and evolution. The support offered by DOCKET is both methodological and practical. From a methodological point of view, its provides guidance on how to capture, integrate and access a variety of knowledge about a software system and its application domain. Its tools support the relevant aspects of knowledge capture and knowledge base consultation (reverse engineering, document processing, knowledge acquisition from human experts, knowledge base browsing). Capturing knowledge during development is fundamental but it often seems to be just an additional overhead with respect to time pressure and deadlines. The major issue here is how to improve the cost-effectiveness of the capturing of these sources of knowledge and combining them with formal sources. This overhead can be minimised by using appropriate methods, tools and notations, and the knowledge capture/storage motivated by offering new re-use facilities for subsequent maintenance and intelligent tutoring/monitoring tools. DOCKET addresses the problems of knowledge elicitation (from experts) and acquisition (from other sources) as a natural fall-out of the development and maintenance process rather than super-imposing a new methodology or new formalisms. It actually integrates different sources of knowledge in an opportunistic way according to whichever sources are most available, reliable and most easily capturable. DOCKET aims to support knowledge elicitation in the form in which it is currently available in the work environment and in the application domain of the users by adopting an opportunistic strategy in that it identifies knowledge sources according their reliability and acquisition cost and it tolerates incomplete and scattered acquisitions. The effective use and expected benefits of reverse engineering tools are greatly limited, or confined to narrow domains, by major semantic gaps. In fact there is a wide disparity between the reading of code or its reverse outputs and the user's ability to do something useful with it (Fisher, 1987; Black et al., 1992). Powerful semantic clues are contained in informal sources, such as documents, and human expertise always plays a fundamental role in software understanding. The cost of software understanding is often an hidden part of development costs, not just of maintenance costs. It includes costs related both to the time required to comprehend software and to that lost due to misunderstanding. The system knowledge sources on which the DOCKET toolset relies are: source code, test cases, documentation (user manuals, programming documentation, specifications, designs, change requests, bug reports, etc.) and human experts (system users, business analysts, IT system experts, operational staff). The analysis of these types of knowledge involves the use of different types of meta-knowledge such as programming language grammars, code cliche and patterns, text heuristics and domain and system knowledge. Any automated knowledge acquisition tool tends to optimise the process by reducing the amount of meta-knowledge required and by cutting down the costs of human intervention for supporting the acquisition process, resulting in a simple updating of knowledge bases with no or little need for extensive (and expensive!) training.

In the framework of the DOCKET project we have also been required to carry out a survey, by means of questionnaires as well as open interviews, on how development, maintenance, and support personnel actually carry out their tasks. The results of this survey have been statistically analised as well as carefully transcribed and analysed (CEC ESPRIT, 1991). We found that the results of the analysis have guided us towards the modelling of a certain type of support to the task. We have also discovered and corrected misconceptions about actual practices so as to align technological solutions to actual needs. In order to locate the support given by the tools in a common shared methodology, a cognitive model of the software understanding and generation process has been developed based on empirical results provided by this wide-ranging survey carried out at the company site.

As pointed out elsewhere"...many of the technical problems are often the result of inadequate management communication and control over the software maintenance process..."(Benedusi & Benvenuto, 1991). Software maintenance basically distinguishes itself from development for the following reason: in development there is no previous constraint on what solution has to be found to the problem; in maintenance, instead, previous solutions which have been already tested should always be taken into account. Brand-new solutions are always worked out on the basis of a trade-off between old and new solutions and on the basis of:

a) the overall complexity of requested intervention,
b) the time devoted to maintenance,
c) the quality of software to be reproduced,
d) the overall cost of testing activity.

The development/maintenance process can be thus modelled as a complex shared task or made up of a set of subtasks, carried out either sequentially or simultaneously by a set of agents by a given time, and consisting of a task environment and a set of preconditions which have to be satisfied.

The DOCKET toolset, once integrated in a software development mainstream, should help a user in gaining a more comprehensive understanding of the overall system. In few years time, due to the massive introduction of CASE tools, corrective maintenance will drastically reduce to a minor percentage (Spurr & Layzell, 1992) and evolutive maintenance will become the major issue.

The whole maintenance task can be divided, for the purpose of DOCKET, into three main phases:

– before a CHANGE,
– during a CHANGE,
– after a CHANGE.

At the beginning the end-user communicates the CHANGE request to the maintainer. This phase is characterised by the construction of a sort of "model" of change. At this stage DOCKET provides knowledge to support the interpretation and location of the change request within the whole system and domain knowledge base. Once the change request has been interpreted, this "model" of change is mapped onto the system and domain knowledge base. This allows change requests to be expressed at any level of representation (business, organizational rules, specifications, design, implementation as well as interface requirements) and to be traced back and forward, up and down to the appropriate (i.e. relevant to the context of task) representational level. This should guarantee a rapid and cost-effective impact analysis including all the areas, modules, programs and files, routines and subroutines which could be affected by the change as it has been modelled. The associated documentation should also be visualised so as to provide an enriched environment for carrying out the change. This characterises the phase of "during a change".

Once the change has been carried out, the same knowledge used for impact analysis is then used for checking the integrity and consistency of the system as well as documentation alignment. This is the "after a change" stage.

We will distinguish three basic ways of using DOCKET:

– an **exploitation** mode,
– an **exploratory** mode,
– a **monitoring** mode.

By **exploitation** mode we intend a modality by which an expert user proposes a maintenance problem to DOCKET and DOCKET helps her to solve it, by providing all the necessary information.

By **exploratory** mode we intend a modality by which a novice user, who has to be trained for the maintenance task, could use DOCKET as an interactive training environment where hypotheses can be tested and ways of solving problems simulated.

By **monitoring** mode we intend the periodical reminder and registration of all the changes on the system so as to monitor alignments between code and documentation.

As regards the *exploitation* mode, we want to concentrate on the following issue: in order to introduce a new tool into an already established work-place we need to know when the potential user will use it and what is its impact on the traditional way of carrying out the task. We need to provide a "need for its use".

In our case the "trigger situation" is an external event, which usually comes in the form of a CHANGE REQUEST.

What is a CHANGE REQUEST? It is a piece of information, usually encoded in a surface form, implicitly full of links to an already existing global system model. It can be interpreted as a complex, multidimensional query to the DOCKET base.

CHANGE REQUEST can be formulated in various ways: it can be informally (either verbally or written down) communicated by final users; it can be based on system "external" outputs such as interface screens or on a system "internal" and intermediate output; it can be supported in its interpretation by means of the user guide etc.

The user, by means of DOCKET, should be helped to formulate the problem "in DOCKET terms" (i.e., in system global understanding terms) so that the tool can start connecting information coming from the change request with the global system knowledge, through a sequence of steps.

There are basically two possible usages of DOCKET. A first possibility is to use DOCKET as an Intelligent Tutoring system which is able to teach and train a novice user by providing different information at different stages of the learning process and even prompting the user so as to have, from his answers, a better monitoring of her stage of learning; in this case DOCKET is not constrained by the final user needs, but its only constraints are the stage of the learning and training process.

A second possibility is that of using DOCKET when the maintainer has received a change request and DOCKET should:

- interpret this change request "in DOCKET terms",
- start from the old interpretation and provide links to the new one by means of the DOCKET model,
- derive new links from the new input; these links are hypotheses in different directions which should be handled by means of heuristics,
- use key-words in the formulated change request to make intelligent complex queries to the DOCKET data-base.

There are several individuals or groups who already knows the application and want either to make a check to be sure, for instance, that the right amount of resources have been allocated to that development task (planning phase) or she wants to find details of certain modules and to access the hierarchy of information at the proper level of detail. Other individuals or group who has been told to maintain the software system but did not participate in the actual development of the system. They will use the DOCKET tool for acquiring the necessary information. Individuals would need more guidance by the system. For instance the system will pose questions to the user so as to identify her basic needs and the level of Abstraction needed. For this reason the system should be able to interpret the user responses so as to classify the user (Kobsa & Wahlster, 1988).

5 Conclusions and Future Directions

The ideas pursued in this paper have been those of designing an "enterprise network" where knowledge is captured, stored and later retrieved as the major requirement for an effective learning and communication throughout the enterprise. We have explored the boundaries of the service enterprise and discovered that, indeed, there are no real boundaries between client and service provider and that mutual communication should be ensured in a fast and efficient way. A major finding is that Organizational Learning is tightly linked to Communication and Visualisation of knowledge and this will shape our future ahead.

This paper has also shown a first attempt for a unifyied, cognitive-based approach to software development. A knowledge based approach to information system development seems the only fruitful way ahead: present and future methodologies should cope with the integration of different human, human-system and human-system-human requirements and different modalities of learning, communication and visualisation.

The relevance of the domain specificities calls for a careful customisation of any general software development methodology. The issue of apprenticeship to a methodology should be seen as a continuous monitoring and control process through the enterprise network as well as a continuous learning and assessment process so as to best capitalise on human resources. The use of a methodology should become an implicit added-value to any creative enterprise in so far as "The really creative person knows that the creation task is accomplished by working according to constraints. If there are not constraints, there is no pure creation" (Fritz, 1989).

Much remains to be done in refining our framework and adapt it to the actual organization we work in. The encouraging results of carefully studied experiment at our site have convinced us of the opportunity of increasing efforts and resources in pilot, controlled experiments and data collection throughout the enterprise can be seen as a viable method for capturing updated requirements, following existing needs and anticipating future desires.

Acknowledgements

The authors wish to thank D. Natale, Head of the Software Engineering Area and E. Pesce, Head of the Research & Innovation Area, both at the Technical Management of SOGEI -FINSIEL S.p.A. for their stimulating and original discussions. They also thank Paul Layzell, DOCKET Project Manager and the UMIST team and all partners of the ESPRIT projects. Ettore Pietrosanti and Paolo Mussetto at the Knowledge Based Systems Department of TECSIEL-FINSIEL SpA,, have provided thoughtful ideas and comments. They also thank Cristina Zucchermaglio and Sebastiano Bagnara for their long-standing scientific and moral support.

References

Atkinson, J. M., & Heritage, J.C. (Eds.) (1984). Structures of social action: Studies in conversation analysis. Cambridge, UK: Cambridge University Press

Bagnara, S., & Failla, A. (1990). Risorse Umane e Informatica d'Ufficio. Economia e Lavoro 3, 101-114

Bannon, L.J. (1986). Computer mediated communication. In: D.A. Norman & S.W. Draper (Eds.), User centered system design. Hillsdale, NJ: Lawrence Erlbaum Assoc.

Bannon, L.J., & Bodker S. (1989). Beyond the interface: Encountering artifacts in use. DAIMI PB 288, Computer Science Dept., Aarhus University, Denmark

Belady, L.A. (1991). From software engineering to knowledge engineering: The shape of the software industry in the 1990s. In: IJSEKE 1, 1

Benedusi, P., & Benvenuto, E. (1991). A theoretical approach to software maintenance. In: Giannetti et al. (Eds.), DOCKET Deliverable SOO6 vol. II. Rome

Black, W.J. et al. (1992). Exploitation of system documentation in design recovery. In: Proc. of International Conference on Software Engineering and Its Applications, Toulouse-92

Bond, A. H., & Gasser, L. (Eds.) (1988). Readings in distributed artificial intelligence. Morgan Kaufmann

Boy, G. (1991). Intelligent assistant systems. Academic Press

Carbonell, J., & Langley, P. (1987). Machine learning. In: S. Shapiro (Ed.), Encyclopedia of A.I. Wiley

Carrière, J.B. (1992). Un modèle stratégique de la technologie pour l'innovation dans l'entreprise. In: Technologies de l'Information et Société, Processus d'apprentissage et d'appropriation, Vol. 4

CEC ESPRIT II Project 5111 DOCKET Technical Annex I, and A. Giannetti et. al. (1991). DOCKET Deliverable: Research Report, SOO6 Vols. I, II, Rome

CEC ESPRIT II Project 5330 NOMOS Technical Annexes I, II, Deliverables

Chignell, M.H., & Hancock, P.A. (1988). Intelligent interface design. In: M. Helander (Ed.), Handbook of human-computer interaction. Amsterdam: North-Holland

Cross, G.R. (1986). Legal knowledge. In: Proceedings of the IEEE, 74(10)

DAFNE Methodology and Tools (1993). FINSIEL

Davis, A. (1990). Software requirements: analysis and specifications. New York: Prentice Hall

Durfee, E.H. (1988). Coordination of distributed problem solvers. Amsterdam: Kluwer Academic Publishers

Ellis, C.A., & Naffah, N. (1987). Design of office information systems. Berlin: Springer-Verlag

Engelbart, D.C. (1988). A conceptual framework for the augmentation of man's intellect. In: I. Grief (Ed.), Computer-supported cooperative work: A book of readings. San Mateo, CA: Morgan Kaufmann

Failla, A., & Bagnara, S. (1989). Managers e Professionals nell'Automazione d'Ufficio. Studi Organizzativi n. 3-4

Fisher, G. (1987). Cognitive view of re-use and re-design. IEEE Software

Fritz, R. (1989). The path of least resistance. New York: Fawcett-Columbine

Gaines, B.R., & Shaw, M.L.G. (1991). Foundations of knowledge acquisition. In: H. Motoda et al. (Eds.), Knowledge acquisition for knowledge based systems. Tokyo: Omsha

Gallino. L. (1988). Menti naturali e menti artificiali: nuove prospettive per la ricerca e i processi formativi. In Quaderni di Sociologia n.10

Giannetti, A. et al. (1992). NOMOS: Knowledge acquisition for normative reasoning systems. In: L. Steels & B. LePape (Eds.), Enhancing the knowledge engineering process. Contributions from ESPRIT. Elsevier

Giannetti, A. (1991). A computational pragmatic theory for business messages structuring and generation. In: R.K. Stamper, P. Kerola, R. Lee, & Lyytinen (Eds.), Collaborative work, social communications and information systems. Amsterdam: North-Holland

Gougen, R. (1991). The dry and the wet. Centre for Requirements Engineering, University of Oxford (personal communication)

Graziadio, B., et. al. (1992). A multi-layered architecture for automatic knowledge acquisition from legal texts. In: Proceedings of Avignon-92 Conference on Expert Systems

Grief, I. (Ed.) (1988). Computer-supported cooperative work: A book of readings. San Mateo, CA: Morgan Kaufmann

Handy, C. (1990). The age of unreason. London: Arrow Books

Hedberg, B. (1991). The role of information systems in imaginary organizations. In R.K. Stamper, P. Kerola, R. Lee, & Lyytinen (Eds.), Collaborative work, social communications and information systems. Amsterdam: North-Holland

Hiltz, S.R., & Turoff, M. (1985). Structuring computer-mediated communication systems to avoid information overload. Comm. ACM 28(7)

IEEE Transactions on Software Engineering (1992). Special Issue on Knowledge Representation and Reasoning in Software Development, 18(6)

Kedzierski, B.I. (1988). Communication and management support in system developments Environments. In: I. Grief (Ed.), Computer supported cooperative work. San Mateo, CA: Morgan Kaufmann

Kobsa, A., & Wahlster, W. (Eds.) (1988). User models in dialog systems. Berlin: Springer-Verlag

Layzell, P., & Champion, R.E.M. (1991). The Esprit DOCKET project. In: Atti del Convegno CNR PF-SICP, Naples

Malone, T.W, Benjamin, R.I., & Yates, J. (1987). Electronic markets and electronic hierarchies. Comm. ACM 30(6)

Mark, W. (1986). Knowledge based interface design. In: D.A. Norman & S.W. Draper (Eds.), User centered system design. Hillsdale, NJ: Lawrence Erlbaum Assoc.

Negrotti, M. (Ed.) (1990). Capire l'artificiale. Torino: Bollati Boringhieri

Prahalad, C.K., & Hamel, G. (1991). The core competence of the corporation. Harvard Business Review 68(3)

Schal, T., & Zeller, B. (1990). A methodological approach to computer supported cooperative work. In: Proceedings of ECCE-5, Urbino, Italy

Senge, P. (1981). La Quinta Disciplina. Sperling & Kupfer

Simon, H.A. (1981). The science of the artificial, 2nd ed. Cambridge, MA: MIT Press

Sperber, D., & Wilson, D. (1986). Relevance: communication and cognition. Cambridge, MA: Harvard University Press

Spurr, K., & Layzell, P. (1992). CASE: current practice and future prospects. Chichester, UK: Wiley

Stamper, R.K., Kerola, P., Lee, R., & Lyytinen (Eds.) (1991). Collaborative work, social communications and information systems. Amsterdam: North-Holland

Suchman, L.A. (1987). Plans and situated actions: the problem of human-machine communication. Cambridge, UK: Cambridge University Press

Sycara, K. (1992). Tutorial on case-based reasoning. ECAI-92, Vienna

Winograd, T., & Flores, F. (1986). Understanding computers and cognition: A new foundation for design. Norwood, NJ: Ablex

Woods, D.D. (1987). Commentary: cognitive engineering in a complex and dynamic world. Int. J. Man-Machine Studies 27, 571-585

Worden, R. (1988). Integrating KBS into information systems – the challenge ahead. In: P.H. Duffin (Ed.), Knowledge based systems. Chicester, UK: Ellis Horwood

Zara, C. (1992). La Strategia delle Competenze chiave nelle grandi aziende diversificate giapponesi. Economia & Management 3

The NATO ASI Series F Volumes Presenting the Results of Activities of the Special Programme on ADVANCED EDUCATIONAL TECHNOLOGY

67: Designing Hypermedia for Learning
76: Multimedia Interface Design in Education
78: Integrating Advanced Technology into Technology Education
80: Intelligent Tutoring Systems for Foreign Language Learning
81: Cognitive Tools for Learning
84: Computer-Based Learning Environments and Problem Solving
85: Adaptive Learning Environments: Foundations and Frontiers
86: Intelligent Learning Environments and Knowledge Acquisition in Physics
87: Cognitive Modelling and Interactive Environments in Language Learning
89: Mathematical Problem Solving and New Information Technologies
90: Collaborative Learning Through Computer Conferencing
91: New Directions for Intelligent Tutoring Systems
92: Hypermedia Courseware: Structures of Communication and Intelligent Help
93: Interactive Multimedia Learning Environments
95: Comprehensive System Design: A New Educational Technology
96: New Directions in Educational Technology
97: Advanced Models of Cognition for Medical Training and Practice
104: Instructional Models in Computer-Based Learning Environments
105: Designing Environments for Constructive Learning
107: Advanced Educational Technology for Mathematics and Science
109: Advanced Educational Technology in Technology Education
111: Cognitive Models and Intelligent Environments for Learning Programming
112: Item Banking: Interactive Testing and Self-Assessment
113: Interactive Learning Technology for the Deaf
115: Learning Electricity and Electronics with Advanced Educational Technology
116: Control Technology in Elementary Education
119: Automating Instructional Design, Development, and Delivery
121: Learning from Computers: Mathematics Education and Technology
122: Simulation-Based Experiential Learning
125: Student Modelling: The Key to Individualized Knowledge-Based Instruction
128: Computer Supported Collaborative Learning
129: Human-Machine Communication for Educational Systems Design
132: Design of Mathematical Modelling Courses for Engineering Education
133: Collaborative Dialogue Technologies in Distance Learning
135: Technology Education in School and Industry
137: Technology-Based Learning Environments
140: Automating Instructional Design:
Computer-Based Development and Delivery Tools
141: Organizational Learning and Technological Change
142: Dialogue and Instruction. Modeling Interaction in Intelligent Tutoring Systems
146: Computers and Exploratory Learning

NATO ASI Series F

Including Special Programmes on Sensory Systems for Robotic Control (ROB) and on Advanced Educational Technology (AET)

Vol. 71: Expert Systems and Robotics. Edited by T. Jordanides and B.Torby. XII, 744 pages. 1991.

Vol. 72: High-Capacity Local and Metropolitan Area Networks. Architecture and Performance Issues. Edited by G. Pujolle. X, 536 pages. 1991.

Vol. 73: Automation and Systems Issues in Air Traffic Control. Edited by J. A. Wise, V. D. Hopkin and M. L. Smith. XIX, 594 pages. 1991.

Vol. 74: Picture Archiving and Communication Systems (PACS) in Medicine. Edited by H. K. Huang, O. Ratib, A. R. Bakker and G. Witte. XI, 438 pages. 1991.

Vol. 75: Speech Recognition and Understanding. Recent Advances, Trends and Applications. Edited by P. Laface and Renato De Mori. XI, 559 pages. 1991.

Vol. 76: Multimedia Interface Design in Education. Edited by A. D. N. Edwards and S. Holland. XIV, 216 pages. 1992. *(AET)*

Vol. 77: Computer Algorithms for Solving Linear Algebraic Equations. The State of the Art. Edited by E. Spedicato. VIII, 352 pages. 1991.

Vol. 78: Integrating Advanced Technology into Technology Education. Edited by M. Hacker, A. Gordon and M. de Vries. VIII, 185 pages. 1991. *(AET)*

Vol. 79: Logic, Algebra, and Computation. Edited by F. L. Bauer. VII, 485 pages. 1991.

Vol. 80: Intelligent Tutoring Systems for Foreign Language Learning. Edited by M. L. Swartz and M. Yazdani. IX, 347 pages. 1992. *(AET)*

Vol. 81: Cognitive Tools for Learning. Edited by P. A. M. Kommers, D. H. Jonassen, and J. T. Mayes. X, 278 pages. 1992. *(AET)*

Vol. 82: Combinatorial Optimization. New Frontiers in Theory and Practice. Edited by M. Akgül, H. W. Hamacher, and S. Tüfekçi. XI, 334 pages. 1992.

Vol. 83: Active Perception and Robot Vision. Edited by A. K. Sood and H. Wechsler. IX, 756 pages. 1992.

Vol. 84: Computer-Based Learning Environments and Problem Solving. Edited by E. De Corte, M. C. Linn, H. Mandl, and L. Verschaffel. XVI, 488 pages. 1992. *(AET)*

Vol. 85: Adaptive Learning Environments. Foundations and Frontiers. Edited by M. Jones and P. H. Winne. VIII, 408 pages. 1992. *(AET)*

Vol. 86: Intelligent Learning Environments and Knowledge Acquisition in Physics. Edited by A. Tiberghien and H. Mandl. VIII, 285 pages. 1992. *(AET)*

Vol. 87: Cognitive Modelling and Interactive Environments. With demo diskettes (Apple and IBM compatible). Edited by F. L. Engel, D. G. Bouwhuis, T. Bösser, and G. d'Ydewalle. IX, 311 pages. 1992. *(AET)*

Vol. 88: Programming and Mathematical Method. Edited by M. Broy. VIII, 428 pages. 1992.

Vol. 89: Mathematical Problem Solving and New Information Technologies. Edited by J. P. Ponte, J. F. Matos, J. M. Matos, and D. Fernandes. XV, 346 pages. 1992. *(AET)*

Vol. 90: Collaborative Learning Through Computer Conferencing. Edited by A. R. Kaye. X, 260 pages. 1992. *(AET)*

Vol. 91: New Directions for Intelligent Tutoring Systems. Edited by E. Costa. X, 296 pages. 1992. *(AET)*

Vol. 92: Hypermedia Courseware: Structures of Communication and Intelligent Help. Edited by A. Oliveira. X, 241 pages. 1992. *(AET)*

Vol. 93: Interactive Multimedia Learning Environments. Human Factors and Technical Considerations on Design Issues. Edited by M. Giardina. VIII, 254 pages. 1992. *(AET)*

NATO ASI Series F

Including Special Programmes on Sensory Systems for Robotic Control (ROB) and on Advanced Educational Technology (AET)

Vol. 94: Logic and Algebra of Specification. Edited by F. L. Bauer, W. Brauer, and H. Schwichtenberg. VII, 442 pages. 1993.

Vol. 95: Comprehensive Systems Design: A New Educational Technology. Edited by C. M. Reigeluth, B. H. Banathy, and J. R. Olson. IX, 437 pages. 1993. *(AET)*

Vol. 96: New Directions in Educational Technology. Edited by E. Scanlon and T. O'Shea. VIII, 251 pages. 1992. *(AET)*

Vol. 97: Advanced Models of Cognition for Medical Training and Practice. Edited by D. A. Evans and V. L. Patel. XI, 372 pages. 1992. *(AET)*

Vol. 98: Medical Images: Formation, Handling and Evaluation. Edited by A. E. Todd-Pokropek and M. A. Viergever. IX, 700 pages. 1992.

Vol. 99: Multisensor Fusion for Computer Vision. Edited by J. K. Aggarwal. XI, 456 pages. 1993. *(ROB)*

Vol. 100: Communication from an Artificial Intelligence Perspective. Theoretical and Applied Issues. Edited by A. Ortony, J. Slack and O. Stock. XII, 260 pages. 1992.

Vol. 101: Recent Developments in Decision Support Systems. Edited by C. W. Holsapple and A. B. Whinston. XI, 618 pages. 1993.

Vol. 102: Robots and Biological Systems: Towards a New Bionics? Edited by P. Dario, G. Sandini and P. Aebischer. XII, 786 pages. 1993.

Vol. 103: Parallel Computing on Distributed Memory Multiprocessors. Edited by F. Özgüner and F. Erçal. VIII, 332 pages. 1993.

Vol. 104: Instructional Models in Computer-Based Learning Environments. Edited by S. Dijkstra, H. P. M. Krammer and J. J. G. van Merriënboer. X, 510 pages. 1993. *(AET)*

Vol. 105: Designing Environments for Constructive Learning. Edited by T. M. Duffy, J. Lowyck and D. H. Jonassen. VIII, 374 pages. 1993. *(AET)*

Vol. 106: Software for Parallel Computation. Edited by J. S. Kowalik and L. Grandinetti. IX, 363 pages. 1993.

Vol. 107: Advanced Educational Technologies for Mathematics and Science. Edited by D. L. Ferguson. XII, 749 pages. 1993. *(AET)*

Vol. 108: Concurrent Engineering: Tools and Technologies for Mechanical System Design. Edited by E. J. Haug. XIII, 998 pages. 1993.

Vol. 109: Advanced Educational Technology in Technology Education. Edited by A. Gordon, M. Hacker and M. de Vries. VIII, 253 pages. 1993. *(AET)*

Vol. 110: Verification and Validation of Complex Systems: Human Factors Issues. Edited by J. A. Wise, V. D. Hopkin and P. Stager. XIII, 704 pages. 1993.

Vol. 111: Cognitive Models and Intelligent Environments for Learning Programming. Edited by E. Lemut, B. du Boulay and G. Dettori. VIII, 305 pages. 1993. *(AET)*

Vol. 112: Item Banking: Interactive Testing and Self-Assessment. Edited by D. A. Leclercq and J. E. Bruno. VIII, 261 pages. 1993. *(AET)*

Vol. 113: Interactive Learning Technology for the Deaf. Edited by B. A. G. Elsendoorn and F. Coninx. XIII, 285 pages. 1993. *(AET)*

Vol. 114: Intelligent Systems: Safety, Reliability and Maintainability Issues. Edited by O. Kaynak, G. Honderd and E. Grant. XI, 340 pages. 1993.

Vol. 115: Learning Electricity and Electronics with Advanced Educational Technology. Edited by M. Caillot. VII, 329 pages. 1993. *(AET)*

NATO ASI Series F

Including Special Programmes on Sensory Systems for Robotic Control (ROB) and on Advanced Educational Technology (AET)

Vol. 116: Control Technology in Elementary Education. Edited by B. Denis. IX, 311 pages. 1993 *(AET)*

Vol. 118: Program Design Calculi. Edited by M. Broy. VIII, 409 pages. 1993.

Vol. 119: Automating Instructional Design, Development, and Delivery. Edited by. R. D. Tennyson. VIII, 266 pages. 1994 *(AET)*

Vol. 120: Reliability and Safety Assessment of Dynamic Process Systems. Edited by T. Aldemir, N. O. Siu, A. Mosleh, P. C. Cacciabue and B. G. Göktepe. X, 242 pages. 1994.

Vol. 121: Learning from Computers: Mathematics Education and Technology. Edited by C. Keitel and K. Ruthven. XIII, 332 pages. 1993. *(AET)*

Vol. 122: Simulation-Based Experiential Learning. Edited by D. M. Towne, T. de Jong and H. Spada. XIV, 274 pages. 1993. *(AET)*

Vol. 123: User-Centred Requirements for Software Engineering Environments. Edited by D. J. Gilmore, R. L. Winder and F. Détienne. VII, 377 pages. 1994.

Vol. 124: Fundamentals in Handwriting Recognition. Edited by S. Impedovo. IX, 496 pages. 1994.

Vol. 125: Student Modelling: The Key to Individualized Knowledge-Based Instruction. Edited by J. E. Greer and G. I. McCalla. X, 383 pages. 1994. *(AET)*

Vol. 126: Shape in Picture. Mathematical Description of Shape in Grey-level Images. Edited by Y.-L. O, A. Toet, D. Foster, H. J. A. M. Heijmans and P. Meer. XI, 676 pages. 1994.

Vol. 127: Real Time Computing. Edited by W. A. Halang and A. D. Stoyenko. XXII, 762 pages. 1994.

Vol. 128: Computer Supported Collaborative Learning. Edited by C. O'Malley. X, 303 pages. 1994. *(AET)*

Vol. 129: Human-Machine Communication for Educational Systems Design. Edited by M. D. Brouwer-Janse and T. L. Harrington. X, 342 pages. 1994. *(AET)*

Vol. 130: Advances in Object-Oriented Database Systems. Edited by A. Dogac, M. T. Özsu, A. Biliris and T. Sellis. XI, 515 pages. 1994.

Vol. 131: Constraint Programming. Edited by B. Mayoh, E. Tyugu and J. Penjam. VII, 452 pages. 1994.

Vol. 132: Mathematical Modelling Courses for Engineering Education. Edited by Y. Ersoy and A. O. Moscardini. X, 246 pages. 1994. *(AET)*

Vol. 133: Collaborative Dialogue Technologies in Distance Learning. Edited by M. F. Verdejo and S. A. Cerri. XIV, 296 pages. 1994. *(AET)*

Vol. 134: Computer Integrated Production Systems and Organizations. The Human-Centred Approach. Edited by F. Schmid, S. Evans, A. W. S. Ainger and R. J. Grieve. X, 347 pages. 1994.

Vol. 135: Technology Education in School and Industry. Emerging Didactics for Human Resource Development. Edited by D. Blandow and M. J. Dyrenfurth. XI, 367 pages. 1994. *(AET)*

Vol. 136: From Statistics to Neural Networks. Theory and Pattern Recognition Applications. Edited by V. Cherkassky, J. H. Friedman and H. Wechsler. XII, 394 pages. 1994.

Vol. 137: Technology-Based Learning Environments. Psychological and Educational Foundations. Edited by S. Vosniadou, E. De Corte and H. Mandl. X, 302 pages. 1994. *(AET)*

Vol. 138: Exploiting Mental Imagery with Computers in Mathematics Education. Edited by R. Sutherland and J. Mason. VIII, 326 pages. 1995. *(AET)*

Vol. 139: Proof and Computation. Edited by H. Schwichtenberg. VII, 470 pages. 1995.

Vol. 140: Automating Instructional Design: Computer-Based Development and Delivery Tools. Edited by R. D. Tennyson and A. E. Barron. IX, 618 pages. 1995. *(AET)*

NATO ASI Series F

Including Special Programmes on Sensory Systems for Robotic Control (ROB) and on Advanced Educational Technology (AET)

Vol. 141: Organizational Learning and Technological Change. Edited by C. Zucchermaglio, S. Bagnara and S. U. Stucky. X, 368 pages. 1995. *(AET)*

Vol. 142: Dialogue and Instruction. Modeling Interaction in Intelligent Tutoring Systems. Edited by R.-J. Beun, M. Baker and M. Reiner. IX, 368 pages. 1995. *(AET)*

Vol. 146: Computers and Exploratory Learning. Edited by A. A. diSessa, C. Hoyles and R. Noss. VIII, 482 pages. 1995. *(AET)*